Sensors: Modeling, Design and Applications

Sensors: Modeling, Design and Applications

Edited by
Sherley Benson

WILLFORD PRESS
www.willfordpress.com

Published by Willford Press,
118-35 Queens Blvd., Suite 400,
Forest Hills, NY 11375, USA

ISBN: 978-1-68285-583-6

Cataloging-in-Publication Data

Sensors : modeling, design and applications / edited by Sherley Benson.
    p. cm.
Includes bibliographical references and index.
ISBN 978-1-68285-583-6
1. Detectors. 2. Detectors--Design and construction. 3. Detectors--Industrial applications.
I. Benson, Sherley.
TK7871.674 .S46 2019
681.2--dc21

For information on all Willford Press publications
visit our website at www.willfordpress.com

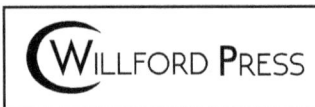

WILLFORD PRESS

Contents

Preface

The main aim of this book is to educate learners and enhance their research focus by presenting diverse topics covering this vast field. This is an advanced book which compiles significant studies by distinguished experts in the area of analysis. This book addresses successive solutions to the challenges arising in the area of application, along with it; the book provides scope for future developments.

Sensors are systems designed to sense change and transfer information to processors for further analysis. The advances in machine design and micromachinery have affected the technological advancements of sensors. The applications of sensors vary across a range of industrial and scientific divices and mechanisms such as medical devices, smartphones, industrial manufacturing units, etc. This book is a compilation of chapters that discuss the most vital concepts and emerging technological trends related to sensor devices and their growing usage. Different approaches, evaluations, methodologies and advanced studies on sensor modeling, design and their applications have been included in this book. Those in search of information to further their knowledge will be greatly assisted by this book.

It was a great honour to edit this book, though there were challenges, as it involved a lot of communication and networking between me and the editorial team. However, the end result was this all-inclusive book covering diverse themes in the field.

Finally, it is important to acknowledge the efforts of the contributors for their excellent chapters, through which a wide variety of issues have been addressed. I would also like to thank my colleagues for their valuable feedback during the making of this book.

Editor

Effects of Operating Temperature on Droplet Casting of Flexible Polymer/Multi-Walled Carbon Nanotube Composite Gas Sensors

Jin-Chern Chiou [1,2], Chin-Cheng Wu [1,*], Yu-Chieh Huang [2], Shih-Cheng Chang [3] and Tse-Mei Lin [1]

[1] Department of Electrical and Computer Engineering, National Chiao Tung University, 1001 University Road, Hsinchu City 30010, Taiwan; chiou@mail.nctu.edu.tw (J.-C.C.); s0450737.eed04g@g2.nctu.edu.tw (T.-M.L.)
[2] Institute of Electrical Control Engineering, National Chiao Tung University, 1001 University Road, Hsinchu City 30010, Taiwan; yuchieh.ece99g@g2.nctu.edu.tw
[3] Institute of Biomedical Engineering, National Chiao Tung University, 1001 University Road, Hsinchu City 30010, Taiwan; scchang1222.iie04g@g2.nctu.edu.tw
*   Correspondence: cjwu028@gmail.com

Academic Editors: Eduard Llobet and Stella Vallejos

Abstract: This study examined the performance of a flexible polymer/multi-walled carbon nanotube (MWCNT) composite sensor array as a function of operating temperature. The response magnitudes of a cost-effective flexible gas sensor array equipped with a heater were measured with respect to five different operating temperatures (room temperature, 40 °C, 50 °C, 60 °C, and 70 °C) via impedance spectrum measurement and sensing response experiments. The selected polymers that were droplet cast to coat a MWCNT conductive layer to form two-layer polymer/MWCNT composite sensing films included ethyl cellulose (EC), polyethylene oxide (PEO), and polyvinylpyrrolidone (PVP). Electrical characterization of impedance, sensing response magnitude, and scanning electron microscope (SEM) morphology of each type of polymer/MWCNT composite film was performed at different operating temperatures. With respect to ethanol, the response magnitude of the sensor decreased with increasing operating temperatures. The results indicated that the higher operating temperature could reduce the response and influence the sensitivity of the polymer/MWCNT gas sensor array. The morphology of polymer/MWCNT composite films revealed that there were changes in the porous film after volatile organic compound (VOC) testing.

Keywords: polymer/multi-walled carbon nanotube composites; droplet casting; operating temperature; impedance spectrum

1. Introduction

Polymer-based sensors are resistive-type gas sensors that are widely used by extant research for gas and vapor sensing owing to their diverse responses to different gases. Polymer composite chemiresistor gas sensor arrays comprise different polymers and carbon particles that interact with an adsorptive analyte and cause electrical property changes. Several previous studies examined the high response and sensitivity of polymer-based sensors for detection of volatile organic compounds [1–3]. Carbon nanotubes (CNTs) have stimulated great interest due to their distinctive electrical, physical, and chemical properties that enable the development of sensitive devices in the field of gas sensing [4,5]. Polymer/MWCNT composites have attracted considerable attention due to fast response and high sensitivity towards environmental gases at room temperature. Recent studies demonstrate feasibility of polymer/MWCNT composites for detection of toxic chemical agents, inorganic vapors, and volatile organic compounds [6–14].

Operating temperature of the sensing layer is a key factor that affects response time, sensitivity, and the baseline for both metal oxide semiconductor (MOS) gas sensors and polymer-based gas sensors with different absorbed gases. Specifically, an MOS gas sensor equipped with an oxide-based layer that operates at high operating temperatures causes a change in charge mobility during chemisorptions of oxygen [15]. The operating temperature of SnO_2-based gas sensors that ranges from 25 °C to 500 °C detects various types of low concentration gases [16,17]. Polymer-based gas sensors are characterized by swelling due to the absorption of a target gas into the polymer layer, and subsequently the variation of an electric signal results in a charge transfer on the surfaces of the CNTs [8,18]. The operating temperature of polymer-based gas sensors corresponds to low temperatures below 80 °C to guarantee a stable response to mitigate the influence of ambient temperature [19–21].

Polymer-carbon black composite gas sensor arrays that are operated at low temperatures were developed and applied in electronic nose systems [19,20]. Extant research has examined resistance changes with respect to varying ambient temperatures in polymer-carbon black films in detail [21,22]. The results indicated that the ambient temperature could influence the resistance and baseline at different molecular weights and different carbon loadings [23]. The results of previous studies indicated that different polymer-carbon black composite gas sensors manipulated at several low operating temperatures could exhibit a decrease in their response to the target gases as the operating temperature is increased [24–26]. Many reports on MWCNT/polymer based gas sensors demonstrated high sensitivity but slow recovery at room temperature to achieve complete desorption of adsorbed gas molecules from the surface of MWCNTs. Thermal treatment is one of the more efficient methods to tackle the poor recovery [27–29]. Nevertheless, our preliminary study had shown that the sensing response of a flexible polymer/MWCNT composite gas sensor was decreased with increasing operating temperature [30]. Extant studies have not examined the effect of temperature on the polymer-carbon black composite sensors, with respect to mechanisms of tunneling, hopping, and thermal expansion. Additionally, it is not fully understood how a variation in temperature affects the electrical properties of polymer/carbon nanotube composite gas sensors and causes different chemical potentials of polymer phase and gas phase.

In this study, a two-layer polymer/MWCNT composite sensing film was fabricated by a droplet casting method, and a flexible printed circuit (FPC) technology was used to fabricate sensing electrodes with embedded heater gas sensor arrays. The operating temperature dependence of electrical characterization and sensor response was investigated. The selected polymers used in a polymer/MWCNT composite sensing film included ethylcellulose (EC), polyethylene oxide (PEO), and polyvinylpyrrolidone (PVP). The effect of different operating temperatures on the electric properties and sensing responses of the polymer/MWCNT composite gas sensor array was tested in the device developed to detect methanol. Furthermore, scanning electron microscopy (SEM) was used to compare differences in morphologies between the sensors before and after the test.

2. Materials

The polymer/MWCNT composite sensing film consisted of two membranes, the top layer and the bottom layer wherein the polymer film and the MWCNT film were deposited, respectively. Both membranes were fabricated via a droplet casting method to form the two-layer structure for gas sensing. Polymers selected for deposition on the MWCNT film included ethylcellulose (200679, Sigma-Aldrich, Saint Louis, MO, USA), polyethylene oxide (43678, Alfa Aesar, Haverhill, MA, USA), and polyvinylpyrrolidone (PVP 10, Sigma-Aldrich, Saint Louis, MO, USA). The selection was based on linear solvation energy relationship (LSER) theory and physical absorption bonding [31,32]. Typically, each of the selected polymers (0.2 g) was dissolved in 20 mL tetrahydrofuran (THF) and was then prepared by sonication for 6 h in an ultrasonic bath at room temperature. The MWCNT used for the composite films were few-walled carbon nanotubes (FWNTs) provided by the XinNano Materials, Inc. (Taoyuan, Taiwan). The approximate dimensions of the MWCNT with 2–5 layers of sidewalls were an average diameter of 4 nm, 10–12 µm average length, and >86% average purity.

The fabrication processes of droplet casting a two-layer sensing film are as follows: first, 1 wt % (2 μL) of MWCNT was deposited on a conductive electrode by a micro jet. The device was then placed in an oven at 70 °C to evaporate THF and furnish the MWCNT film. The selected polymers were then deposited by adding a droplet of 1 wt % (2 μL) solution (1 mg/mL THF) on the MWCNT layer to form the film. Finally, the device was dried for 24 h at 60 °C, and the solvent was completely evaporated prior to use. The sensor resistance after each casting step was monitored to limit the value within a range of 10 kΩ–200 kΩ to guarantee the reproducibility. The morphology of all polymer/MWCNT composite films was confirmed by the SEM image as shown in Figure 1a–d. The morphology of a polymer sensing film was examined using an SEM (NOVA NANO SEM 450, FEI Co., Hillsboro, OR, USA) with 10 kV acceleration voltage. The pore sizes of EC/MWCNT film remained in a range of 0.7–1.1 μm. The pores with the largest diameter were in the range from 1.2 μm to 1.4 μm.

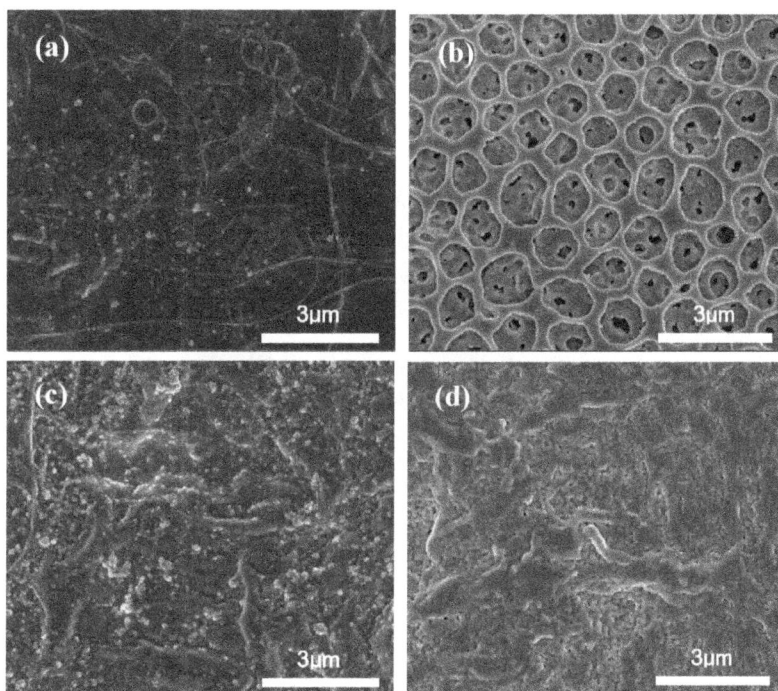

Figure 1. SEM morphology of polymer/MWCNT composite films before the test (**a**) MWCNT film; (**b**) EC/MWCNT film; (**c**) PEO/MWCNT film; and (**d**) PVP/MWCNT film.

3. Design and Fabrication

A cost-effective gas sensor array was fabricated by flexible printed circuit industry technologies. The flexible gas sensor array was comprised of three different types of polymer/MWCNT composite sensing films arranged in a 3 × 3 matrix pattern. Each type of the selected polymer was arranged in one of the rows in the matrix. The fabricated flexible gas sensor array exhibited excellent flexibility, as shown in Figure 2a. The insets indicate the sensing electrode of a single sensor element and the heater.

The sensing electrode was composed of copper with 35 μm thickness, 220 μm line width, and 220 μm line spacing. The through hole-machined well with 130 μm thickness was positioned and then adhered to the upper side of the sensing electrode to guarantee a filled polymer composite film placed in a specific area [33]. The configuration of the fabrication and cross-section view of the flexible gas sensor array is shown in Figure 2b.

(a)

Polymer	MWCNT	Cu
Polyimide	SUS304	RTD

(b)

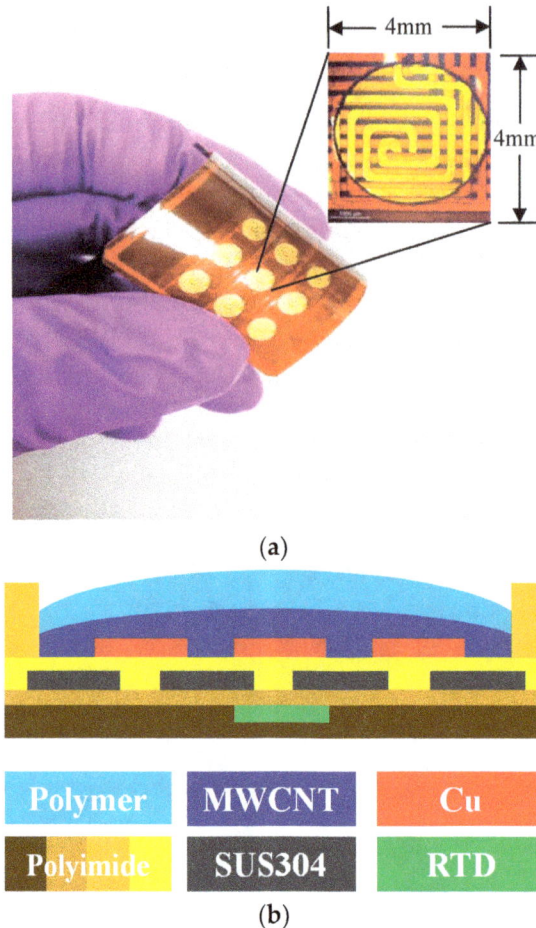

Figure 2. (**a**) configuration of the flexible gas sensor array. The inset shows the sensing electrode (**top** electrode) and the heater (**bottom** electrode); and (**b**) the cross-sectional schematic structure of the single gas sensor.

The heater had a 50 μm thickness and a geometry corresponding to 20 mm × 20 mm. In contrast, the width and spacing of the single heater line was 220 μm and 280 μm, respectively. The heater was made of stainless steel (SUS304) to provide a thermostat operating temperature. These temperatures included room temperature, 40 °C, 50 °C, 60 °C, and 70 °C. The platinum resistance temperature detector (RTD) was embedded in a polyimide substrate to enable feedback control at the operating temperature. To prevent heat loss from the substrate, 130 μm polyimide films were adhered to the bottom side of the sensor substrate. Both the heater as well as the sensing electrode were designed in double-spiral shapes in a square area [30].

The architecture of the gas sensor array control system was comprised of a flexible gas sensor array sensor, an interface circuit, a micro control unit, and a human–machine interface. The system was designed to drive the sensor array and the heater, and to control the operating temperature and collect response data from each sensor. Figure 3 shows a block diagram of the proposed gas sensor array control system.

When a flexible gas sensor array was operated at a specific temperature for target gas detection, the varied resistance of each sensor was obtained through a multiplexer (MUX). The resistance was then converted to voltage signals by a sensor interface circuit (SIC). The multichannel signals were recorded through a micro control unit (MCU, C8051F120, Silicon Laboratories, Inc., Austin, TX, USA) and then synchronized display was obtained on the human–machine interface (HMI, see Figure 3b).

(a)

(b)

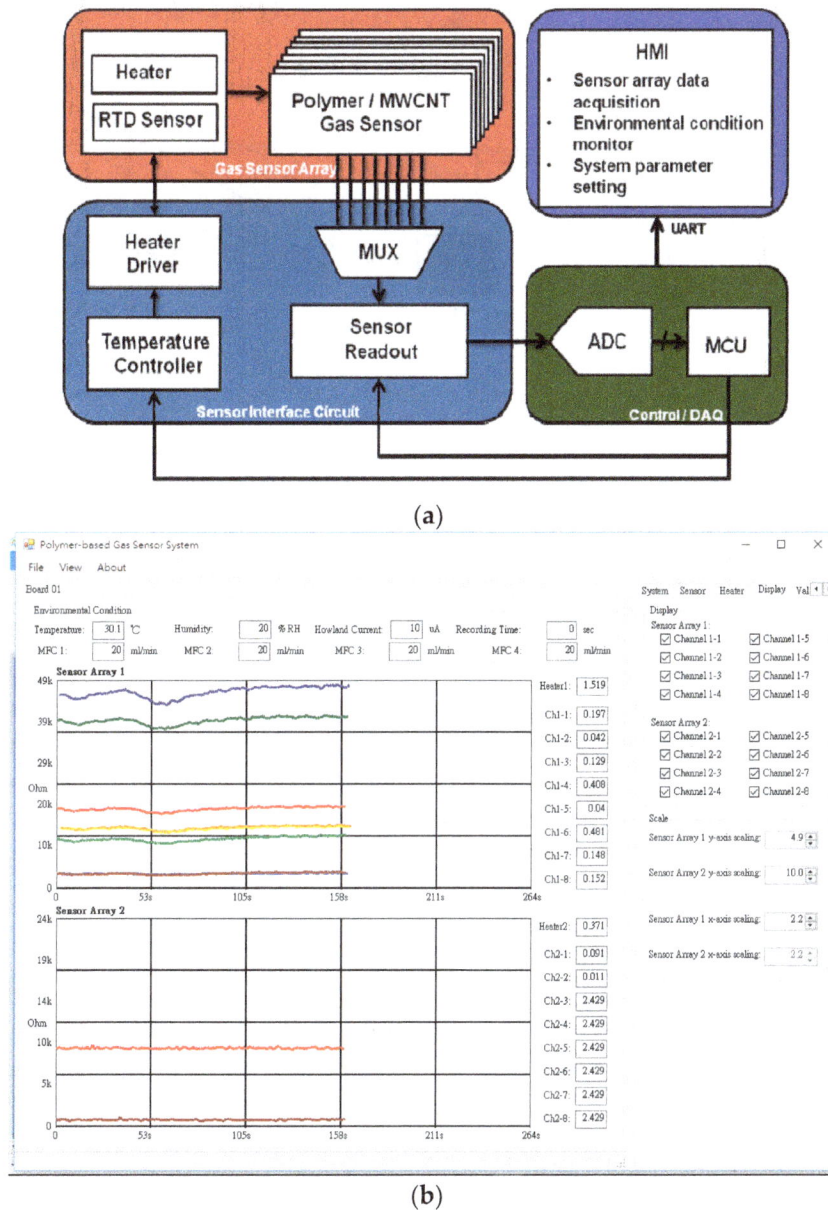

Figure 3. (**a**) system block of the gas sensor array control system; and (**b**) human–machine interface software.

The driver and feedback control circuit of the heater are shown in Figure 4a. The operational amplifier was connected to a voltage source and operated a bipolar junction transistor that functioned as a switch for the current to the heater [34,35]. The 1 kΩ platinum RTD sensor that measured the change in the operating temperature of the heater was driven by a constant current source. The signal of the RTD sensor was obtained by a voltage follower and was then connected to one of the inputs in the differential amplifier in the compensator circuit. This output signal was compared to the reference temperature set-point voltage for driving the heater. This feedback control system for the heater was used to obtain the steady-state electrical power consumption curves of the heater under 500 mL/min airflow conditions given the existence of the composite-sensing layer, as shown in Figure 4b. The electrical power consumption of the heater was a function of the operating temperature range (35.36–84.03 °C) in a flexible gas sensor array.

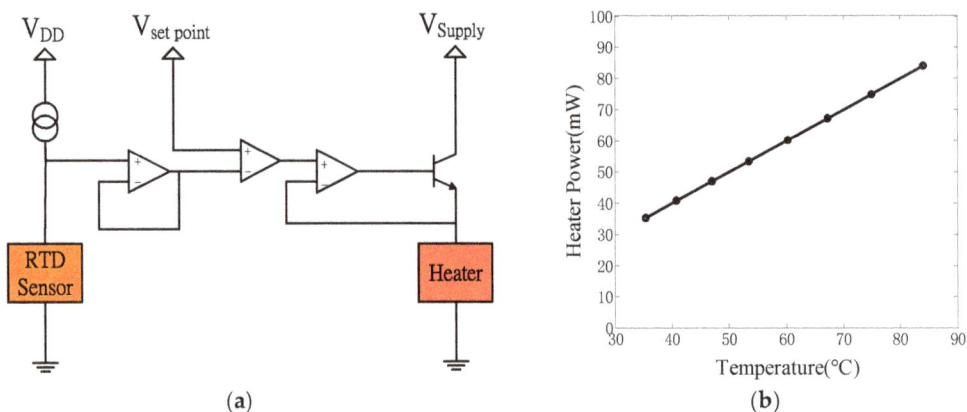

(a)   (b)

Figure 4. (**a**) temperature control circuit of the heater; and (**b**) power consumption vs. operating temperature of the microheater for the flexible gas sensor array.

4. Experiments and Discussion

4.1. Transient Response of the Polymer/MWCNT Composite Film

The transient response experiment included four separate heating stages to operate the microheater by heating the flexible gas sensor array to temperatures of 40 °C, 50 °C, 60 °C, and 70 °C under a 500 mL/min airflow condition. In each stage, the flexible gas sensor array was first maintained at room temperature to obtain the recovery baseline for 10 min. The heater was then used to heat the sensor array to the specific operating temperature for 10 min. Figure 5a shows the transient response of the heater to the relative operating temperature. The profile display indicated that the time taken to heat the array to the operating temperature range (±0.5 °C) using the heater was less than 2 min.

The transient responses of three different polymer/MWCNT composite films are shown in Figure 5b. Evidently, the recovery baselines of the EC/MWCNT and PEO/MWCNT composite films were slightly shifted and PVP/MWCNT was heavily shifted. Three different polymer/MWCNT composite films exhibited a negative temperature coefficient resistance (NTC) inclination as the operating temperature increased.

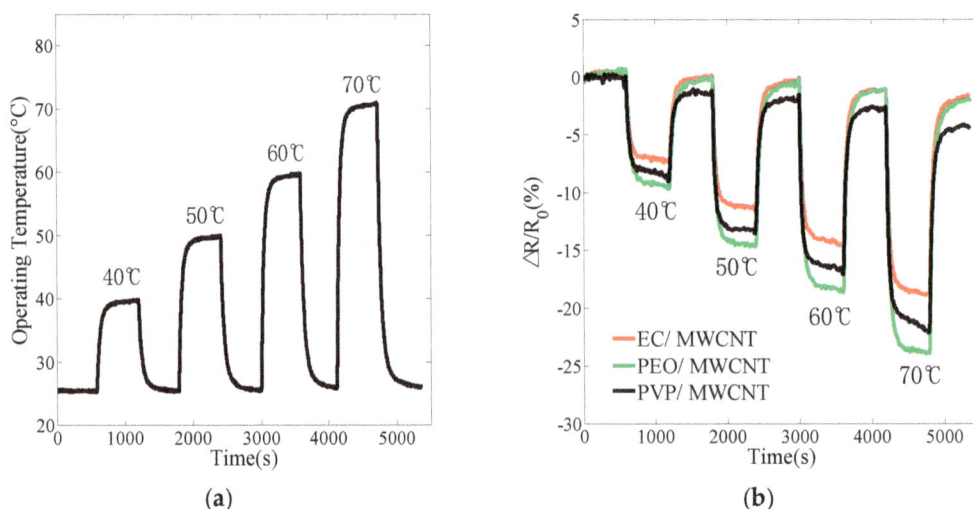

(a)   (b)

Figure 5. Cyclical heating to operating temperatures of 40 °C, 50 °C, 60 °C, and 70 °C. (**a**) the transient response of the heater to the relative operating temperature; and (**b**) the responses of normalized resistance of the polymer/MWCNT composite sensor.

With respect to different operating temperatures, the EC/MWCNT and PEO/MWCNT film revealed a better immunity to temperature influence with a variation in resistance The PEO/MWCNT film revealed the widely transient response with respect to the operating temperature that was considerably more stable than other films. The resistance of the PVP/MWCNT film was stable at 40 °C and 50 °C, but unstable at 60 °C and 70 °C.

4.2. Impedance Spectrum Property

Impedance measurement was performed using an Agilent 4292A impedance analyzer (Agilent Technologies, Santa Clara, CA, USA)in the frequency range of 100 Hz to 1 MHz using a modulation voltage of 500 mV (peak to peak) [4,36]. The impedance measurements were measured at different operating temperatures under the 500 mL/min airflow conditions. The impedance spectrum of the MWCNT film and three different polymer/MWCNT composite films are shown in Figure 6. The sensor responses were comparable, as shown in the figure.

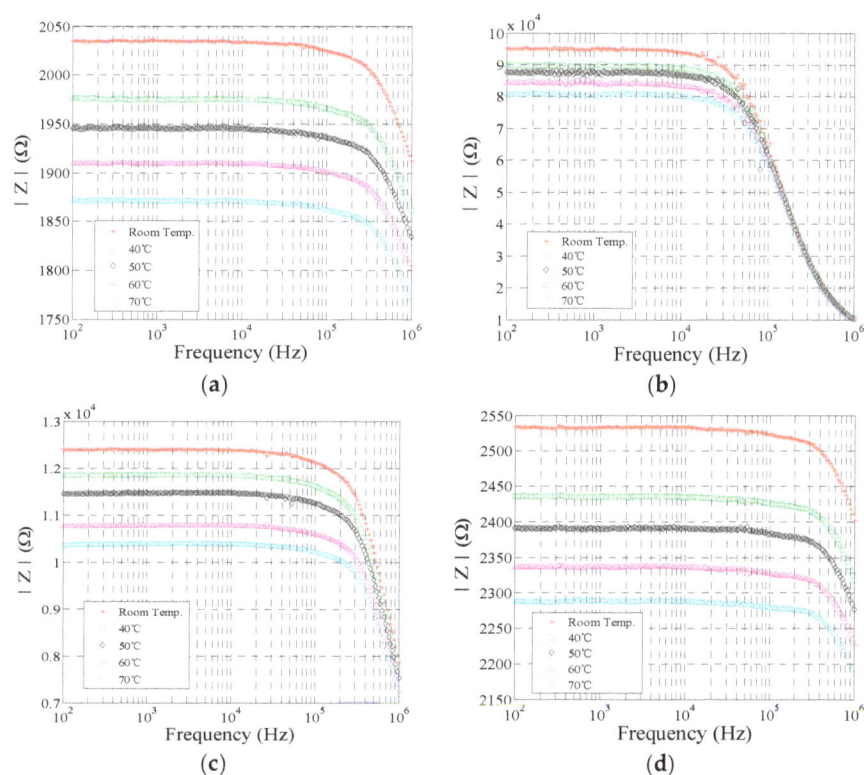

Figure 6. Impedance spectrum of the MWCNT film and polymer/MWCNT composite film at different operating temperatures. (**a**) MWCNT film; (**b**) EC/MWCNT film; (**c**) PEO/MWCNT; and (**d**) PVP/MWCNT.

As the measurements indicate, the resistance behavior of MWCNT film decreased with increases in operating temperatures at frequencies below 100 kHz, corresponding to the operating temperatures (the following operating temperatures: room temperature, 40 °C, 50 °C, 60 °C, and 70 °C). The equivalent circuit model of the polymer/MWCNT composite film was examined. It included two components, namely the resistance and capacitance effects [4]. Significant differences in the behavior of the impedance spectrum were not observed in the other polymer/MWCNT composite films. The impedance spectra of the other polymer/MWCNT composite films revealed a circuit model equivalent to that of the MWCNT film. However, the capacitance effect of each polymer/MWCNT composite film occurred in a different frequency range. As observed in Figure 6, the resistance effects of the EC/MWCNT and PEO/MWCNT composite films were observed below 10 kHz. Additionally, the resistance effect of the PVP/MWCNT composite film was observed below 100 kHz.

4.3. The Response of the Sensor Array

The polymer composite film absorbed the target gas when it was introduced into the reaction chamber. As the gas was introduced, the film swelled up slightly, and this induced the change in the distance between nanoparticles. The change in the resistance of the film could then be measured by an instrument [8,18]. The experimental setup used in the measurements is shown in Figure 7.

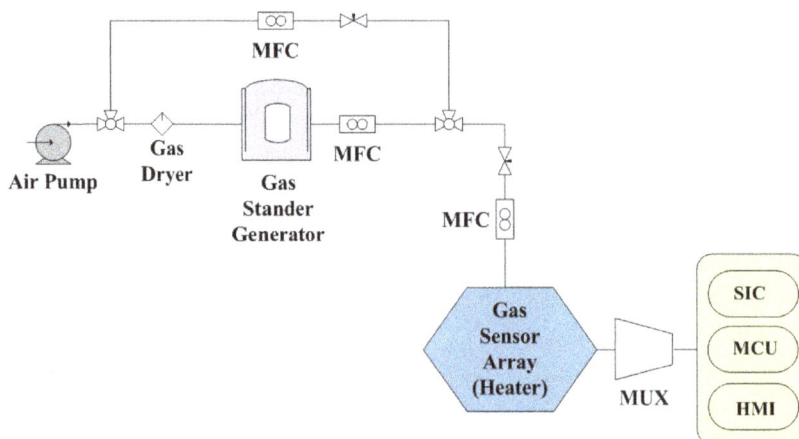

Figure 7. Experimental setup of the gas sensing system.

The 1.5% of ethanol gas was controlled by a mass flow controller under a flow rate of 500 mL/min. Dry air (25 °C, 45% relative humidity (RH)) was used as background gas, and the flow rate was set at 500 mL/min. The flexible polymer/MWCNT composite gas sensor array was placed inside a reaction chamber with 60 mL capacity. The gas sensing response measurement consisted of several steps in each gas-testing cycle. First, the heater was heated to the operating temperature, and then dry air was introduced into the reaction chamber for 10 min to obtain a reference resistance baseline. When the temperature of the heater was stable, the ethanol gas was introduced into the reaction chamber for 5 min. The polymer films were adsorbed and swollen due to gas molecules. Following this, dry air was introduced for 10 min to enable desorption from the polymer film.

Normalized resistance changes ($\Delta R/R_0\%$) of the polymer/MWCNT composite films were determined using $\Delta R/R_0\% = [(R_{max} - R_0)/R_0] \times 100$, where R_0 denotes the mean value of sensor resistance from t = 1~100 s when the sensor was exposed to dry air in equilibrium, and R_{max} denotes the maximum resistance when the sensor was exposed to ethanol. In order to obtain sufficient response information to analysis, the polymer/MWCNT composite sensing film was exposed to 1.5% ethanol with different operating temperatures. Figure 8 shows the response patterns of the normalized data when the polymer/MWCNT composite sensing film was exposed to ethanol with different operating temperatures. The response patterns exhibited that EC/MWCNT and PVP/MWCNT sensors show a decreased response of sensitivity with an increase in operating temperature. Increasing operating temperature could result in increased polymer chain mobility to form percolation networks for sensing response, but simultaneously provide the electrons more energy to overcome the potential barrier and cause more tunneling contribution to decrease the sensing response [37]. The glass transition temperature of PEO is very low and an increase in temperature could result in increased polymer chain mobility at higher operating temperatures [23,37,38]. Hence, the sensitivites of the PEO sensor at 60 °C and 70 °C have better responses than at 50 °C.

All three polymer/MWCNT composite films showed a decrease in sensitivity response with an increase in the operating temperature. The results indicated that the polymer chain mobility increased with an increase in operating temperature with respect to sensing response. A suitable operating temperature could provide a flat baseline for target gas recognition. The baseline shift was severe in the PEO/MWCNT film because this film involves a lower glass transition temperature material that could be sensitive to the operating temperatures.

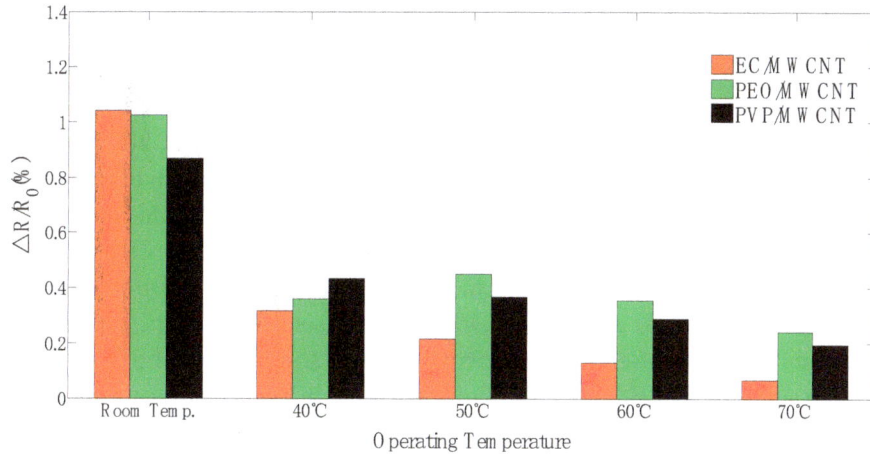

Figure 8. Normalized resistance response of polymer/MWCNT composites films exposed to 1.5% of ethanol gas with respect to different temperatures.

4.4. SEM Morphology of Polymer/MWCNT Composite Films

The morphology of each polymer/MWCNT composite film after the aforementioned test was investigated via SEM and is shown in Figure 9. The significant differences of the porous EC/MWCNT film indicated that the pores evidently expanded and became larger when compared to the initial pore sizes. The pore size range corresponded to 2.3–2.8 μm with cavity sizes in the range of 0.3–2.1 μm. The surface of the PVP/MWCNT film could shrink after the application of a series of thermal cycles. There were no obvious changes in the other two films after the aforementioned test.

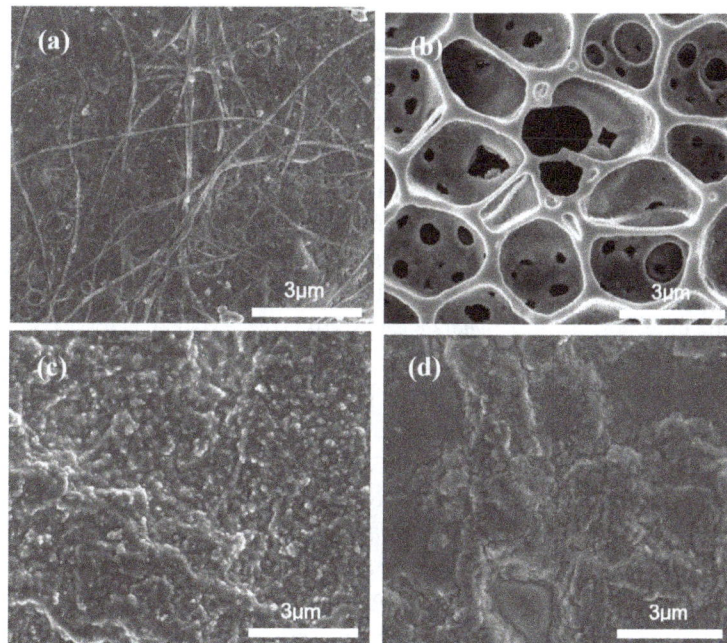

Figure 9. SEM morphology of polymer/MWCNT composite films after the test (**a**) MWCNT film; (**b**) EC/MWCNT film; (**c**) PEO/MWCNT film; and (**d**) PVP/MWCNT film.

5. Conclusions

A flexible polymer/MWCNT gas sensor offers several advantages including cost effectiveness, lower power consumption, reproducibility, lightweight, and flexibility given its potential integration in electronic noses and portable consumer products. However, the ambient environments of these applications involve several variables that influence sensor performance. Temperature is one such variable in which temperature variations pose a critical problem for reducing the sensitivity of the sensor. To date, extant research has not focused on the effect of operating temperature on a polymer-based gas sensor. Current studies examine environments with a constant operating temperature.

The gas absorptions and interaction mechanisms of polymer/MWCNT composite films are dominated by two principles: namely, physisorption and chemisorption. Both of these principles could change with respect to different operating temperatures. The experiment in this study investigated the effect of operating temperature on the responses of a flexible polymer/MWCNT gas sensor. The results indicated that higher operating temperature could mitigate the influence of ambient temperature but reduce the response. Both of these effects could influence the sensitivity of the polymer/MWCNT gas sensor array. The morphology after the aforementioned test showed that the pores of EC/MWCNT expanded, but the surface of PVP/MWCNT film started to shrink. The reusability and the life cycle of each polymer/MWCNT composite film should be considered at a suitable operating temperature to prevent thermal expansion and subsequent destruction of the pores. A future study will examine the effect of other ambient variables and the performance under mechanical strain on the flexible polymer/MWCNT gas sensor array.

Acknowledgments: This work was supported in part by the Ministry of Science and Technology, Taiwan, R.O.C. (under Contract Number: MOST 103-2221-E-009-192-MY3 and MOST 105-2218-E-009-018). The authors would like to thank the National Chip Implementation Center for chip fabrication. The authors thank the team of the Chemistry Division on the National Chung-Shan Institute of Science & Technology, especially Yu-Ping Wang, Li-Chun Wang, Jen-Chin Wu and Ching-Lung Lin for valuable discussions about polymer and MWCNT characterization.

Author Contributions: These authors contributed equally to this work.

Conflicts of Interest: The authors declare no conflict of interest.

References

1. Ryan, M.A.; Shevade, A.V.; Zhou, H.; Homer, M.L. Polymer-carbon black composite sensors in an electronic nose for air-quality monitoring. *MRS Bull.* **2004**, *29*, 714–719. [CrossRef] [PubMed]

2. Lonergan, M.C.; Severin, E.J.; Doleman, B.J.; Beaber, S.A.; Grubbs, R.H.; Lewis, N.S. Array-based vapor sensing using chemically sensitive, carbon black-polymer resistors. *Chem. Mater.* **1996**, *8*, 2298–2312. [CrossRef]

3. Ryan, M.A.; Lewis, N.S. Low power, lightweight vapor sensing using arrays of conducting polymer composite chemically-sensitive resistors. *Enantiomer* **2000**, *6*, 159–170.

4. Hafaiedh, I.; Elleuch, W.; Clément, P.; Llobet, E.; Abdelghani, A. Multi-walled carbon nanotubes for volatile organic compound detection. *Sens. Actuators B Chem.* **2013**, *182*, 344–350. [CrossRef]

5. Wang, Y.; Yang, Z.; Hou, Z.; Xu, D.; Wei, L.; Kong, E.S.W.; Zhang, Y. Flexible gas sensors with assembled carbon nanotube thin films for DMMP vapor detection. *Sens. Actuators B Chem.* **2010**, *150*, 708–714. [CrossRef]

6. Wang, F.; Gu, H.W.; Swager, T.M. Carbon nanotube/polythiophene chemiresistive sensors for chemical warfare agents. *J. Am. Chem. Soc.* **2008**, *130*, 5392–5393. [CrossRef] [PubMed]

7. Chang, C.P.; Yuan, C.L. The fabrication of a MWNTs–polymer composite chemoresistive sensor array to discriminate between chemical toxic agents. *J. Mater. Sci.* **2009**, *44*, 5485–5493. [CrossRef]

8. Philip, B.; Abraham, J.K.; Chandrasekhar, A. Carbon nanotube/PMMA composite thin films for gas-sensing applications. *Smart Mater. Struct.* **2003**, *12*, 935–939. [CrossRef]

9. Li, J.; Lu, Y.; Ye, Q.; Cinke, M.; Han, J.; Meyyappan, M. Carbon nanotube sensors for gas and organic vapor detection. *Nano Lett.* **2003**, *3*, 929–933. [CrossRef]

10. Abraham, J.K.; Philip, B.; Witchurch, A.; Varadan, V.K.; Reddy, C.C. A compact wireless gas sensor using a carbon nanotube/PMMA thin film chemiresistor. *Smart Mater. Struct.* **2004**, *13*, 1045–1049. [CrossRef]

11. Zhang, B.; Fu, R.W.; Zhang, M.Q.; Dong, X.M.; Lan, P.L.; Qiu, J.S. Preparation and characterization of gas-sensitive composites from multi-walled carbon nanotubes/ polystyrene. *Sens. Actuators B Chem.* **2005**, *109*, 323–328. [CrossRef]

12. Niu, L.; Luo, Y.; Li, Z. A highly selective chemical gas sensor based on functionalization of multi-walled carbon nanotubes with poly(ethylene glycol). *Sens. Actuators B Chem.* **2007**, *126*, 361–367. [CrossRef]

13. Spitalsky, Z.; Tasis, D.; Papagelis, K.; Galiotis, C. Carbon nanotube–polymer composites: Chemistry, processing, mechanical and electrical properties. *Prog. Polym. Sci.* **2010**, *35*, 357–401. [CrossRef]

14. Llobet, E. Gas sensors using carbon nanomaterials: A review. *Sens. Actuators B Chem.* **2013**, *179*, 32–45. [CrossRef]

15. Barsan, N.; Weimar, U. Understanding the fundamental principles of metal oxide based gas sensors; the example of CO sensing with SnO_2 sensors in the presence of humidity. *J. Phys. Condens. Matter* **2003**, *15*, R813. [CrossRef]

16. Chiou, J.C.; Tsai, S.W.; Lin, C.Y. Liquid Phase Deposition Based SnO_2 Gas Sensor Integrated with TaN Heater on a Micro-hotplate. *IEEE Sens. J.* **2013**, *13*, 2466–2473. [CrossRef]

17. Berger, F.; Sanchez, J.; Heintz, O. Detection of hydrogen fluoride using SnO_2-based gas sensors: Understanding of the reactional mechanism. *Sens. Actuators B Chem.* **2009**, *143*, 152–157. [CrossRef]

18. Bai, H.; Shi, G. Gas sensors based on conducting polymers. *Sens. J.* **2007**, *7*, 267–307. [CrossRef]

19. Ryan, M.A.; Homer, M.L.; Buehler, M.G.; Manatt, K.S.; Zee, F.; Graf, J. Monitoring the Air Quality in a Closed Chamber Using an Electronic Nose. In Proceedings of the 27th International Conference on Environmental Systems, Lake Tahoe, NV, USA, 14–17 July 1997.

20. Kim, Y.S.; Ha, S.C.; Yang, Y.; Kim, Y.J.; Cho, S.M.; Yang, H.; Kim, Y.T. Portable Electronic Nose System Based on the Carbon Black-Polymer Composite Sensor Array. *Sens. Actuators B Chem.* **2005**, *108*, 285–291. [CrossRef]

21. Xie, H.F.; Yang, Q.D.; Sun, X.X.; Yang, J.J.; Huang, Y.P. Gas sensor arrays based on polymer-carbon black to detect organic vapors at low concentration. *Sens. Actuators B Chem.* **2006**, *113*, 887–891. [CrossRef]

22. Buehler, M.G.; Ryan, M.A. Temperature and humidity dependence of a polymer-based gas sensor. *Proc. SPIE* **1997**, *3082*, 40–48.

23. Homer, M.L.; Lim, J.R.; Manatt, K.; Kisor, A.; Manfreda, A.M.; Lara, L.; Jewell, A.D.; Yen, S.P.S.; Zhou, H.; Shevade, A.V.; et al. Temperature effects on polymer-carbon composite sensors: Evaluating the role of polymer molecular weight and carbon loading. *Proc. IEEE Sens.* **2003**, *2*, 877–881.

24. Ha, S.C.; Yang, Y.S.; Kim, Y.S.; Kim, S.H.; Kim, Y.J.; Cho, S.M. Environmental temperature-independent gas sensor array based on polymer composite. *Sens. Actuators B Chem.* **2005**, *108*, 258–264. [CrossRef]

25. Ha, S.C.; Kim, Y.S.; Yang, Y.; Kim, Y.J.; Cho, S.M.; Yang, H.; Kim, Y.T. Integrated and microheater embedded gas sensor array based on the polymer composites dispensed in micromachined wells. *Sens. Actuators B Chem.* **2005**, *105*, 549–555. [CrossRef]

26. Kim, Y.S. Microheater-integrated single gas sensor array chip fabricated on flexible polyimide substrate. *Sens. Actuators B Chem.* **2006**, *114*, 410–417. [CrossRef]

27. Santucci, S.; Picozzi, S.; Gregorio, F.D.; Lozzi, L.; Cantalini, C.; Valentini, L.; Kenny, J.M.; Delley, B. NO and CO gas adsorption on carbon nanotubes: Experiment and theory. *J. Chem. Phys.* **2003**, *119*, 10904–10910. [CrossRef]

28. Mangu, R.; Rajaputra, S.; Singh, V.P. MWCNT–polymer composites as highly sensitive and selective room temperature gas sensors. *Nanotechnology* **2011**, *22*, 215502. [CrossRef] [PubMed]

29. Sharma, S.; Hussain, S.; Singh, S.; Islam, S.S. MWCNT-conducting polymer composite based ammonia gas sensors: A new approach for complete recovery process. *Sens. Actuators B Chem.* **2014**, *194*, 213–219. [CrossRef]

30. Wu, C.C.; Chiou, J.C.; Wang, Y.P.; Wang, L.C. Flexible polymer/multi-walled carbon nanotube composite sensor array equipped with microheater for gas sensing. In Proceedings of the 2016 International Conference on Manipulation, Automation and Robotics at Small Scales (MARSS), Paris, France, 18–22 July 2016.

31. Hierlemann, A.; Zellers, E.T. Use of linear solvation energy relationships for modeling responses from polymer-coated acoustic-wave vapor sensors. *Anal. Chem.* **2001**, *73*, 3458–3466. [CrossRef] [PubMed]

32. Chuang, P.K.; Wang, L.C.; Kuo, C.T. Development of a high performance integrated sensor chip with a multi-walled carbon nanotube assisted sensing array. *Thin Solid Films* **2013**, *529*, 205–208. [CrossRef]

33. Zee, F.; Judy, J.W. Micromachined polymer-based chemical gas sensor array. *Sens. Actuators B Chem.* **2001**, *72*, 120–128. [CrossRef]

34. García-Guzmán, J.; Ulivieri, N.; Cole, M.; Gardner, J.W. Design and simulation of a smart ratiometric ASIC chip for VOC monitoring. *Sens. Actuators B Chem.* **2003**, *95*, 232–243. [CrossRef]

35. Kim, J.C.; Chung, J.T.; Lee, D.J.; Kim, Y.K.; Kim, J.W.; Hwang, S.W.; Ju, B.K.; Yun, S.K.; Park, H.W. Development of temperature feedback control system for piezo-actuated display package. *Sens. Actuators A Phys.* **2009**, *151*, 213–219. [CrossRef]

36. Amrani, M.E.H.; Persaud, K.C.; Payne, P.A. High-Frequency Measurements of Conducting Polymers—Development of a New Technique for Sensing Volatile Chemicals. *Meas. Sci. Technol.* **1995**, *6*, 1500–1507. [CrossRef]

37. Li, Q.; Xue, Q.Z.; Gao, X.L.; Zheng, Q.B. Temperature dependence of the electrical properties of the carbon nanotube/polymer composites. *Express Polym. Lett.* **2009**, *3*, 769–777. [CrossRef]

38. Zhou, Y.X.; Wu, P.X.; Cheng, Z.-Y.; Ingram, J.; Jeelani, S. Improvement in electrical, thermal and mechanical properties of epoxy by filling carbon nanotube. *Express Polym. Lett.* **2008**, *2*, 40–48. [CrossRef]

Relative Estimation of Water Content for Flat-Type Inductive-Based Oil Palm Fruit Maturity Sensor

Norhisam Misron [1,2,*], Nor Aziana Aliteh [1], Noor Hasmiza Harun [3], Kunihisa Tashiro [4], Toshiro Sato [4] and Hiroyuki Wakiwaka [4]

[1]   Faculty of Engineering, Universiti Putra Malaysia, 43400 Serdang, Selangor, Malaysia; aziana.teh@gmail.com
[2]   Institute of Advance Technology (ITMA), Universiti Putra Malaysia, 43400 Serdang, Selangor, Malaysia
[3]   Medical Engineering Section, Universiti Kuala Lumpur-British Malaysia Institute, Batu 8, Jalan Sg Pusu, 53100 Gombak, Selangor, Malaysia; noorhasmiza@unikl.edu.my
[4]   Faculty of Engineering, Shinshu University, Wakasato 4-17-1, Nagano 380-8553, Japan; tashiro@shinshu-u.ac.jp (K.T.); labyam1@shinshu-u.ac.jp (T.S.); wakiwak@shinshu-u.ac.jp (H.W.)
*   Correspondence: norhisam@upm.edu.my

Academic Editor: Simon X. Yang

Abstract: The paper aims to study the sensor that identifies the maturity of oil palm fruit bunches by using a flat-type inductive concept based on a resonant frequency technique. Conventionally, a human grader is used to inspect the ripeness of the oil palm fresh fruit bunch (FFB) which can be inconsistent and inaccurate. There are various new methods that are proposed with the intention to grade the ripeness of the oil palm FFB, but none has taken the inductive concept. In this study, the resonance frequency of the air coil is investigated. Samples of oil palm FFB are tested with frequencies ranging from 20 Hz to 10 MHz and the results obtained show a linear relationship between the graph of the resonance frequency (MHz) against time (Weeks). It is observed that the resonance frequencies obtained for Week 10 (pre-mature) and Week 18 (mature) are around 8.5 MHz and 9.8 MHz, respectively. These results are compared with the percentage of the moisture content. Hence, the inductive method of the oil palm fruit maturity sensor can be used to detect the change in water content for ripeness detection of the oil palm FFB.

Keywords: inductive concept; air coil; resonance frequency; oil palm; maturity classification; moisture content

---

1. Introduction

The *Elaeis guineensis* species of oil palm tree are the most common type used for palm oil production. Oil palm fruit will undergo a process through crude palm oil milling to get the crude palm oil for various products. The ripeness of the oil palm is commonly determined based on visual inspection which is manually done by a human grader. The manual grading proposed by the Malaysian Palm Oil Board (MPOB) is used as the main reference during the inspection for harvesting the oil palm fruit bunches. The inspections are based on the surface's visual color of the oil palm fruit and the number of fruit loose from the bunch [1]. Furthermore, it is important to pluck the oil palm fresh fruit bunch (FFB) at the right time, combined with the accurate ripeness assessments of the oil palm fruit, in order to maintain the optimum quality of palm oil.

To increase the efficiency as well as the accuracy of the oil palm FFB ripeness detection, researchers proposed a new non-destructive detection that can replace the conventional visual method. Various methods for oil palm fruit maturity detection have been developed and tested.

The color vision system is one of the popular methods used for this application. Osama et al. [2] proposed a method of categorizing oil palm FFB maturity through the use of a portable four-band system. The classification system uses a portable and active optical sensor which consists of four spectral bands. The classification of the sensor uses an advanced digital camera and a computer set that is required for the data analysis [2,3]. However, this method requires continuous light intensity monitoring during the image acquisition process for it to be accurately determined by the computer.

The next method is done by using a microwave moisture sensor which utilizes the moisture content of the oil palm fruitlets. Yeow et al. [3,4] proposed an application of the microwave moisture sensor for the identification of the ripeness of oil palm fruit where the content of moisture ranged from 30% to 80% on the basis of wet weight as the main objective of the study. The fabricated sensor operates at the frequency between 1 GHz and 5 GHz, and the overall system is comprised of the fabricated sensor and a vector network analyzer (VNA) that is computer-controlled. The procedure to conduct the measurement is complicated and time-consuming, and the equipment must be used indoors.

Shariffudin et al. [5] presented a monitoring system for oil palm fruit development and ripeness using magnetic resonance imaging (MRI) along with bulk nuclear magnetic resonance (NMR). The proposed monitoring system focused on tracking the development of intact fruit. The entire measurements of spin-spin relaxation times (T2 values) were conducted at the level of 2.35 Tesla and at 200 °C. This method requires skilled personnel to operate complicated and expensive machines, which limit it to indoor testing.

Non-destructive near-infrared (NIR) spectroscopy uses two spectrometers to examine the oil palm fruit with different modes and their chemical content is analyzed using the partial least square regression (PLSR) model [6]. This method again requires indoor testing, in addition to the complicated sample preparation as well as the expensive equipment.

This paper is about a new grading method which uses a simple inductive technique, where the focus is based on the resonance frequency of the air coil as well as the change in the air coil's stray capacitance.

2. Design Basic Principles

2.1. Electrical Diagram

The single flat-type coil representation is shown in Figure 1a. The flux linkages flow through the fruit sample and the inductance of the air coil is measured by using an impedance analyzer. The whole system of the inductive oil palm fruit sensor can be further explained as illustrated in Figure 1b. This inductive oil palm fruit sensor requires operating at the high-frequency response and thus the effect of inductance is very high. The value of the capacitance is predicted from the value of the stray capacitance which came from both the air coil and fruitlet samples.

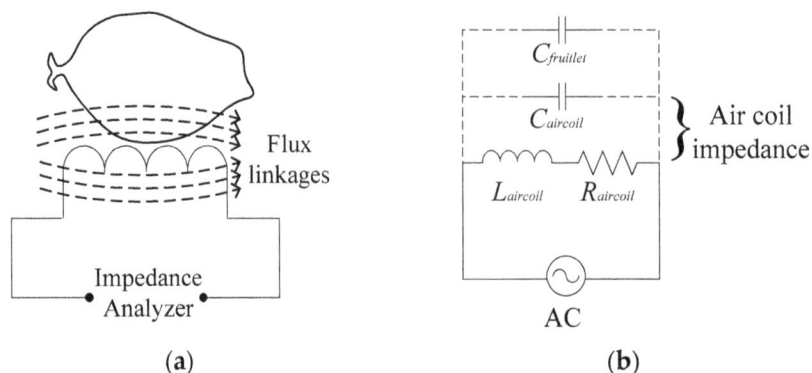

Figure 1. (a) Single flat-type shape air coil electrical diagram; (b) Series RLC circuit representation.

For each non-ideal inductor, it has a special frequency point named as a natural frequency where the impedance of the circuit is dominated by inductive properties at low frequency, while at high frequency it is dominated by capacitive properties. At the intermediate point in between two properties, the impedance transition becomes purely resistive. This is the resonance frequency of an RLC circuit, also known as a resonant circuit, in radians per second and every non-ideal inductor has it. This resonance point is called the self-resonance frequency of the inductor. The self-resonance frequency is where the inductor will start to behave like a capacitor beyond the self-resonance point. The resonant frequency is the frequency where the inductive reactance and capacitive reactance are the same and cancel one another. This response is shown in Figure 2. The reactance, up to the resonant frequency, is positive, but beyond resonance, the reactance becomes negative. From basic circuit theory, it is known that a negative reactance is associated with a capacitor. Therefore, above the self-resonance frequency (SRF), the inductor behaves like a capacitor.

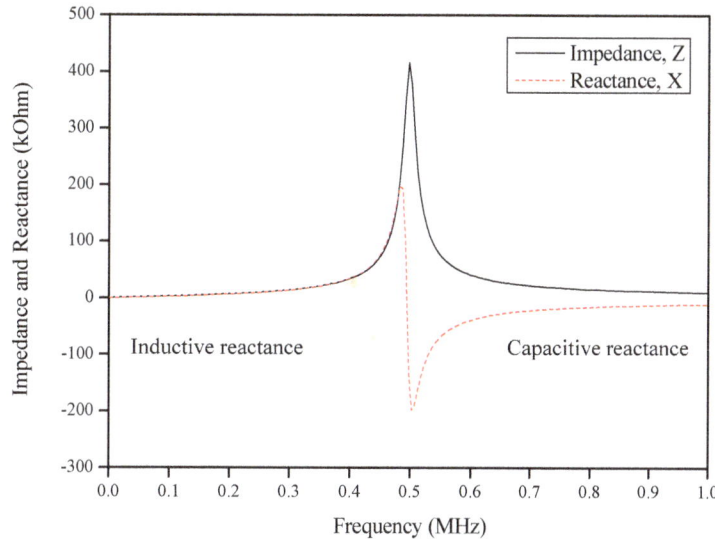

Figure 2. Example of the air core inductor impedance and reactance.

The resonant frequency of the inductive oil palm fruit sensor is assumed to have a constant by considering the effects of both the air coil structure and coil casing. Therefore, it can be derived as follows:

$$f_r = \frac{1}{2\pi\sqrt{LC}} \tag{1}$$

where f_r, L and C are the resonant frequency (Hz), inductance (H) and capacitance (F), respectively.

The values of the inductance and capacitance are based on the estimation calculation of the air coil's structure. Through this empirical analysis, the value of the capacitance is estimated based on the existence of stray capacitance in the air coil. The capacitance value of this flat-type air coil uses the simplified formula which estimates the capacitance to determine the self-resonance frequency of the inductor [7].

This proposed inductive concept of the oil palm fruit sensor involves the moisture content which is also related to the changes in the permittivity of the fruitlet. Since the oil palm fruitlet is a non-conducting material, the permeability value of the fruitlet is extremely small. It is identified that the permeability value of water is 1.2566270×10^{-6} Hm^{-1} and the permeability of air, μ_0, is $1.256637061 \times 10^{-6}$ Hm^{-1}. As for the capacitance, the permittivity value for air, ε_0, is 8.854×10^{-12} Fm^{-1}, while the relative permittivity of water and oil are 80 and 2~3, respectively [8].

2.2. Sensor's Structure

The flat-type oil palm fruit sensor is rectangular in shape as shown in Figure 3. The main parameters of the air coil consist of the height (inner height, h_{in} = 1 mm), width (inner width, w_{in} = 6 mm) and length (inner length, l_{in} = 5 mm) of the sensor. The sensor's fabrication uses a plastic called Perspex which is a non-conducting material that minimizes the flux disturbance in the sensor.

With the coil wire diameter D_o of 0.12 mm, the sample of fruitlets from the oil palm fresh fruit bunch is tested with a frequency ranging between 20 Hz to 10 MHz for a total duration of 16 weeks. The air coil's number of turns is fixed at 170 turns. The set-up and parameters shown in Table 1 are constant throughout the experiment.

Table 1. Specifications for field testing experimental setup.

Parameter/Part	Value/Type
Type of Measurement Setup	Series (L_s–R_s)
Voltage (mV)	500
Frequency (MHz)	0.02–10
Sweep (points)	200
Coil wire diameter, D_o (mm)	0.12
Number of turns, N	170

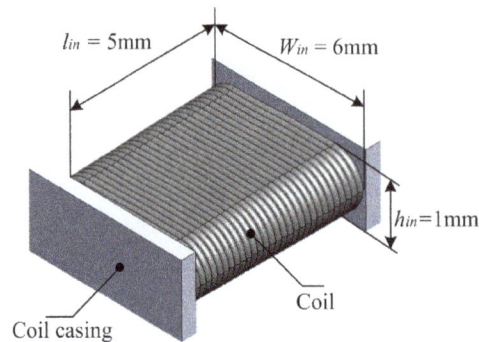

Figure 3. Flat-type air coil 3D structure.

2.3. Resonance Characteristics

Through analyzing the inductance characteristics for air, and ripe and unripe fruitlets, the graph shows different resonance frequency values with similar curves for all samples. The resonance frequency from this experiment is identified where the inductance is at its maximum value. From Figure 4, the inductance characteristics of each sample are represented by colors; that is, the black line represents air, the red line represents ripe fruit and the blue line represents unripe fruit. The sensor portrays a similar inductance characteristic curve for all samples throughout the experiments.

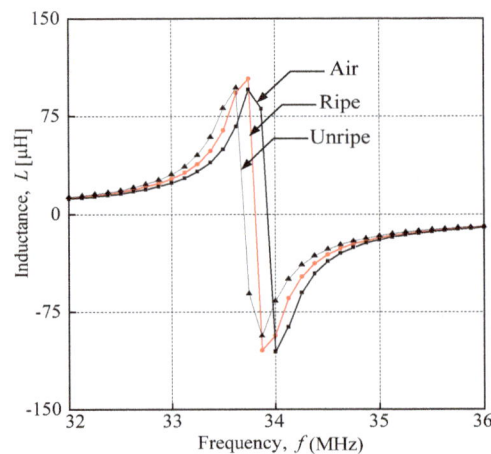

Figure 4. Oil palm fruit sensor inductance characteristics for air, ripe and unripe fruits.

2.4. Fruit Sample

Sample selection and the category are standardized throughout the research. Each sample was taken from the same bunch; each sample was measured three times to ensure the consistency of the measurement for each sample. Standard specifications outlined by MPOB are used in sample selection, such as the fruitlet surface's color and its age. The ripe fruitlet is orange whereas the unripe fruitlet sample is dark purple in color. Unripe fruitlet samples were selected around the seventh weeks after anthesis (WAA) and at 18th WAA for the ripe fruitlet [9]. Each sample was freshly taken from the same oil palm fresh fruit bunch (FFB) on the testing day. The test had to be completed on same day that samples were taken in order to avoid inconsistency and contamination. The characteristics of the samples selected for this research are summarized in Table 2. Figure 5 shows the percentage content of the ripe and unripe oil palm FFB that was tested for its content.

Table 2. Characteristics of the selected samples.

Category	Surface Color	Age (WAA)
Unripe	Dark purple	After 7
Ripe	Red orange	18–21

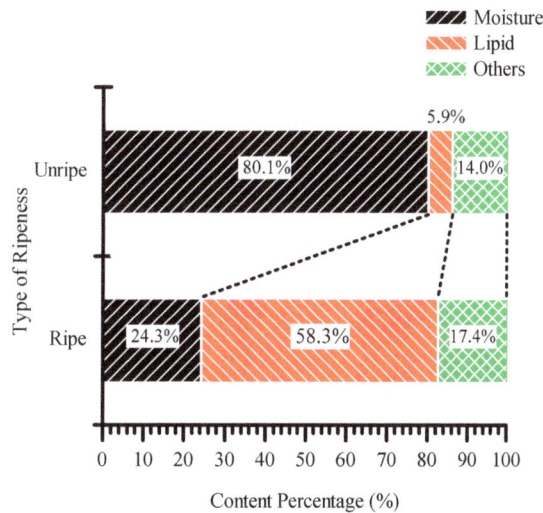

Figure 5. The percentage of content for unripe and ripe oil palm fruit.

3. Mathematical Analysis

3.1. Inductance Estimation

Based on the parameters set up in Table 1, inductance for the single flat-type air coil was deduced from the air coil's formula, proposed in Reference [10]. For a finite length of coil wire, the inductance of the flat-type air coil can be calculated using Nagaoka's coefficients as in the following equation:

$$L_{aircoil} = C_{nagaoka} \, L_0 \tag{2}$$

where $L_{aircoil}$ (H) is the inductance of the air coil, L_0 (H) is the inductance of the ideal inductor and $C_{nagaoka}$ is the Nagaoka's coefficient. The formula for the ideal coil L_0 is as follows:

$$L_0 = \frac{\mu_0 \pi N^2 (W_{in} + D_o)^2}{4 \, l_{in}} \tag{3}$$

where μ_0 is the permeability in vacuum (H/m), N is number of turns of the coil windings, W_{in} is the inner width of the air coil (m), D_o is the diameter of the coil wire (m), and l_{in} is the inner length of the air coil (m). The Nagaoka coefficient can be found by the following equation:

$$C_{nagaoka} = \frac{4}{3\pi}\frac{1}{k'}\left[\frac{k'^2}{k^2}(K-E) + E - k\right] \tag{4}$$

$$k = \sqrt{\frac{(W_{in}+D_o)^2}{(W_{in}+D_o)^2 + l_{in}^2}} \tag{5}$$

where k is the elliptic modulus, k' is the complementary elliptic module, K is the complete elliptic integral of the first type and E is the elliptic integral of the second type. The complete elliptic integral of the first type K and the second type E are calculated to estimate the value of inductance [10]. The definitions for both K and E are described by the following equations:

$$K = K(k) = \int_0^{\pi/2} \frac{1}{\sqrt{1-k^2\sin^2\theta}}\,d\theta \tag{6}$$

$$E = E(k) = \int_0^{\pi/2} \sqrt{1-k^2\sin^2\theta}\,d\theta \tag{7}$$

Since the calculation of the coefficient requires the estimation of the complete elliptic integral, an approximation is needed for practical and easy approximation calculation. Therefore, C. Hastings' approach is used for the inductance calculation for this flat-type air coil.

$$K(k) = \left(1.3862944 + 0.1119723k'^2 + 0.0725296k'^4\right)$$
$$+ \left(\frac{1}{2} + 0.1213478k'^2 + 0.0288729k'^4\right)\ln\left(\frac{1}{k'^2}\right) \tag{8}$$

$$E(k) = \left(1 + 0.4630151k'^2 + 0.1077812k'^4\right) + \left(0.2452727k'^2 + 0.0412496k'^4\right)\ln\left(\frac{1}{k'^2}\right) \tag{9}$$

3.2. Self-Capacitance Estimation

For the air coil's self-capacitance, the turn-to-turn capacitance is formed from the basic cell shown in Figure 6. It is observed that two adjacent turns of different layers and two adjacent turns of the same layer for the basic cell are identical. Thus, the winding's inner part can be divided into identical basic cells.

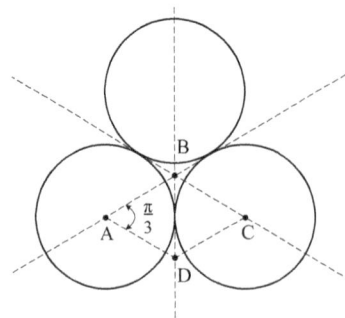

Figure 6. Representation of turn-to-turn capacitance for basic ABCD.

The total stray capacitances of a single basic cell of ABCD from Figure 6 can be calculated using the simplified formula by considering the turn-to-air gap capacitance, the turn-to-turn capacitance and the turn-to-insulation coating capacitance.

A near-accurate simplifying approach obtained from Reference [7] where $\theta = 0°$ and the plot of Equations (11) and (12) resulted in the path of the electric field in the air gap becoming zero and the corresponding basic cell elementary capacitance given by Equation (11) remaining constant. Therefore, for a small value of angle θ, the basic cell air gap's elementary capacitance is larger than the series combination of the coating's elementary capacitance.

Due to the coil's geometrical symmetry, the electric field lines need to be equally shared between the adjacent conductors. Therefore, by considering two adjacent conductors, the basic cell elementary capacitance dC between two opposite corresponding elementary surfaces of the conductor dS is given by:

$$dC = \epsilon \frac{dS}{x} \text{ (F)} \tag{10}$$

where $\epsilon = \epsilon_0 \epsilon_r$ is the permittivity of the medium with $\epsilon_r = 5.1$ for enamel, and x is the length of the electric field line connecting two elementary surfaces. Each basic cell elementary surface location can be described by an angular coordinate, θ. Hence, the basic cell elementary capacitance dC depends on the angular coordinate, θ.

The insulating coating equivalent capacitance in the basic cell C_{ttc} is given by:

$$C_{ttc} = \epsilon_0 \epsilon_r \frac{\pi W_{in} \theta}{\ln \frac{D_o}{D_c}} \text{ (F)} \tag{11}$$

From the geometrical consideration, the length of the assumed path and the elementary surface of the wire which includes the coating are in the form of the elementary ring of the length, πW_{in}. The air gap capacitance as the elementary capacitance per unit angle C_{ttg} is:

$$C_{ttg} = \epsilon_0 \pi W_{in} \left[\cot\left(\frac{\theta}{2}\right) \cot\left(\frac{\pi}{12}\right) \right] \text{ (F)} \tag{12}$$

Angle θ^* corresponds to the crossing point at which Equations (11) and (12) can be equated, which yields:

$$\theta^* = \cos^{-1}\left(1 - \frac{\ln\frac{D_o}{D_c}}{\epsilon_r}\right) \tag{13}$$

Therefore, the total capacitance of the basic cell is the parallel combination of the capacitance, C_{tt}:

$$C_{tt} = C_{ttc} + C_{ttg} = \epsilon_0 \pi W_{in} \left[\frac{\epsilon_r \theta^*}{\ln \frac{D_o}{D_c}} + \cot\left(\frac{\theta^*}{2}\right) - \cot\left(\frac{\pi}{12}\right) \right] \text{ (F)} \tag{14}$$

Assuming that the overall stray capacitance of the coil with N turns is in sequence, it converges to:

$$C_s = 1.83 C_{tt} \text{ (F) for N} \geq 10 \tag{15}$$

4. Results and Discussions

4.1. Calculated Results

Based on specification in Table 1 and Figure 2, the value of the inductance, the self-capacitance and the resonance frequency is calculated.

The inductance value is calculated with Equation (2). The value of the Nagaoka coefficient, $C_{nagaoka}$, is 0.642941286 as acquired by Equation (4) together with Equations (5)–(9). The value of the ideal coil L_0, obtained from Equation (3), is 213.7 µH. Therefore, the estimated inductance of the flat-type air coil $L_{aircoil}$ is 137.4 µH.

The self-capacitance approximation of this flat-type air coil is obtained from the total capacitance of the basic cell C_{tt} from Equation (14), which is 0.920 pF, with the angular coordinate of the intersection angle for Equations (11) and (12), which is $\theta^* = 0.4326$ rad or $24.79°$. From Equation (15) we can obtain the approximate value of the stray capacitance, $C_s = 1.68397$ nF. From both values of $L_{aircoil}$ and C_s, we can predict that the resonance frequency from Equation (1) is 10.5 MHz.

It is expected that the calculated value of the resonance frequency exceeds the experimental value. The calculation does not include the capacitance value of the fruitlet and it just considers the capacitance of the air coil only. Furthermore, it is also noted that the approximation does have an error percentage that affects the calculated resonance frequency to be higher than the actual value.

4.2. Field Testing Results and Analysis

Figure 7 shows the results from three bunches of the oil palm FFB at different levels of maturity according to the number of weeks. All graphs show a similar curve where the resonant frequency increases as the week of the testing increase. The field test was conducted simultaneously on three different fruit bunches until they fully ripened. The number of weeks it took for each bunch to reach its optimum maturity differed for each bunch, as these bunches were initially at different stages of ripeness. From Figure 7, the blue-colored symbol represents the bunch that was predicted to be at eight weeks of anthesis when the field testing started and it ended after 10 weeks as the fruitlet started to loosen from the bunch. The red-colored symbol represents the bunch predicted to be at the age of four weeks of anthesis while the black-colored symbol represents the bunch at two weeks of anthesis. All three bunches provided a similar pattern throughout the testing period and the range of the resonant frequencies for all these bunches was recorded within 6.3 MHz to 9.9 MHz.

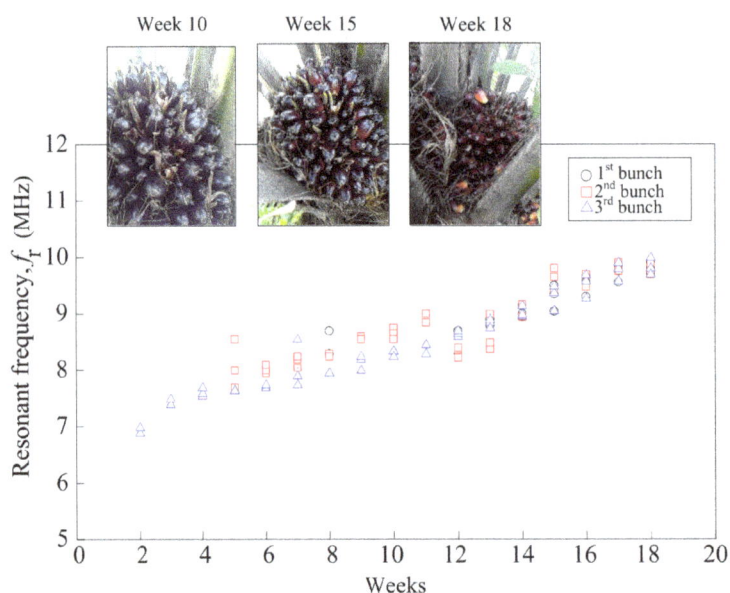

Figure 7. Oil palm fruit sensor frequency characteristics.

The resonant frequencies of these three bunches were recorded at 9.8 MHz to 9.9 MHz when the fruitlet started to loosen from the bunch and the color of the mesocarp turned to yellowish-red as described in the grading manual published by the Malaysian Palm Oil Board. Figure 7 also shows the picture of the fruit bunches at different levels of maturity corresponding to the weeks the field testing was conducted. The oil palm fruit bunches is predicted to be at 10th WAA, 15th WAA and 18th WAA, respectively. The prediction was based on the week when the fruitlet started to loosen from the bunch and was assigned as 18th WAA.

In order to find the relationship between the maturity of the oil palm fruitlets and the moisture content, we observed the value of capacitance which affects the resonance frequency from the frequency stated in Equation (1).

Based on the measured resonance frequency obtained, a graph of capacitance and moisture content against the week is plotted and the significance of the relationship is modeled. An approximation line is plotted and the relation between the moisture content is approximated between the number of weeks and the capacitance. The capacitance values are obtained with Equation (1) using $L_{aircoil} = 137.4$ μH as calculated from Equation (2) and the resonance frequency obtained through field measurement. It is observed that as the number of weeks progressed, the capacitance value obtained through the measured resonance decreased. This is due to the loss of moisture in the fruitlet and the increase in lipid content as the fruitlet in the fruit bunches ripened. Referring to the data obtained from Figure 5, the percentage of moisture content for unripe and ripe oil palm fruitlets is marked near the approximation line as shown in Figure 8.

It is deduced that the approximation equation capacitance value, C (F) for respective weeks of the curve is as follows:

$$C = -C_{tt}\ \ln(Week) + 4.8 \times 10^{-12} \tag{16}$$

where C_{tt} is 0.920 pF, Week represents the weeks after anthesis (WAA) of the FFB, followed by the experimental correction for the capacitance approximation curve. This equation is applied only for the moisture content range of 20% to 85%, as a value beyond this range is not possible.

Therefore, from the approximate moisture content of the unripe fruitlet curve, the unripe fruitlet with an 80.1% moisture content, from data in Figure 5, falls near Week 3, and the ripe fruitlet with a 24.3% moisture content falls near Week 18, as shown in Figure 8.

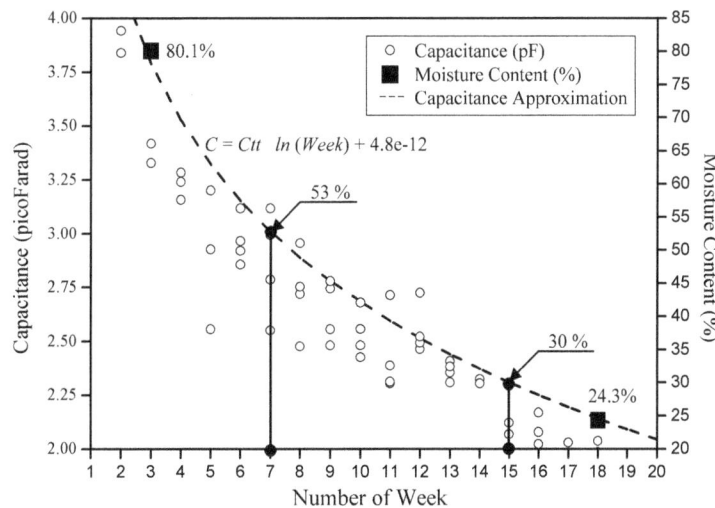

Figure 8. Graph of capacitance and moisture content of the fruitlet against week.

From the graph obtained from Figure 8, the relationship between the capacitance from the resonance and the moisture content percentage can be predicted using the equation below:

$$Percentage\ of\ Moisture\ Content = -MC_{ripe}\ \ln(Week) + 111 \tag{17}$$

where MC_{ripe} is the moisture content of the ripe fruitlet which is approximately 30% [3] and Week refers to the weeks after anthesis (WAA), followed by the experimental correction value for the moisture content approximation equation.

The maturity process of the oil palm FFB starts from the seventh WAA until the 15th WAA [1]. From Figure 8, it is approximated that at the seventh WAA, the water content of the fruitlet of the FFB is estimated at around 53%, and it further decreased at the 15th WAA to 30% which is the recommended time for harvesting [1].

The unripe sample is stated to be selected before the 12th WAA with the initial 80.1% water content, which is coinciding with the estimation curve around the third WAA, which is a plausible estimation. For the ripe sample, it is found that 24.3% moisture content is overlapping the estimation curve at the 18th WAA, which is plausible as the ripe fruitlet is estimated to be of age between the 16th WAA to the 20th WAA [1].

5. Conclusions

The flat-type oil palm maturity sensor functions as a device to detect the maturity of the oil palm fruitlet which corresponds with the increase of the measured resonance frequency. It is observed that the ready-to-harvest fruitlet on Week 18 obtains a frequency of 9.8 MHz in comparison to the frequency of 8.5 MHz at Week 10. Further analysis shows that the stray capacitance of the sensor changes as the fruit matures in relation with the moisture content of the fruit. Through the estimation curve, the moisture content of the fruitlet decreases in a negative logarithm function. The experimental correction value in the estimation curve is due to the shape, position and experimental setup for that particular measurement. Thus, there is a potential for this inductive concept to be developed as an oil palm fruit maturity sensor.

Author Contributions: This article is written and analyzed by Nor Aziana Aliteh with the data collected by Noor Hasmiza Harun; supervised by Norhisam Misron, Kunihisa Tashiro, Toshiro Sato and Hiroyuki Wakiwaka for the manuscript improvement.

Conflicts of Interest: The authors declare no conflict of interest.

References

1. Malaysian Palm Oil Board. *Oil Palm Fruit Grading Manual*, 2nd ed.; Malaysian Palm Oil Board (MPOB): Bangi, Malaysia, 2003.
2. Osama, M.B.S.; Sindhuja, S.; Abdul, R.M.S.; Helmi, Z.M.S.; Reza, E.; Meftah, S.A.; Mohd, H.M.H. Classification of oil palm fresh fruit bunches based on their maturity using portable four-band sensor system. *Comput. Electron. Agric.* **2012**, *82*, 55–60.
3. Yeow, Y.K.; Abbas, Z.; Khalid, K. Application of microwave moisture sensor for determination of oil palm fruit ripeness. *Meas. Sci. Rev.* **2010**, *10*, 7–14. [CrossRef]
4. Abbas, Z.; Yeow, Y.K.; Shaari, A.H.; Khalid, K.; Hassan, J.; Saion, E. Complex permittivity and moisture measurements of oil palm fruits using an open-ended coaxial sensor. *IEEE Sens. J.* **2005**, *5*, 1281–1287. [CrossRef]
5. Sharifudin, M.S.; Cardenas-Blanco, A.; Gao Amin, M.H.; Soon, N.G.; Laurance, D.H. Monitoring development and ripeness of oil palm fruit (*Elaeis guineensis*) by MRI and bulk NMR. *Int. J. Agric. Biolog.* **2010**, *12*, 101–105.
6. Bonnie, T.Y.P.; Mohtar, Y. Determination of palm oil residue in palm kernel and palm oil methyl ester using near infrared spectroscopy. *J. Oil Palm Res.* **2011**, *23*, 1055–1059.
7. Massarini, A.; Kazimierczuk, M.K. Self-capacitance of inductors. *IEEE Trans. Power Electron.* **1997**, *12*, 671–676. [CrossRef]
8. Jakoby, B.; Vellekoop, M.J. Physical sensors for water-in-oil emulsions. *Sens. Actuators A. Phys.* **2004**, *110*, 28–32. [CrossRef]
9. Harun, N.H.; Misron, N.; Sidek, R.M.; Aris, I.; Ahmad, D.; Wakiwaka, H.; Tashiro, K. Investigations on a novel inductive oil palm fruit sensor. *Sensors* **2014**, *14*, 2431–2448.
10. Tashiro, K.; Wakiwaka, H.; Mori, T.; Nakano, R.; Harun, N.H.; Misron, N. Experimental confirmation of cylindrical electromagnetic sensor design for liquid detection application. In *Sensing Technology: Current Status and Future Trends IV*, 1st ed.; Mason, A., Mukhopadhyay, S., Eds.; Springer International Publishing: Basel, Switzerland, 2015; Volume 12, pp. 119–137.

3

A Visual Analytics Approach for Station-Based Air Quality Data

Yi Du [1], **Cuixia Ma** [2], **Chao Wu** [3], **Xiaowei Xu** [1], **Yike Guo** [3], **Yuanchun Zhou** [1] and **Jianhui Li** [1,*]

[1]   Department of Big Data Technology and Application Development, Computer Network Information Center, Chinese Academy of Sciences, Beijing 100190, China; duyi@cnic.cn (Y.D.); xuxiaowei@cnic.cn (X.X.); zyc@cnic.cn (Y.Z.)

[2]   Intelligence Engineering Laboratory, Institute of Software, Chinese Academy of Sciences, Beijing 100190, China; cuixia@iscas.ac.cn

[3]   Department of Computing, Imperial College London, London SW7 2AZ, UK; chao.wu@imperial.ac.uk (C.W.); y.guo@imperial.ac.uk (Y.G.)

*   Correspondence: lijh@cnic.cn

Abstract: With the deployment of multi-modality and large-scale sensor networks for monitoring air quality, we are now able to collect large and multi-dimensional spatio-temporal datasets. For these sensed data, we present a comprehensive visual analysis approach for air quality analysis. This approach integrates several visual methods, such as map-based views, calendar views, and trends views, to assist the analysis. Among those visual methods, map-based visual methods are used to display the locations of interest, and the calendar and the trends views are used to discover the linear and periodical patterns. The system also provides various interaction tools to combine the map-based visualization, trends view, calendar view and multi-dimensional view. In addition, we propose a self-adaptive calendar-based controller that can flexibly adapt the changes of data size and granularity in trends view. Such a visual analytics system would facilitate big-data analysis in real applications, especially for decision making support.

Keywords: visual analytics; spatio-temporal visualization; time series visualization; multi-dimensional visualization; air pollution

1. Introduction

Air pollution is becoming a pressing issue. A recent study [1] showed that approximately 3.2 million people died from air pollution-related causes in 2010 worldwide, 2.1 million of whom were from Asia. Additionally the number of paediatric patients in China with pneumonia has increased dramatically [2]. Air quality in China has become a hotly debated issue, and people want to be more informed about it. As a result, the Chinese government has been providing the public with air quality data from across the country.

Air quality data can be collected by different means, including monitoring stations and remote sensing satellites. Now, with pervasive sensing capability and deployment of large scale sensor infrastructure, we now have the capabilities to build "big" air quality datasets. Regardless of the method used, the collected data are usually spatio-temporal, and contain the location and time at which they were recorded. The data collected by monitoring stations are considered more accurate. Although each station can detect the air quality continuously, the published data often have different time granularities. For example, some stations may publish the data on an hourly basis, whereas others only release data a certain number of times per day. The stations are also distributed at different locations to detect air quality in specific areas. Further, the collected data are multi-dimensional and include values of NO_2, SO_2, $PM_{2.5}$, PM_{10} and many other parameters.

The richness of the data collected by these stations offers opportunities for people to better understand air quality. In addition to knowing the current air quality, people can also analyze the trends, abnormalities and other interesting patterns about air quality. Researchers can conduct in-depth analysis to understand the causes and consequences of bad air quality.

However, the diversity of sensed data posts challenges for data analysis. In this paper, we propose a visual analysis system called AirVis to support the analysis of multi-dimensional spatio-temporal air quality data. With improved time series visualization methods and comprehensive interaction techniques, the system is aimed at helping people find more interesting patterns, which is very important to support decision making in real application, especially when the dataset has high volume and dimension. The main contributions of this paper are as follows:

1. AirVis, a visual analysis system, is proposed. This system can help the public and domain scientists find interesting patterns easily.
2. The new mechanism has comprehensive interactions and combines multi-dimensional visualization, spatio-temporal analysis, and multi-scale methods. This provides a general approach for big-data air pollution analysis.
3. A new adaptive development method is supported by multi-scale time series visualization and interaction.

The paper is organized as follows: Section 2 reviews related work from two different aspects, Section 3 describes the data source and the pre-processing of the dataset, and Section 4 is a system overview of AirVis. Then, we introduce the adaptive techniques of time series visualization in detail in Section 5. In Section 6, we give use cases to prove the usability of the system. Finally, we discuss some use cases and conclude the paper.

2. Related Work

2.1. Environment Related Visual Analytics

The environment is an essential facet in the development of society. It concerns many research areas [3,4], such as geography and ecology. Among this research, we can summarize some of the analysis tasks of environment-related domains: trends, abnormalities, cause, impact, and policies. A visual analytical system on these domains can help complete such tasks. Most of those systems are based on spatio-temporal datasets, and focus on different areas of the environment. EarthSystemsVisualizer (ESV) [5] and the systems proposed by Compieta [6] are two visualization systems developed to address large weather datasets. Both systems are task-based, and can help researchers complete analysis. HydroQual [7] is a system for the visual analysis of river water quality. It uses data collected by water quality stations. Compieta and HydroQual both incorporate data mining into their system, which can display mining results and spatio-temporal visualization. Vismate [8] is another visual analysis system for visualizing climate change. The system uses land surface observation data collected by meteorological observation stations. Similar to HydroQual, Vismate uses station-based data. In this system, three different visualization techniques are used to help analyze the long-term changes in climate. All of the visual analysis systems above can help in analyzing the environment. However, many of them cannot address multi-dimensional or multi-scale datasets well. Qu [9] proposed a visual analysis system for analyzing the air pollution problem in Hong Kong, which has a very similar domain to AirVis. AirVis uses a similar dataset as Qu, but the former is on a much larger scale and has different spatial and temporal granularities. It uses data collected by air quality monitoring stations. The dataset is updated hourly, and every data item contains multiple parameters.

2.2. Multi-Dimensional Spatio-Temporal Data Visualization

Because AirVis aims to address multi-dimensional and large-scale datasets, we investigate the related visual analytics techniques and interaction methods. The literature describes several spatio-temporal visualization techniques that help analyze spatio-temporal data [10–14]. Although they use different visual encodings, all these visualization techniques use a map as a basis. Qu [9] used parallel coordinates to analyze the multiple attributes of the air quality dataset so people can easily find the relationships between different attributes. Guo [15] proposed VIS-STAMP, a visualization system for space-time and multivariate patterns. Although both systems consider multi-dimensional analysis, the analyses are independent of spatial relationships. We cannot easily find the patterns behind the multi-dimensional data and spatial location.

There are also many temporal data visualization techniques used in spatio-temporal visual analytics. Aigner [16] reviewed some time series data visualization methods, and used that, some new techniques to visualize time series data [17,18]. Many of these methods do help in spatio-temporal visual analysis, however, when they are used in spatio-temporal visual analysis, they are usually used as a display control, and the interactions of the visual techniques and the interactions between spatial and temporal visualization are limited.

2.3. Multi-Scale Techniques in Spatial and Temporal Visualization

Multi-scale analysis is an important method for spatio-temporal visual analysis. It is similar to an interactive interface [19] that allows zooming, and the scalable analysis is reflected in both spatial and temporal facets. In spatial visualizations, we can analyze data by different continents [5], countries, regions [8] and stations [7]. Taking advantage of these systems, we let users select different spatial scales. In a time series visualization, as Aigner described [16], the time series data can be seen as linear time and cyclic time. The multi-scale analysis can reflect both facets. In addition to using different scales, time series data can be displayed at different levels of detail. The literature [20,21] discusses two time series visualization techniques, which use rectangular view for visualizing large time series data.

However, when time series visualization techniques are used in spatio-temporal analysis, we cannot pay attention on both points of view. Yuan [22] used cyclic rectangular view to visualize the pattern from different time of day, which is a similar techniques to [20,21]. However, the scale of time can be much more flexible, not just to analyze the cyclic pattern of different hours in days and weeks. [23,24], Landesberger [24] proposed a set of algorithms to help find the time steps, which is similar to our methods. However, the algorithm proposed here can be used in trend view, which has multiple time series charts. The methods fit Focus + Context well.

3. Data Source Materials and Methods

The method proposed in this paper is general for other air quality data from various sensor networks. As a case study, we adopted a dataset from collected at stations operated by the China National Environmental Monitoring Center (CNEMC) as a case study. CNEMC has 1437 stations across China. The number of stations of each province is shown in Table 1. CNEMC updates the air quality data every hour. The published data contains the value of SO_2, NO_2, CO, O_3, $PM_{2.5}$ and PM_{10}. However, CNEMC does not provide past data. To analyze the air quality situation, we developed a web crawler, which can grab the hourly updated data and store them in a database. We began collecting data in December 2013. In this system, we use the air quality data from 1 January 2014, to 31 December 2014. In total, we have approximately12 million data items. Based on the names of the monitoring stations, we found all of the coordinates in the stations, which can be used on map-based visualization.

As described by CNEMC, the value of an item can occasionally be "null" when hardware or network problems occur. We also had several crashes of the crawler. To analyze the dataset efficiently, we pre-processed the raw data before the analysis. First, we cleaned the raw data. We found some null or obviously wrong (zero for example) monitoring values in the raw data and we removed such item from the raw dataset. Then, we mapped each data items with the space coordinates of its monitoring station. Finally, we generate a new dataset based on the processed dataset.

Table 1. Provinces and number of stations.

Province	Number of Stations	Province	Number of Stations	Province	Number of Stations
Beijing	12	Tianjin	14	Chongqing	17
Shanghai	10	Inner Mongolia	44	Liaoning	78
Jilin	33	Heilongjiang	57	Shanxi	58
Jiangsu	72	Zhejiang	47	Anhui	68
Fujian	37	Jiangxi	60	Shandong	74
Henan	75	Hubei	51	Hunan	78
Guangdong	102	Guangxi	50	Hainan	7
Hebei	53	Sichuan	94	Guizhou	33
Yunnan	40	Tibet	18	Shanxi	50
Gansu	34	Qinghai	11	Ningxia	19
Sinkiang	41				

4. System Overview

We integrate three types of visualization techniques in this system. A map-based metaphor is used to visualize the distribution of stations and overall air quality situation. Trend view is used to visualize the trend of air quality items. Finally, calendar view is used to visualize the detail of different day and time circularly. When analyzing multi-dimensional data, we use multiple map-based visualizations and calendar views with one trend view. In each dimension, the color mappings of the map-based view and calendar view are the same.

4.1. Map-Based Metaphors

We use Google Maps as our base map (Figure 1). Inside the map, there are two types of map-based metaphors for visualizing data. First, the points on the map indicate the location of stations. The colors of the points denote the average value during the year. Second, a heat map is designed to visualize the overall air quality of an area. Based on those two visualization techniques, we add interactions to support multi-scale-ey find.

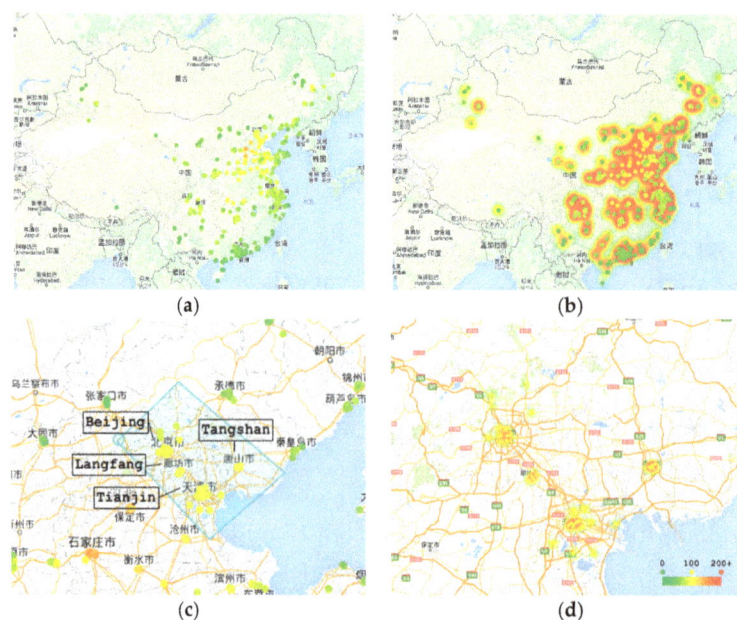

(a)   (b)

(c)   (d)

Figure 1. Map based views. Color from green to red means the air quality from good to bad. (**a**) Location of all the stations; (**b**) Overall situation of air quality in China; (**c**) Polygon selection tools used on the map. Beijing, Tianjin and several cities of Hebei Province are selected; (**d**) Detail of selected area.

4.2. Trends View

Line chart is a classic visualization method to display trends of different times. However, when the size of time varying data increases, effectively displaying all of the data in one line chart becomes a challenge. We developed a trend view based on the Focus + Context approach. The trend view is organized in two connected parts: On the top, there is an overall line chart that is used to visualize all data at a raw granularity; on the bottom, a detailed line chart is displayed. The overall line chart supports a brush to select a time period range, and the detailed line chart shows the detail of this period. In contrast to the traditional Focus + Context line charts, we select the granularity of both the top and the bottom charts automatically, which is useful for the dataset with different time granularities. The algorithm of the selection method is shown in Section 5. Figure 2 shows trend view at different time granularity.

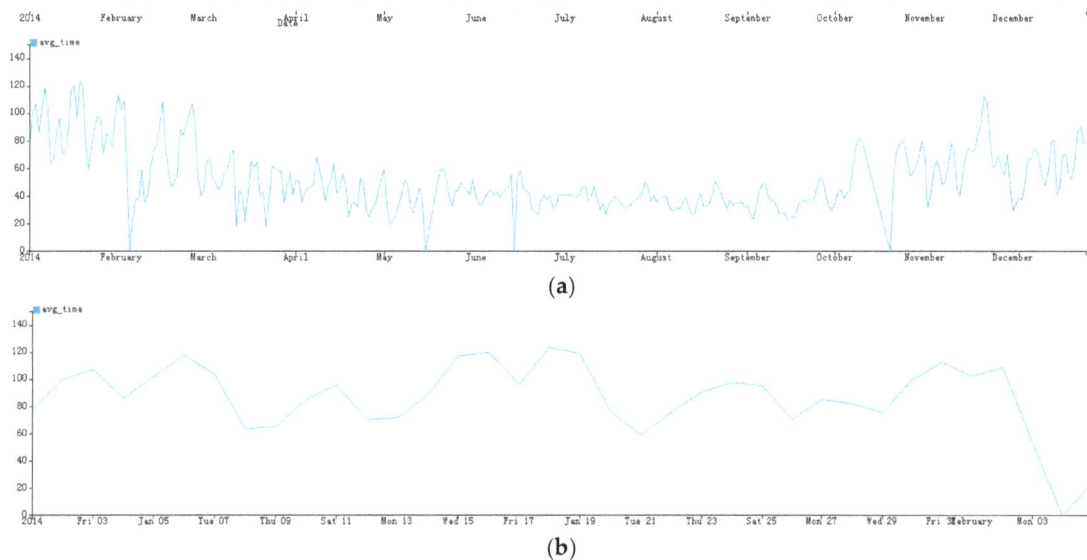

Figure 2. Trend view in AirVis. Plot (**a**) shows the initial trends view where the top and the bottom line chart have the same granularity; Plot (**b**) shows the focus view of the bottom line chart, which of it has the scale of hour, instead of day in the top line chart.

4.3. Calendar View

A calendar view is often used to visualize periodical time series data. However, the periodical time series patterns must also reflect in hours, minutes or even seconds, which cannot be visualized in traditional calendar views.

Therefore, we implemented a new calendar view that integrates the traditional calendar view and rectangular view to help finding multi-scale periodical timer series patterns. In this new view, a calendar is displayed to show the daily recorded air quality. With this calendar, we can find the average air condition of a given month, week and day. As shown in the top figure in Figure 3, there are 12 large blocks, each of which represents a month. When the time scale becomes small, the granularity of the calendar changes. A periodical rectangular view is used to discover the patterns of different days and hours. When the scale changes, the granularity of the time changes as well. As shown in the bottom figure in Figure 3, each line indicates a day of the selected time extent. The text on the left of each line shows the date, and the color shows whether the day is a weekday or weekend.

(a)

(b)

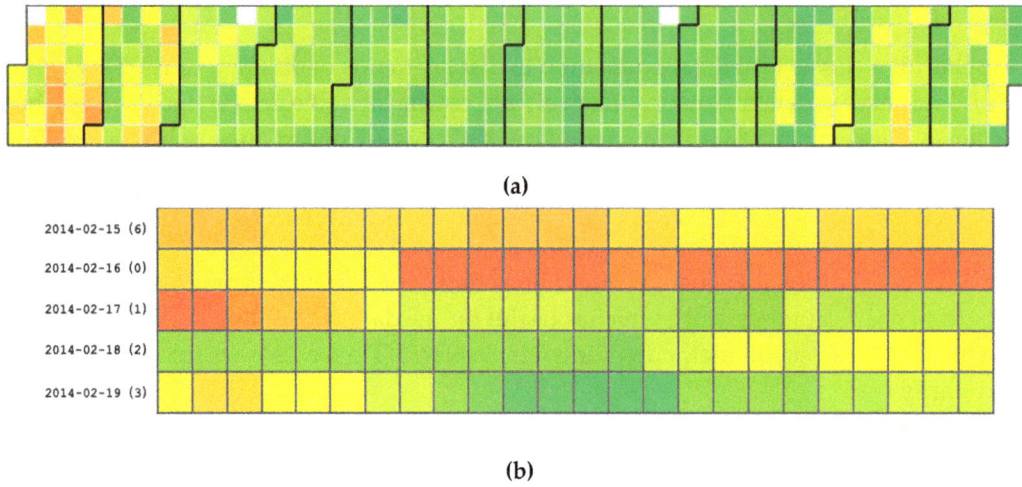

Figure 3. Calendar view in AirVis. Color from green to red means the air quality from good to bad. (**a**) Daily value of $PM_{2.5}$ in several stations of China; (**b**) Hourly value of $PM_{2.5}$ recorded by the same stations.

4.4. Multi-Dimensional View

To support the visual analysis of multiple air quality parameters, we developed a multi-dimensional view. We did not choose ordinary multi-dimension visualization techniques such as scatter plots and parallel coordinates for several reasons. First, those techniques are independent of spatial relationships, so it is difficult find the patterns behind the multi-dimensional data and spatial locations. Second, there are at most six attributes except for the spatial and temporal attributes, and the amount of data is not sufficiently large to justify using high-dimensional visualization techniques. When analyzing using a multi-dimensional view, each parameter of the data uses a separate map-based chart and calendar view. Trend view uses multiple lines to display different attributes. As shown in Figure 4, the difference and relationship among the three parameters in both the spatial and temporal dimensions can be observed.

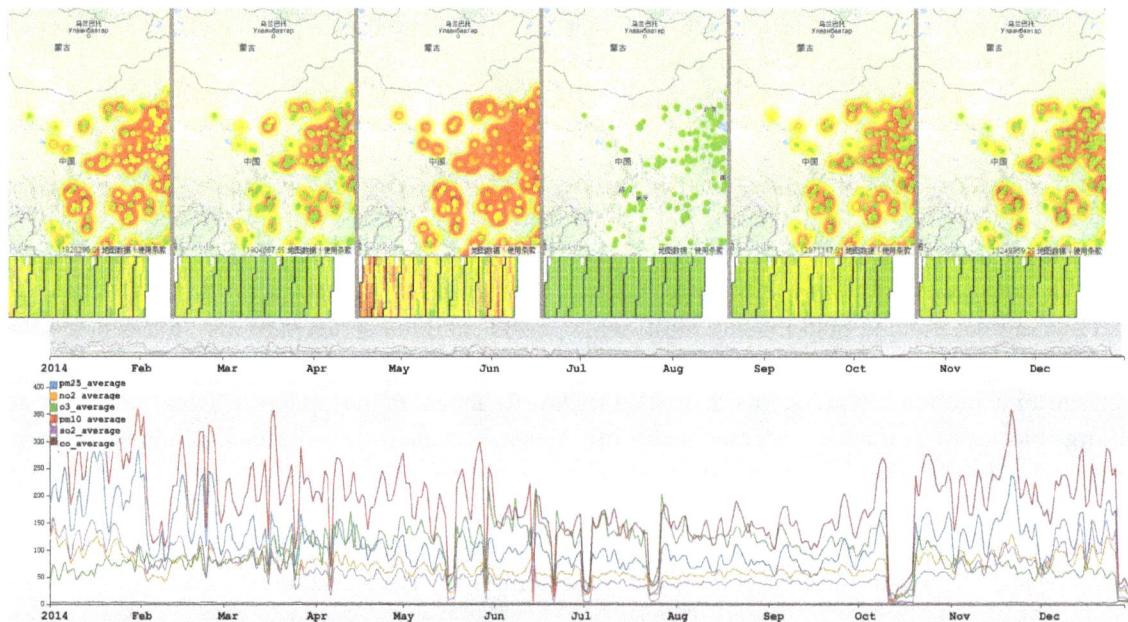

Figure 4. Multi-dimensional view in AirVis.

4.5. Interactions

Interaction is very important when analyzing multi-dimensional data. AirVis incorporates some interaction tools to facilitate an analysis. First, a polygon selection tool is provided. As shown in Figure 1, this tool is used to select stations on the map, and users can find station patterns of interest. They can select stations by province, city, or even terrain. A brush tool on the context line chart is also available. When brushing on the context line chart, the adaptive algorithms find the best data of the time interval, and choose the best visualization techniques. Another tool is tooltips on plots. When analyzing the relationship and differences, users can click plots on the map. After clicking, a detailed calendar view appears to aid the analysis. Figure 5 shows two different stations in the calendar view of a specified time period.

Figure 5. Tooltips on plots. Chinese words in the figure are POIs (point of interests) of a region in Beijing.

5. Adaptive Multi-Scale Trend View

Time series data analysis requires two types of tasks: linear time analysis and periodical time analysis. In addition, there is a relationship between the granularities of time series data. For example, if the scale of the data is one year, users can analyze the linear pattern of seasons, months and days by different requirements. They can also analyze the cyclic patterns of a fixed time frame of a given day in a year, the cyclic patterns between weekdays and weekends, and so on. If the granularity of the dataset is much smaller, we can analyze patterns of different minutes, seconds, and even milliseconds, which is useful for analyzing different time series dataset.

Motivated by the date and time structures in programming languages, we propose a novel design guideline of time series data. This guideline can address linear and cyclic time analysis. First, we define the granularity of time in Table 2. We also define nine time levels as year, season, month, week, day, hour, minute, second, and millisecond, as shown in Figure 6. For each granularity of time, there is a level where it belongs. For example, in level "Day", we have Sd, E, D, F, W and d, which are all used to define "Day", but have different granularities than the top level.

Table 2. 9 Levels and Their Descriptions.

Level	Granularity	Description
Year	y	Year
Season	Se	Season in year
Month	M	Month in year
Week	W	Week in month
	w	Week in year
	D	Day in year
	Sd	Day in season
Day	d	Day in month
	F	Day of week in month
	E	Day in week
Hour	H	Hour in day
Minute	m	Minute in hour
Second	s	Second in minute
Millisecond	S	Millisecond in second

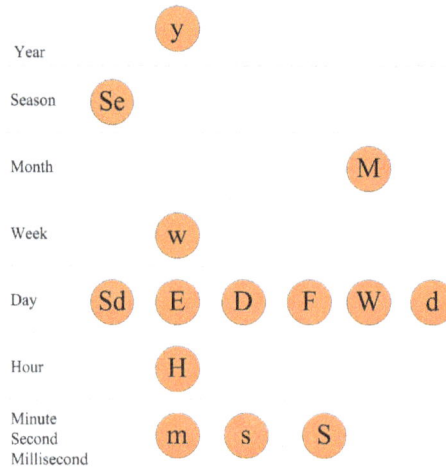

Figure 6. Level and granularity.

5.1. Linear Analysis Determination

We showed the granularity and level definition in Figure 6. For linear visual analysis, we should follow the nine levels as the granularity of the time series data. We define the scale between each level as an array S. The value of S is (1000, 60, 60, 24, 7, 4, 3, 4). Before designing the time series visualization, we define the following variables:

(1)   Count the overall number of data items C.
(2)   Define the minimum granularity of the time series data g_{min}, which belongs to one of the nine levels.
(3)   Determine the approximate display resolution of the screen R.

A trend view similar to that we propose for AirVis is very common in many visualization systems. The Focus + Context visualization approach is integrated with two line charts. Using Algorithm 1, we calculate the property granularity of the overview visualization. As described, the algorithm calculates the proper level of the result chart. The algorithm traverses from the minimum granularity of the time series data to the year, which is the maximum granularity of the data. During this period, the algorithm compares the overall number of data items in the visited level with the resolution of the display to find the property granularity of the overview view visualization. After running Algorithm 1, g_{max} will return. Then, we can calculate the moment when the scale of the detail line chart changes, as shown below.

$$\frac{\sqrt{R}}{\prod_{i=g_{min}}^{i<g_{max}} s[i]} \tag{1}$$

In the algorithm, A is a constant used to determine the threshold of the display.

Algorithm 1

---

for $i \leftarrow level(g_{min})$ to $level(y)$ do
if$((C \leftarrow (C/S[i]))/\sqrt{R}) < A$
return$i$ as g_{max}
end if
end for
return $level(y)$ as g_{max}

We apply the algorithm to the air quality dataset. As described above, the minimum level of the dataset is "H". Assume that we want to display the trends chart with two line charts on a

900×500 space, and apply Algorithm 1 to A. After running the algorithm, the result maximum level of the dataset is day. Thus, we can display the trend chart of each day. Additionally, the threshold of switching between the normal Focus + Context trend chart and the zoom-enabled one is 29, which means that if the extent of the brush is less than 29 days, the hour dataset will display on the focus view. As shown in Figure 7, when the brush extent is smaller than the threshold, the hourly data will display on the detail line chart.

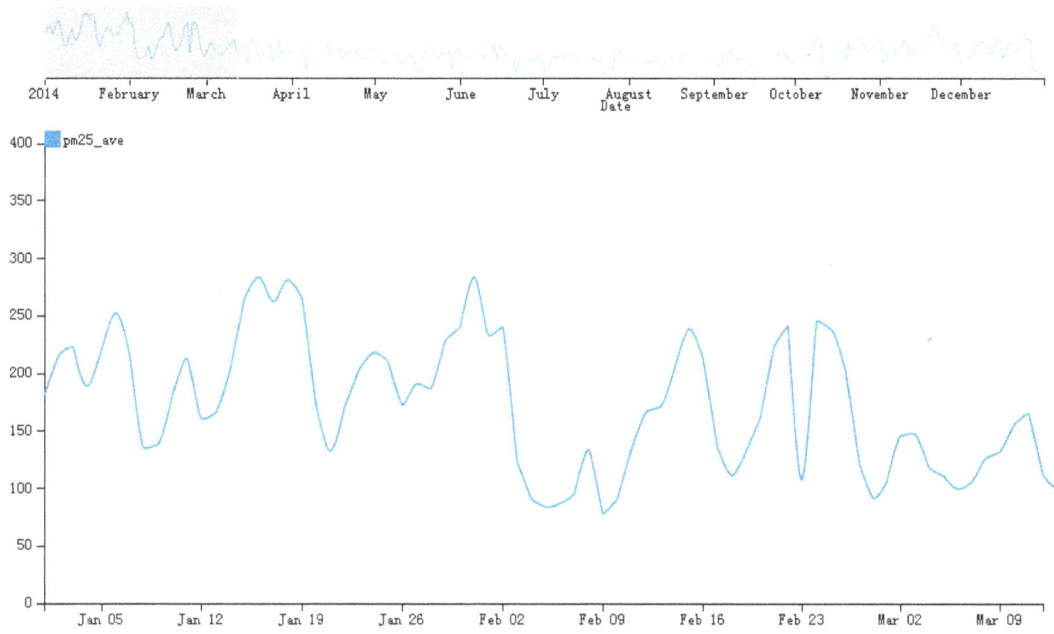

(a)

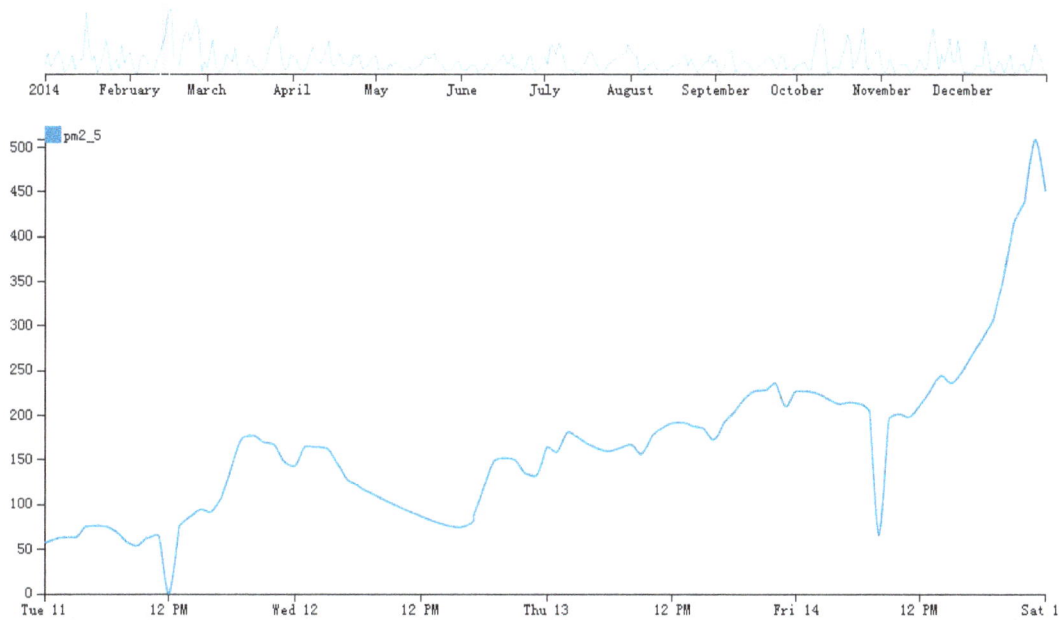

(b)

Figure 7. Trend view before and after running Algorithm 1. (**a**) Traditional Focus + Context view of trend view; (**b**) Focus + Context view using Algorithm 1.

The above algorithm and determination method are suitable when the time series visualization is integrated with two line charts with different granularities. We then extend the algorithm to a more common scenario, in which the number of line chart is k, and we use a recursive method to choose the best integration. First, we define the "best integration" as choosing the best integration from all the available levels. In this algorithm, we use the minimum variance of the scale of all selected levels as the "best integration". The inputs of the algorithm are the minimum and maximum granularities calculated by Algorithm 1, and the number of line charts. In one recursion, the algorithm estimates whether the number of charts is equal to k. If so, the variance is calculated to determine whether it is the best integration. If not, a loop from the level of small granularity to level of large granularity is executed, in which the recursion method is invoked and the scale of each selection is calculated. The algorithm is shown as Algorithm 2.

Algorithm 2

define *min* as a large number
end define
define $cal(g_{small}, g_{big}, n)$
if n equals k
cal the variance of r
if $r < min$
$min \leftarrow r$
end if
end if
if $n < k$
for i $\leftarrow level(g_{small})$ to $level(g_{big})$
$r[n] \leftarrow \prod_{i=g_{small}}^{i<g_{big}} S[i]$
$nLevel[n] \leftarrow i$
$cal(i, g_{big}, n)$
end if
end define
call cal(level(g_{min}, g_{max}, 0))
return nLevel

We give several examples of the algorithm's usage. Table 3 gives several examples of the nLevel of different groups of number of charts and granularity boundary.

Table 3. 9 Levels and Their Descriptions.

g_{min}	g_{max}	Number of Charts	nLevel
m	M	3	[m, H, M]
s	Y	3	[s, h, Y]
s	Y	4	[s, m, d, Y]
s	Y	5	[s, m, h, w, Y]

5.2. Cyclic Analysis Determination

Based on the linear method, we give the cyclic analysis determination. We add arrows to Figure 6 when the relationship of each level is known. For example, we add an arrow from "y" to "D", which means that we can analyze the cyclic pattern of the same day. Then, a directed acyclic graph is constructed, as shown in Figure 8a. An arrow indicates that the cycle is allowed, such as when the minimum granularity of the time series data is hour, and the scale of the data is larger than one year. A sample of the cyclic patterns we can analyze is as follows:

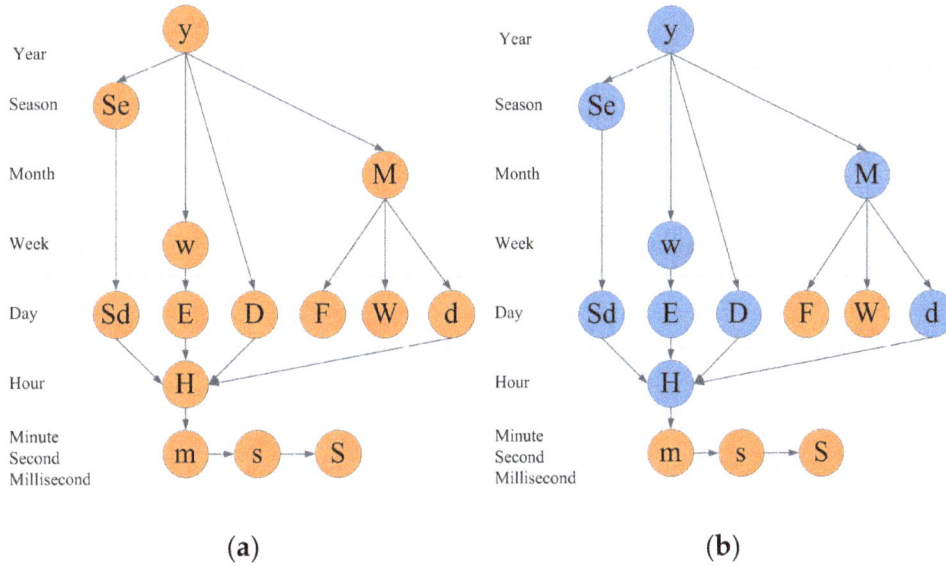

Figure 8. Cyclic analysis graph. (**a**) Level and granularity with analysis relationships; (**b**) All the cyclic patterns in the Level and granularity graph.

"First, we can analyze the cyclic pattern of weeks in several years. Then, we can also analyze the cyclic pattern of days in several weeks. Finally, we can analyze the cyclic pattern of hours in several days".

Similar patterns of the first step are months in several years, seasons in several years, days in several years, while similar patterns of the second step are days in several months, days in several seasons, hours in several days. Figure 8b is all the cyclic patterns we can analyze. The blue circles of the graph construct all the cyclic patterns.

If we take the air quality dataset, for example, the scale of the data is one year, so the cyclic patterns that we can analyze are shown in Figure 9a. The patterns that we support in AirVis are shown in Figure 9b.

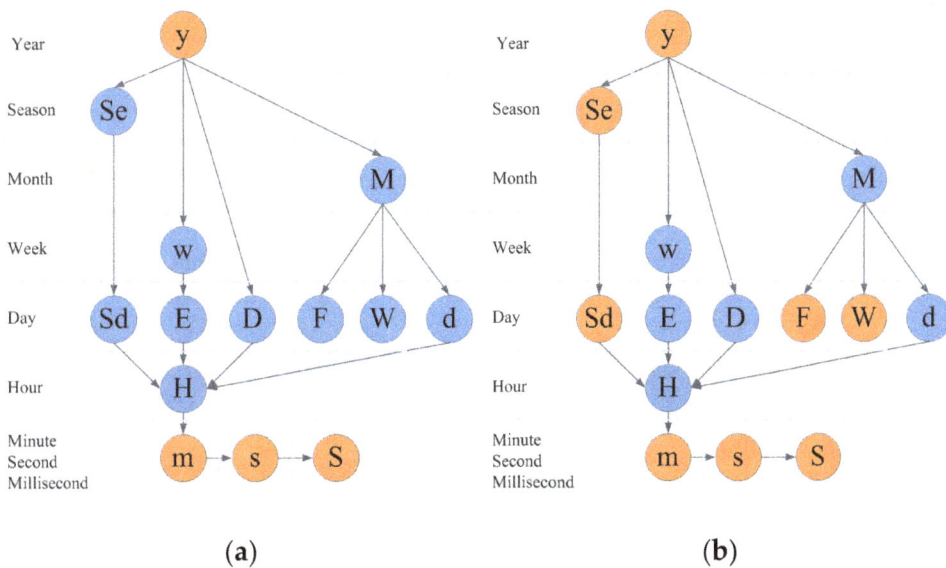

Figure 9. Cyclic analysis graph in AirVis. (**a**) All supported cyclic patterns of the air quality dataset; (**b**) All supported cyclic patterns in AirVis.

6. Case Study

6.1. PM$_{2.5}$ Analysis

First, we analyze one of the most important indices of air quality, PM$_{2.5}$. As shown in Figure 10, the interface of the system includes a map-based view, a calendar view and a trend view. Using this system, we can find some general information:

(1) The overall distribution of the stations. We can see that Eastern China has more stations than Western China and that most of the stations are placed in large and medium-sized metropoli, such as Beijing.

(2) The initial visualization shows the overall situation of PM$_{2.5}$ in 2014. From this, we find that northern China has a higher average value of PM$_{2.5}$ than other areas of China.

(3) From the calendar view, we find that spring and winter have notably higher concentration of PM$_{2.5}$ than summer and autumn. From the line chart, we find some interesting patterns in addition to the seasonal differences. We find that although the overall trends exist, the values of neighboring days change significantly.

(4) AirVis supports the flexible selection of areas and stations. By analyzing the daily trend of different areas in China, shown in Figure 11, we find that the daily value of PM$_{2.5}$ of Beijing is higher than in the Shandong Province and the Yangtze River Delta. The situation in the Yangtze River Delta during spring time is slightly better than that in the Shandong Province.

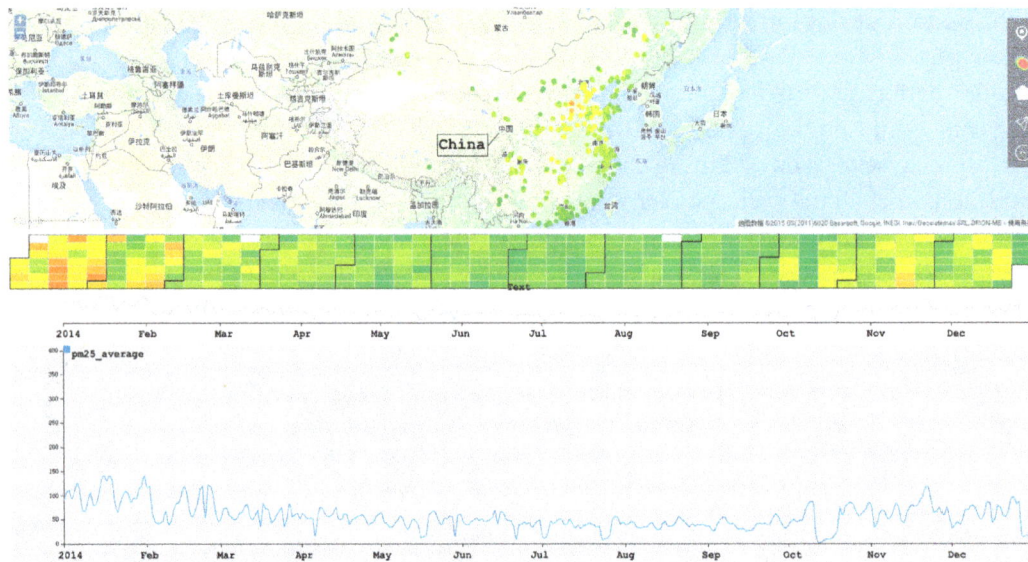

Figure 10. Daily trends of different area in China. Top: Daily trends of Beijing, Tianjin and Hebei province. Middle: Daily trends of Shandong Province. Bottom: Daily trends of Yangtze River Delta.

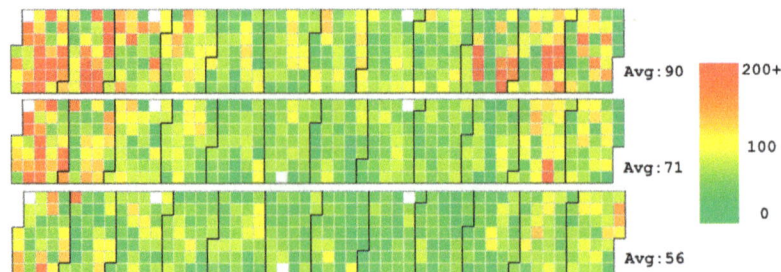

Figure 11. The overall view of AirVis. Chinese words in the figure are country and city names in this regions, which will not affect the understanding of the figure.

Then, we focus observation to Beijing. There are 12 stations in Beijing. By selecting all 12 stations on the map, we can analyze the air quality in Beijing. We also find some interesting patterns:

(1)   There are similar trends between Beijing and the rest of China. Stations record higher concentration of $PM_{2.5}$ during spring and winter than during summer and autumn. However, Beijing has much higher averages.

(2)   There is a very interesting patterns of the value of $PM_{2.5}$ from 13 February to 28 February. As we can see in Figure 12, the value of $PM_{2.5}$ changes from a high value to a low value. After three days of low value (17th–19th), the value rises again (20th–26th). After the 26th, the value reduces to a low value again. To our knowledge, the reason for the rise and fall of $PM_{2.5}$ value because of wind in the city. However, the value reduces following the 17th, and there is no wind during the period. The detail of the hourly value of those days can be seen in Figure 13, and we can see that the change of the value is a gradual process.

Figure 12. The daily trends of $PM_{2.5}$ in Beijing. The selected area shows one of the interesting patterns we found.

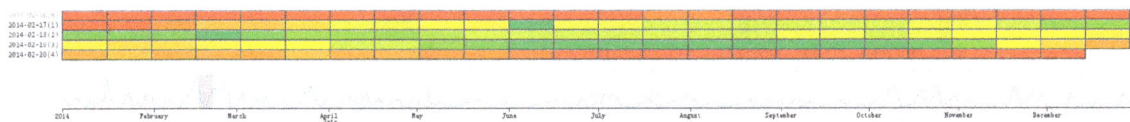

Figure 13. The hourly trends of $PM_{2.5}$ in Beijing from 16 February to 20 February. This shows the detail change of the interesting pattern.

6.2. Multi-Dimensional Analysis

Similar to the $PM_{2.5}$ analysis process, we first provide a general overview of the parameters in the air quality dataset. As shown in Figure 14, the overall trend of the selected six attributes are similar, especially for the value of $PM_{2.5}$ and PM_{10}. When analyzing the six attributes in Beijing, we found the average daily trends of $PM_{2.5}$ and PM_{10} are similar. As described in Section 5.1, we found that the value of $PM_{2.5}$ decreased starting on 17 February, which was very strange. As shown in the multi-dimensional view, we found that the value of PM_{10} also had a similar decreasing trend. By comparing the values of $PM_{2.5}$ and PM_{10} using AirVis, we found that there are two significant high values of PM_{10} on 24 February to 25 February, and on 17 March to 18 March. However, the value of $PM_{2.5}$ from 24 February to 25 February was high, whereas that from 17 March to 18 March was low.

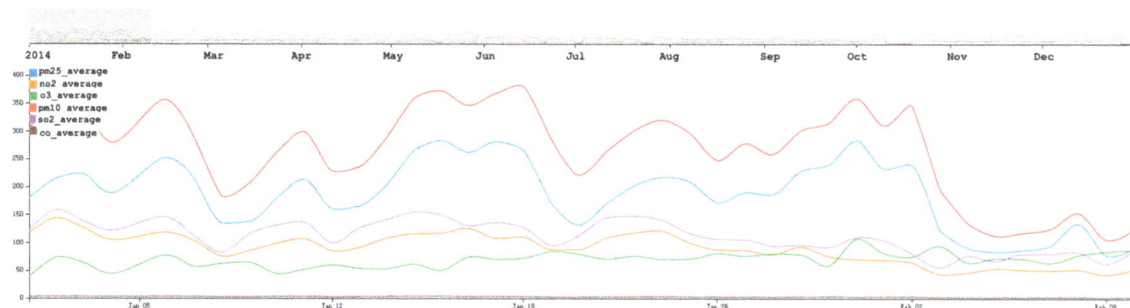

Figure 14. Daily trends of six attributes in China. As shown in the figure, the blue line represents the value of $PM_{2.5}$, the orange line represents the value of NO_2, the green line represents the value of O_3, the red line represents the value of PM_{10}, the purple line represents the value for SO_2, and the brown line represents for the value for CO.

6.3. Domain Experts Feedback

Our system was assessed by two domain experts of Chinese Academy of Sciences. Their expertise included in areas of regional air pollution, indoor and urban air pollutants. During the consultation, we first discussed the domain requirements. Then we provided our system to them and allowed them to explore the air quality data using our system. We collected the feedback on both the usability of the system and the explanation of the result.

For the system itself, both the two researchers gave positive feedback about the exploration process of our system. In their traditional way of doing research on air quality, they have to use different tools to help them generate different visualizations. One of a time consuming tasks is to convert data from one tool to another. They found our system to be particularly helpful as it integrates multiple visualizations that can generate visualizations at different time ranges and spatial regions in real time. One expert mentioned that the linked view is very helpful for exploring the data, especially for the link between the calendar view and the trend view. The other expert was interested in the zoomable map with heat map, which provided a flexible interaction. One researcher was particularly eager to explore his own data using our system. However, our system did not support the capability to interactively ingest new data sources. We leave this as future work. The researchers also gave some suggestions on optimizing the system. They said there were many clutters when they analyzed in multi-dimensional view. They commented that the overall patterns and anomalies found using the system were helpful clues for future analysis. They also suggested integrating other data sources into our system (e.g., traffic, atmospheric variables) to further explore potential patterns of air pollution.

7. Discussion

After finding the patterns using AirVis, we attempted to analyze their causes. Some of the changes in air quality are obvious, such as the seasonal changes of $PM_{2.5}$ concentration. Some findings further prove the theories that we previously established, such as the functions of the wind on the air quality.

The $PM_{2.5}$ values in Northern China are very high. The reasons are complex, but the industrial structure and terrain are important possible explanations. When digging into one specific area, we found that the value of $PM_{2.5}$ has increased gradually since 20 February. This is mostly due to the working of the city. When there is not sufficient wind to blow $PM_{2.5}$ out of the city, and the city itself generates $PM_{2.5}$ gradually, this causes an increase of $PM_{2.5}$.

We also observed that the recorded values of $PM_{2.5}$ from 17 to 18 March were low, whereas the multi-dimensional data analysis shows that the value of PM_{10} during those two days was very high. By looking at the weather during those days, we found that there was a strong north wind during those days. The wind blew $PM_{2.5}$ away and brought in PM_{10} from the northern part of China.

Similarly, we found that the value of $PM_{2.5}$ started to decrease on 17 February. However, when analyzing the multi-dimensional view, we found that the value of PM_{10} was also low, which is different from the days in the previous case. The weather of those days was very calm. The experts were also confused by this pattern. They suggested several possible explanations: (1) the source of $PM_{2.5}$ decreased in the city (e.g., some events caused the number of the moving cars to decrease); (2) the data proposed by CNEMC were wrong; or (3) some other undiscovered factor affecting $PM_{2.5}$ caused the situation.

By analyzing the data multi-dimensionally, we found that the values of $PM_{2.5}$, PM_{10} and other air quality attributes have some relationships. Wind occasionally caused an increase of PM_{10} and a decrease of $PM_{2.5}$, but at other times, the value of $PM_{2.5}$, PM_{10} and other air quality attributes were positively correlated.

8. Conclusions and Future Work

In brief, this paper present AirVis, a visual analytical system for air quality analysis. In the system, three different views are integrated. A map-based view is used to analyze the spatial distribution of stations and the situation of different areas. A calendar view gives users insight into the cyclic situation of air quality. This view is designed as an Overview + Detail calendar view. When interesting patterns are discovered in the map-based or trend views, we can dig into the data to find the cyclic trends for every hour per day. A trend view can display the quality trend by two line charts. Similar to the calendar view, trend view is also connected by two relevant parts. These two parts are designed with a Focus + Context approach.

Motivated by programming language, we improve the trend view by providing nine levels to describe time. There are several granularities in each and based on the methods and the graphs generated by the levels, we give design guidelines to help design visualizations to indicate linear and cyclic trends of time-varying datasets. In addition, we also give algorithms to determine the granularities and the moment when scale should change. However, although we proposed algorithms and provided some use cases and the results, we did not note when the description of level changes in scale and granularity. In our discussion, we believe that only the definition of the level and the scale array change can solve the problem. The algorithms proposed need not change. In future works, we will find additional datasets to prove the hypothesis.

Using AirVis, we gave two use cases: an analysis of the $PM_{2.5}$ situation in China and a study of six attributes of air quality data. During the analysis, we found some interesting patterns that were not easy to find otherwise. These findings can help scientists analyze changes in air quality. When using AirVis to analyze multi-dimensional air quality datasets, we use separate map-based views and calendar views, instead of using parallel coordinate or other multi-dimension data visualization methods. The reason is that the maximum number of dimensions of the air quality dataset is six. However, evaluation on those two methods is also necessary in future work. In Section 7, we discussed some causes for the patterns we found using AirVis. Among those causes, we found that air quality is correlated to natural factors such as wind patterns and temperature. Additionally, in some research on the cause and impact of air pollution, many other types of datasets such as land usage, economic development and car ownership data, are used. However, AirVis does not support visual analysis of these causes. In future work, we will extend AirVis' support for visual analysis of correlation between air quality data and other datasets.

As mentioned by the domain experts, the exploration process of the system could do great help to find patterns. The analysis results are also very interesting, and could be a good supplement to current research on air quality. However, the cause and influence of air pollution is related to many factors [25–27], such as wind, temperature, land use, emission of pollutants. In order to increase the flexibility, we will support the capability to interactively ingest new data sources. We will also try to include additional data sources to further explore potential causal relationships of air quality (e.g., traffic) in the future work.

Acknowledgments: This work was supported by the National Key Research Program of China under Grant No. 2016YFB1000600 and No. 2016YFB0501900, Natural Science Foundation of China under Grant No. 61402435 and No. U1435220, Knowledge Innovation Program of Chinese Academy of Sciences under Grant No. CNIC_QN_1507.

Author Contributions: Yi Du and Cuixia Ma developed the whole system; Yi Du also wrote the draft of the paper. Chao Wu wrote and revised the introduction, abstract and related works; Yuanchun Zhou and Jianhui Li wrote and revised the discussion and conclusion. Xiaowei Xu and Yike Guo revised the writing of this paper.

Conflicts of Interest: The authors declare no conflict of interest.

References

1. Lim, S.S.; Vos, T.; Flaxman, A.D.; Danaei, G.; Shibuya, K.; Adair-Rohani, H.; Amann, M.; Anderson, H.R.; Andrews, K.G.; Aryee, M.; et al. A comparative risk assessment of burden of disease and injury attributable to 67 risk factors and risk factor clusters in 21 regions, 1990–2010: A systematic analysis for the Global Burden of Disease Study 2010. *Lancet* **2013**, *380*, 2224–2260. [CrossRef]
2. Xu, P.; Chen, Y.; Ye, X. Haze, air pollution, and health in China. *Lancet* **2013**, *382*, 2067. [CrossRef]
3. Zhang, J.; Mauzerall, D.L.; Zhu, T.; Liang, S.; Ezzati, M.; Remais, J.V. Environmental health in China: Progress towards clean air and safe water. *Lancet* **2010**, *375*, 1110–1119. [CrossRef]
4. Hoek, G.; Beelen, R.; de Hoogh, K.; Vienneau, D.; Gulliver, J.; Fischer, P.; Briggs, D. A review of land-use regression models to assess spatial variation of outdoor air pollution. *Atmos. Environ.* **2008**, *42*, 7561–7578. [CrossRef]
5. Harrower, M.; MacEachren, A.; Griffin, A.L. Developing a geographic visualization tool to support earth science learning. *Cartogr. Geogr. Inf. Sci.* **2000**, *27*, 279–293. [CrossRef]
6. Compieta, P.; Di Martino, S.; Bertolotto, M.; Ferrucci, F.; Kechadi, T. Exploratory spatio-temporal data mining and visualization. *J. Vis. Lang. Comput.* **2007**, *18*, 255–279. [CrossRef]
7. Accorsi, P.; Fabrègue, M.; Sallaberry, A.; Cernesson, F.; Lalande, N.; Braud, A.; Fabrègue, M.; Cernesson, F.; Braud, A.; Teisseire, M. HydroQual: Visual Analysis of River Water Quality. In Proceedings of the IEEE Symposium on Visual Analytics Science and Technology, Paris, France, 25–31 October 2014; pp. 123–132.
8. Li, J.; Zhang, K.; Meng, Z.P. Vismate: Interactive Visual Analysis of Station-Based Observation Data on Climate Changes. In Proceedings of the IEEE Symposium on Visual Analytics Science and Technology, Paris, France, 25–31 October 2014; pp. 133–142.
9. Qu, H.; Chan, W.Y.; Xu, A.; Chung, K.L.; Lau, K.H.; Guo, P. Visual analysis of the air pollution problem in Hong Kong. *IEEE Trans. Vis. Comput. Graph.* **2007**, *13*, 1408–1415. [CrossRef] [PubMed]
10. Wood, J.; Dykes, J.; Slingsby, A.; Clarke, K. Interactive Visual Exploration of a Large Spatio-temporal Dataset: Reflections on a Geovisualization Mashup. *IEEE Trans. Vis. Comput. Graph.* **2007**, *13*, 1176–1183. [CrossRef] [PubMed]
11. Tominski, C.; Schulze-Wollgast, P.; Schumann, H. 3D information visualization for time dependent data on maps, in Information Visualisation. In Proceedings of the Ninth International Conference on Information Visualisation (IV'05), Washington, DC, USA, 6–8 July 2005; pp. 175–181.
12. Peuquet, D.J.; Kraak, M.-J. Geobrowsing: Creative thinking and knowledge discovery using geographic visualization. *Inf. Vis.* **2002**, *1*, 80–91. [CrossRef]
13. Tominski, C.; Schumann, H.; Andrienko, G.; Andrienko, N. Stacking-Based Visualization of Trajectory Attribute Data. *IEEE Trans. Vis. Comput. Graph.* **2012**, *18*, 2565–2574. [CrossRef] [PubMed]
14. Scheepens, R.; Willems, N.; Van de Wetering, H.; Andrienko, G.; Andrienko, N.; van Wijk, J.J. Composite Density Maps for Multivariate Trajectories. *IEEE Trans. Vis. Comput. Graph.* **2011**, *17*, 2518–2527. [CrossRef] [PubMed]
15. Diansheng, G.; Jin, C.; MacEachren, A.M.; Liao, K. A Visualization System for Space-Time and Multivariate Patterns (VIS-STAMP). *IEEE Trans. Vis. Comput. Graph.* **2006**, *12*, 1461–1474. [CrossRef] [PubMed]
16. Aigner, W.; Miksch, S.; Muller, W.; Schumann, H.; Tominski, C. Visual methods for analyzing time-oriented data. *IEEE Trans. Vis. Comput. Graph.* **2008**, *14*, 47–60. [CrossRef] [PubMed]
17. Krstajic, M.; Bertini, E.; Keim, D.A. CloudLines: Compact Display of Event Episodes in Multiple Time-Series. *IEEE Trans. Vis. Comput. Graph.* **2011**, *17*, 2432–2439. [CrossRef] [PubMed]

18. Ziegler, H.; Jenny, M.; Gruse, T.; Keim, D.A. Visual market sector analysis for financial time series data. In Proceedings of the IEEE Symposium on Visual Analytics Science and Technology, Salt Lake City, UT, USA, 24–29 October 2010; pp. 83–90.

19. Hornbæk, K.; Bederson, B.B.; Plaisant, C. Navigation patterns and usability of zoomable user interfaces with and without an overview. *ACM Trans. Comput. Hum. Interact.* **2002**, *9*, 362–389. [CrossRef]

20. Hao, M.C.; Dayal, U.; Keim, D.A.; Schreck, T. Multi-resolution techniques for visual exploration of large time-series data. In Proceedings of the EuroVis, Norrköping, Sweden, 23–25 May 2007; pp. 27–34.

21. Hao, M.C.; Dayal, U.; Keim, D.A.; Schreck, T. Importance-driven visualization layouts for large time series data. In Proceedings of the IEEE Symposium on Information Visualization, Minneapolis, MN, USA, 23–25 October 2005; pp. 203–210.

22. Wang, Z.C.; Ye, T.Z.; Lu, M.; Yuan, X.R.; Qu, H.M.; Yuan, J.; Wu, Q.L. Visual Exploration of Sparse Traffic Trajectory Data. *IEEE Trans. Vis. Comput. Graph.* **2014**, *20*, 1813–1821. [CrossRef] [PubMed]

23. Nocke, T.; Schumann, H.; Bohm, U.; Flechsig, M. Information visualization supporting modelling and evaluation tasks for climate models. In Proceedings of the 2003 Winter Simulafion Conference, New Orleans, LA, USA, 7–10 December 2003; pp. 763–771.

24. Von Landesberger, T.; Bremm, S.; Andrienko, N.; Andrienko, G.; Tekusova, M. Visual analytics methods for categoric spatio-temporal data. In Proceedings of the IEEE Symposium on Visual Analytics Science and Technology, Seattle, WA, USA, 14–19 October 2012; pp. 183–192.

25. Sun, Y.; Zhuang, G.; Tang, A.; Wang, Y.; An, Z. Chemical characteristics of $PM_{2.5}$ and PM_{10} in haze-fog episodes in Beijing. *Environ. Sci. Technol.* **2006**, *40*, 3148–3155. [CrossRef] [PubMed]

26. Zhou, X.; Cao, Z.; Ma, Y.; Wang, L.; Wu, R.; Wang, W. Concentrations, correlations and chemical species of $PM_{2.5}/PM_{10}$ based on published data in china: Potential implications for the revised particulate standard. *Chemosphere* **2016**, *144*, 518–526. [CrossRef] [PubMed]

27. Lai, S.; Zhao, Y.; Ding, A.; Zhang, Y.; Song, T.; Zheng, J.; Ho, K.F.; Lee, S.; Zhong, L. Characterization of pm 2.5 and the major chemical components during a 1-year campaign in rural Guangzhou, southern China. *Atmos. Res.* **2016**, *167*, 208–215. [CrossRef]

4

Feasibility Test of a Liquid Film Thickness Sensor on a Flexible Printed Circuit Board Using a Three-Electrode Conductance Method

Kyu Byung Lee [1,2], Jong Rok Kim [3], Goon Cherl Park [1] and Hyoung Kyu Cho [1,*]

[1] Department of Nuclear Engineering, Seoul National University, Seoul 08826, Korea;
     kblee@kins.re.kr (K.B.L.); parkgc@snu.ac.kr (G.C.P.)

[2] Department of Nuclear Safety, Korea Institute of Nuclear Safety, Daejeon 34142, Korea

[3] Thermal Hydraulics Safety Research Division, Korea Atomic Energy Research Institute,
     Daejeon 34057, Korea; jongrok@kaeri.re.kr

* Correspondence: chohk@snu.ac.kr

Academic Editors: Remco J. Wiegerink and Luis J. Fernandez

Abstract: Liquid film thickness measurements under temperature-varying conditions in a two-phase flow are of great importance to refining our understanding of two-phase flows. In order to overcome the limitations of the conventional electrical means of measuring the thickness of a liquid film, this study proposes a three-electrode conductance method, with the device fabricated on a flexible printed circuit board (FPCB). The three-electrode conductance method offers the advantage of applicability under conditions with varying temperatures in principle, while the FPCB has the advantage of usability on curved surfaces and in relatively high-temperature conditions in comparison with sensors based on a printed circuit board (PCB). Two types of prototype sensors were fabricated on an FPCB and the feasibility of both was confirmed in a calibration test conducted at different temperatures. With the calibrated sensor, liquid film thickness measurements were conducted via a falling liquid film flow experiment, and the working performance was tested.

Keywords: liquid film thickness; liquid film sensor; three electrode conductance method; two-phase flow experiment

1. Introduction

Understanding the characteristics of liquid films in gas-liquid two-phase flows is an important part of the safety and performance analyses of nuclear power plant systems. For this reason, numerous studies have been conducted in an effort to measure local liquid film thicknesses under two-phase flow conditions. There are various methods that can be used to measure film thicknesses. Several widely applied methods are the ultrasonic, optical, neutron and electrical methods. The ultrasonic method measures the liquid film thickness using the time difference between signals reflected from the boundary interface of the medium. As the velocity of sound is affected by its medium, however, the ultrasonic method has rather low precision [1]. Therefore, this method is limited when used to measure the thicknesses of thin films. Furthermore, multiple-point measurements with the ultrasonic method are confined due to the high cost of the device [1]. Optical methods require a high-speed camera, diffraction tools, and X-ray tomography. Though most optical methods have high spatial and time resolutions, they are limited when applied to complicated flow conditions, as the light is distracted at the boundary interface [2]. Neutron-based methods usually measure the film thickness with a tomography-based approach. Though the spatial resolution of this method is sufficiently high, only time-averaged data is provided. This method also incurs high costs when setting up the measurement device. The electrical method is widely used in two-phase experiments given that it can measure not only the liquid film thickness but the void fraction. Most electrical methods use the electric conductivity of the liquid to measure the liquid film thickness. For non-conductive liquids, the capacitance method is applied to measure the film thickness. Generally, electrical methods for liquid film measurements have high time resolutions due to their electrical characteristics. However, the electrode size and geometry cause relatively low spatial resolutions in multiple-point measurements, representing a limitation of the electrical methods.

Damsohn et al. [3] proposed an electrical method on a printed circuit board (PCB) in order to measure local liquid film thicknesses and to overcome the above-mentioned limitation of the electrical method regarding the spatial resolution. A PCB, which allows the elaborate fabrication of electrodes, enabled high spatial resolutions of the liquid film thickness field to be realized. Furthermore, by coupling the sensor with a wire-mesh circuitry system, high time resolutions are achieved simultaneously. The wire-mesh sensor principle is based on a matrix-like arrangement of the measuring points. Two sets of wire electrodes are stretched perpendicularly to each other, with a small axial separation distance between them. The transmitter electrodes are sequentially activated while all receiver electrodes are parallel-sampled in such a way that the electrical conductivity of the fluid at each crossing point can be evaluated. Based on these measurements, the sensor is thus able to determine the instantaneous fluid distribution across the cross-section [3].

In other work, the electrode used in the electrical method also was fabricated on a flexible printed circuit board (FPCB) for application onto a curved surface. Arai et al. [4] and D'Aleo et al. [5] also used this method to measure the liquid film thickness in an annulus channel and in a microchannel, respectively. Coupling a PCB or FPCB with wire-mesh circuitry provides high time and space resolutions simultaneously, but the experimental condition is restricted to an isothermal condition, as the electrical conductivity of the liquid is affected by its temperature. According to Hayashi [6], the conductivity of water increases by approximately 2% when the water temperature increases by 1 °C.

The objective of this study is to develop the concept of a liquid film sensor fabricated on an FPCB, which works accurately under temperature-varying flow conditions. The motivation behind this arose from an experiment by Yang et al. [7] on the two-dimensional liquid film behavior. They measured reductions in the liquid film thickness after impingement on a rectangular duct with a transverse air flow using the ultrasonic method; the experimental data were then used to improve the wall and interfacial friction factor models of the nuclear reactor safety analysis code. However, under actual flow conditions, the working fluids are saturated steam and subcooled water. Therefore, condensation occurs at the interaction between the two phases. The condensation increases the water temperature and changes the electrical conductivity of the fluid. As a result, the conventional electrical method cannot be applied under this flow condition. Furthermore, the plate duct in their experiment simulated the unfolding downcomer annulus of a nuclear reactor pressure vessel. However, the geometry of interest in practical situations is the annulus channel rather than the plate duct. Therefore, a liquid film sensor which can be used with a curved geometry is required in order to extend the findings of Yang et al. to more realistic flow conditions. The fabrication of a liquid film sensor on an FPCB was selected for this reason.

For liquid film measurements under varying conductivity conditions, a three-electrode method was proposed initially by Coney [8] using parallel rectangular electrodes. According to him, if the receiving electrode is segmented into two parts perpendicularly to the current direction, the liquid film thickness can be measured with the current ratio acquired by the two segmented receiving electrodes. This works because the current flowing to the farther and closer parts of the segment will be proportional to the conductivity, and their ratio will depend only upon the film thickness. Kim et al. [9] conducted experimental research on the three-electrode conductance method for liquid film measurements under varying temperature conditions. They fabricated a sensor on a PCB and assessed the temperature compensation capability of the sensor.

In the present study, the three-electrode conductance method for liquid film thickness measurements by Kim et al. [9] was extended to apply the sensor to an FPCB and increase the degree of integration of the measurement points. To do this, two types of liquid film sensors were fabricated on an FPCB, one with the design from Kim et al. [9] and the other with a new design of electrode arrangement for denser integration of the measurement points. Prototype sensors were fabricated on an FPCB after sensitivity studies of the sensor performance with the electrode geometry via an electrical potential analysis. The prototype sensors were calibrated at different temperatures and the feasibility of temperature compensation was investigated. The calibrated sensor was applied in a falling liquid film experiment in varying temperature conditions, and its working performance was tested. This paper presents the procedure of the design of the sensor, the calibration results, and the feasibility test results and discusses required future improvements.

2. Design and Fabrication of the Prototype Sensor

2.1. Fabrication of the Sensor on an FPCB

The three-electrode method uses the current ratio (I_1/I_2) to measure the film thickness, as presented in Figure 1, rather than the absolute value of the current. Coney [8] proposed the concept of the three-electrode method with conformal transformation as follows. In this theory, the current ratio can be determined by Equation (1).

$$\frac{I_1}{I_2} = \frac{F(\beta, m_1)}{F(\pi/2, m_1) - F(\beta, m_1)}, \tag{1}$$

where F(β,m_1) is an incomplete elliptic integral of the first kind defined by Equation (2),

$$F(\beta, m_1) = \int_0^\beta \left(1 - m_1 sin^2\alpha\right)^{-0.5} d\alpha, \tag{2}$$

where β and m_1 are factors defined by Equations (3) and (4):

$$sin^2\beta = \frac{sinh\frac{\pi}{2h}(\lambda_2 - 1)sinh\frac{\pi}{2h}(\lambda_1 - \lambda_s)}{sinh\frac{\pi}{2h}(\lambda_1 - 1)sinh\frac{\pi}{2h}(\lambda_1 + \lambda_s)}, \tag{3}$$

$$m_1 = \frac{sinh\frac{\pi}{2h}(\lambda_2 - 1)sinh\frac{\pi}{2h}(\lambda_1 - 1)}{sinh\frac{\pi}{2h}(\lambda_2 + 1)sinh\frac{\pi}{2h}(\lambda_1 + 1)}, \tag{4}$$

where π is a pi (3.14159 ...), λ_s and h are the dimensionless distance and dimensionless film thickness, respectively, and both are normalized by a. Following the equations above, the current ratio is independent of temperature changes [8]. However, the equations above were derived from simplified geometry. Therefore, electrical potential analysis was conducted for precise analysis in the present study. Because the current ratio can compensate for conductivity or temperature changes, the measurement error is expected to be minimized under a temperature-varying condition. The theoretical background of the three-ring method assumes that the impedance of the liquid film has some resistance and negligible reactance. Thus, the impedance of the liquid film is a function of the temperature and the thickness. Given that the electric conductivity is a function of the temperature, an impedance ratio of A:B and A:C only depends on the film thickness if the liquid temperature in the A–B region is identical to that in the A–C region.

Based on the design proposed by Kim et al. [9], the electrodes were fabricated on an FPCB in the present study, as shown in Figure 2a. The specific dimensions of the electrodes are described in Figure 2b. The transmitter electrode is located on one side and is rectangular in shape, and the receiver-1 and receiver-2 electrodes are placed on the other side with a 0.1 mm gap between them. A ground electrode exists on both the top and bottom of the receivers to prevent an end effect, leaving a 0.1 mm gap between the receivers and the ground electrode in each case. The end effect refers to a bypass current generated around the edge of the electrode. In order to confirm the geometry effect, two electrodes of different sizes (a = 1.0 mm, λ = 3, a = 1.0 mm, and λ = 5) were manufactured.

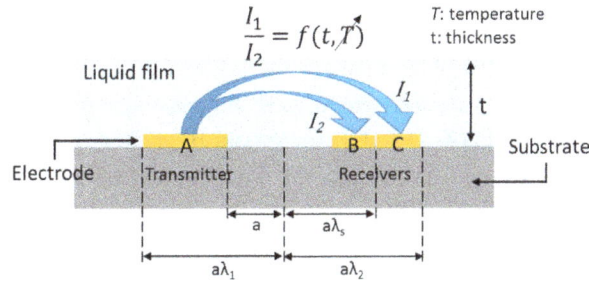

Figure 1. Principle of the three-electrode conductance method [9].

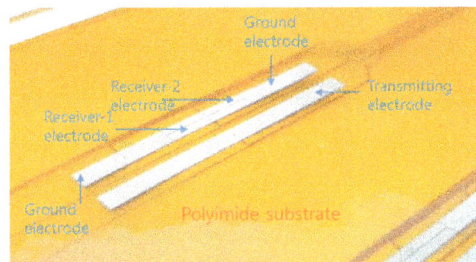

(a)

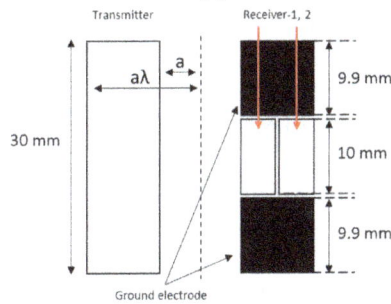

(b)

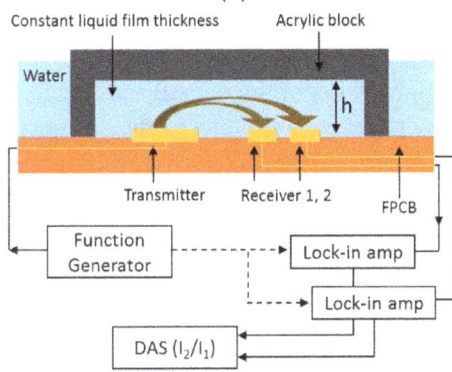

(c)

Figure 2. (**a**) configuration of the electrodes; (**b**) specific dimensions of the electrodes; and (**c**) calibration experiment and circuitry setup.

As shown in Figure 2c, calibration was carried out using insulated acrylic blocks, which can create different liquid film thicknesses (h = 0.3, 0.4, 0.5, 0.8, 1.4, 2.5, 3.8, and 5.1 mm). The liquid film thickness is held constant because the acrylic block confines the liquid film thickness structurally. An AC voltage signal of 0.1 V is induced from the function generator to the transmitter at a frequency of 1 kHz, and the current signals generated by the receivers are transmitted to a DAS (NI-9205, 16-bit) through a lock-in amp. Figure 3 presents the acquired experimental data with various film thicknesses of 0.5 mm–6 mm and water temperatures of 30 °C–60 °C. The increase in the current ratio with the liquid film thickness is shown in Figure 3a. When the liquid film thickness was 2.6 mm, the absolute values of the current outputs from the two receiving electrodes and their ratio with various water temperatures were compared, as shown in Figure 3b. The absolute values of the current output increase with the water temperature as the water conductivity increases. As the water temperature was increased from 30 °C to 60 °C, the output signals of I_1 and I_2 increased by 47% and 45%, respectively. However, the current ratio remained at a constant value of 0.757 within $\pm1.0\%$ of error. This calibration result clearly demonstrates the advantages of the three-electrode method and confirms the feasibility of fabricating the sensor on an FPCB for liquid film thickness measurements.

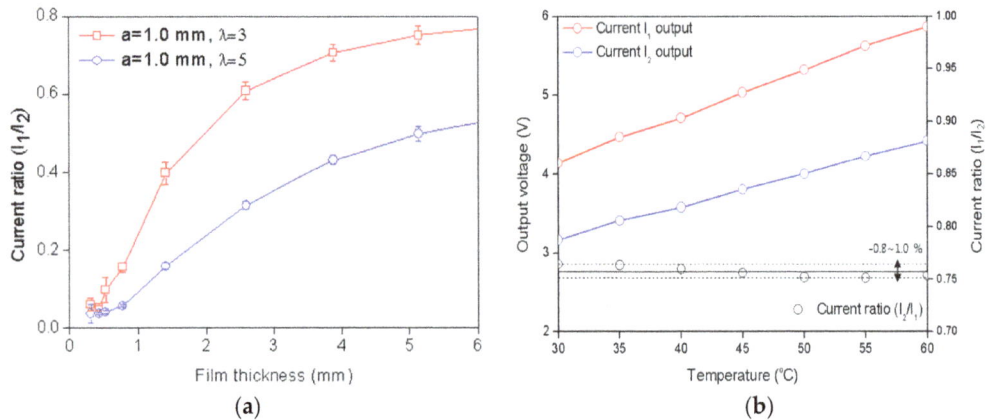

(a)   (b)

Figure 3. (**a**) current ratio with changes in the film thickness and electrode geometry; and (**b**) current output signal and current ratio at varying temperatures.

2.2. Design of a Prototypic Sensor for Higher Integration of the Measurement Points

The electrode design by Kim et al. [9] in the three-electrode method presents a limitation when fabricating an integrated electrode configuration to achieve a high spatial resolution with several measurement points owing to the crosstalk effect. If the sensor elements are too close to each other, the transmitting signal can reach the neighboring sensor elements' receiving electrodes, lowering the measurement accuracy when this occurs. This motivated the design of a new type of electrode configuration that can achieve a higher integration degree and a high spatial resolution. In the present study, a three-electrode sensor with a ring-type electrode arrangement is proposed, as shown in Figure 4, indicating a part (3 × 3 array) of the whole sensor.

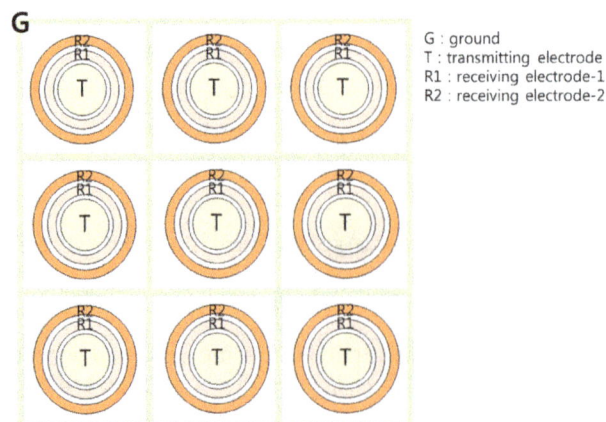

Figure 4. Disposition of multiple probes.

The transmitter electrode is located in the center of a concentric circle, and receiver-1 and receiver-2 are positioned and enclosed within the transmitter. Given the enclosed geometry, the ring-type arrangement can prevent the end effect. The ground electrode is positioned between each sensor element to minimize the crosstalk effect. The electric current flow at each sensor element has a radial symmetric pattern such that it is easy to arrange multiple sensors, as presented in Figure 4. Thus, this configuration is advantageous for achieving higher spatial resolutions in liquid film thickness field measurements as compared to previous sensor configurations.

In order to design and optimize the geometry of the newly proposed sensor electrodes, an electrical potential field simulation was conducted using the COMSOL Multiphysics code [10]. The electrical potential field developed in a liquid film can be analyzed by solving the Maxwell equation. In order to confirm the relationship between the electrode geometry and the current ratio, an electrical potential field simulation was conducted while changing the radius of the electrode. The target measurement thickness was 0.5–3.0 mm, which is in the range used in the experiment by Yang et al. [7].

Figure 5a shows the calculation domain of the COMSOL simulation. A plane water film with electric conductivity covered the ring-type electrodes, and the other boundary of the bottom plane was set to simulate an insulated plane. Electric potential of 1V was applied to the transmitter electrodes, with the receivers set to the ground potential. The electrical potential field was developed, as presented in Figure 5b, and the electric flux diffused from the center transmitter and converged to the two receivers. The predicted value of the current ratio depending on the liquid film thickness is indicated in Figure 6 with various radii of the outmost ring (R_{r1}). The current ratio increases with the liquid film thickness and becomes saturated to a certain value as the liquid film becomes thicker. According to the calculation, a larger outmost diameter of the sensor element produces better characteristics when measuring the target liquid film thickness. Contrary to this, a large sensor element reduces the degree of integration of the sensor array. For this reason, a radius of 4.5 mm and a pitch of 15 mm were selected for the present sensor. Afterwards, a series of calculations were conducted to determine the specific dimensions of the electrodes. According to numerical simulations, increasing the outer ring width increases the slope of the current ratio along with the liquid film thickness, but the signal becomes saturated at thinner thicknesses if the width is too large. In contrast, increasing the inner ring width increases the saturation thickness but makes the current ratio signal insensitive. Thus, the prototype sensor design was determined considering these two effects and restrictions during the manufacturing step. Finally, the dimensions of the ring-type electrodes were determined, as presented in Figure 7.

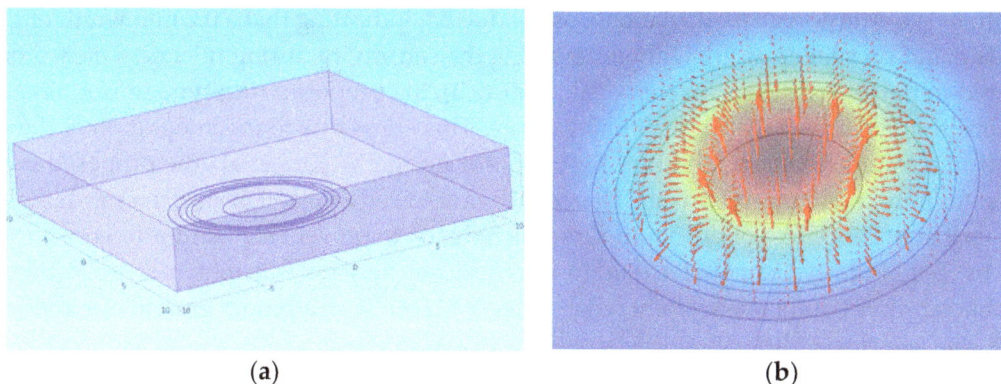

(a)   (b)

Figure 5. (a) Calculation domain of the electrical potential analysis; and (b) electrical potential field around the electrodes.

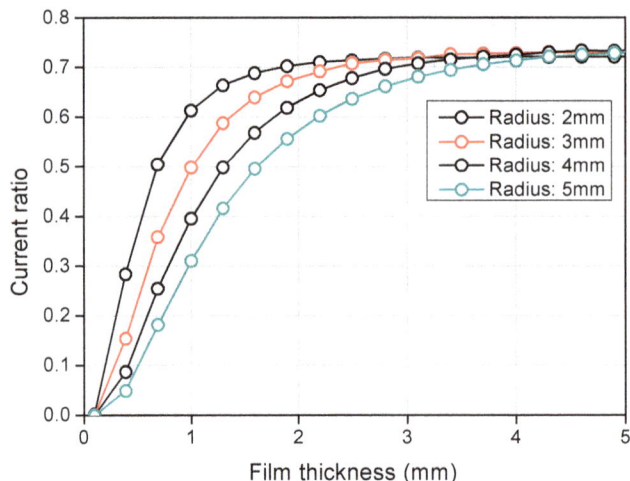

Figure 6. Electrical potential field calculation result according to the radius size.

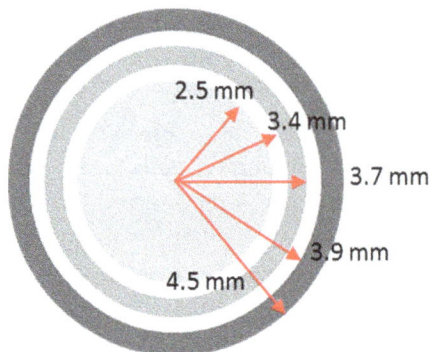

Figure 7. Specific dimensions of the electrodes.

As crosstalk distorts the signal of the sensor, a ground electrode is required to absorb the leakage current. In this study, the sensor elements were uniformly distributed with a 15 mm pitch, and a ground electrode was added between the elements, as shown in Figure 4. The ground electrode had a lattice shape with a narrow width. In order to determine the width of the ground electrode, an additional electric potential calculation was conducted, indicating that a 0.2 mm width of ground electrode could prevent crosstalk without lowering the sensitivity during thickness measurements. Compared to the sensor design by Kim et al. (Figure 3), the proposed ring-type sensor has slightly lower sensitivity and a smaller measurement range. This arises due to the integration of the sensor elements in a confined region. Contrary to the sensor design by Kim et al., the proposed sensor is designed for multi-point measurements. Due to the reduced sensor element size, the signal sensitivity with regard to the thickness of the liquid film was reduced to prevent signal interference (crosstalk) caused by neighboring elements. Although electrical interference was unavoidable in this study, reasonable sensitivity and measurement ranges were achieved with a proper ground electrode design.

2.3. Fabrication of a Prototypic Sensor and Data Acquisition Setup

Based on the electrode design as determined by electrical potential calculations, a prototype sensor was fabricated on an FPCB with a 24 × 12 array of the sensing element, as shown in Figure 8. The dimension of the measurement part is 360 mm × 180 mm (height and width, respectively). A cross-sectional view of the prototype sensor is presented in Figure 9. The substrate material of the FPCB is polyimide and the thickness of each layer is approximately 13 μm, with electrode and circuit wire consisting of copper with a thickness of 18 μm. The electrode has an initial layer of nickel on the copper surface, followed by a thin protective layer of gold. The total thickness of the FPCB sensor is approximately 330 μm, and there are grooves (30 μm) between the electrodes and polyimide film. As the FPCB is composed of four layers, each group of signal lines was separated to prevent a short circuit.

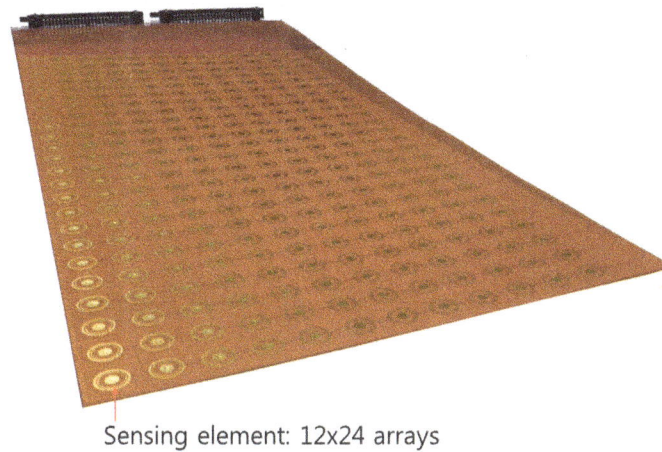

Sensing element: 12x24 arrays

Figure 8. Configuration of the prototype sensor.

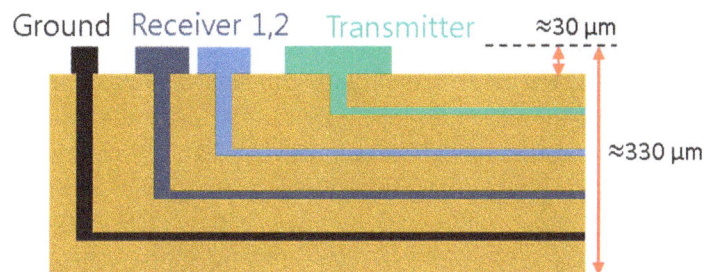

Figure 9. Cross-sectional view of the FPCB.

Figure 10 shows the overall signal processing unit of the FPCB sensor. Twelve transmitter electrodes located in the same row are connected to each other, thus giving the sensor 24 transmitting lines. Then, using a switching device, a single transmitting line is activated, moving electrically from the top to the bottom to cover all 24 transmitting lines. In addition, 24 receiver-1 electrodes and 24 receiver-2 electrodes positioned in the same column are connected in parallel; the sensor in this case has 12 receiver-1 lines and 12 receiver-2 lines. The input voltage was imposed by a function generator, and an AC sine wave was used to transmit the signal. The receiver lines are connected to a converter circuit board where the output currents are converted to voltage signals. These voltage signals are then transferred to a data acquisition system. As presented in Figure 10, the switching device changes the channel of the transmitter line such that the measurements of the liquid film thickness are conducted sequentially. Channel switching can be conducted both manually and automatically. For automatic switching, the switching board changes the channel using a trigger signal from the function generator. In this experiment, the frequency of the switching trigger signal is confined to 50 Hz due to a performance limitation of the data acquisition module. At this frequency, it takes about 0.5 s to switch all 24 transmitting lines, and the entire sensor domain can be accounted for by approximately 2 Hz. The data acquisition system has 24 channels of receivers, and each channel receives data at a sampling rate of 500 Hz.

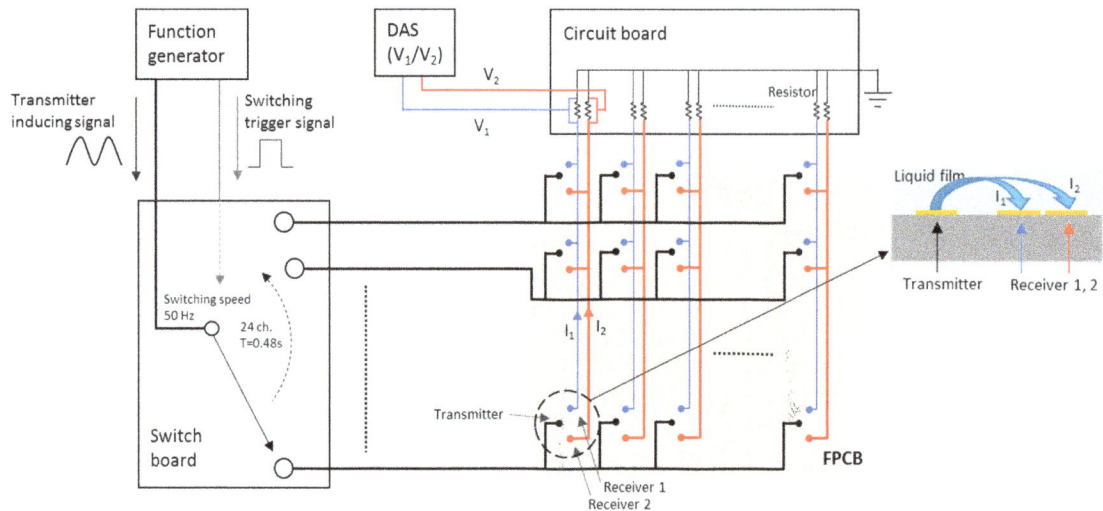

Figure 10. Signal transfer system with the FPCB sensor.

2.4. Calibration of the Prototypic Sensor

Because every probe composing the FPCB sensor has different characteristics with regard to the others owing to fabrication tolerance, calibration should cover all probes for application to liquid film flow measurements. Given that the size of the modified FPCB sensor is too large for calibration in one trial, calibration was conducted by dividing it into four sub-sections. A schematic diagram of the calibration device and its setup are presented in Figure 11. The calibration range was 0–3.0 mm with 0.5 mm steps and the water condition for calibration was 18 °C and 22 µS/cm. Figure 12 illustrates a typical example of the calibration process after three repeated runs. With an increase in the liquid film thickness, the current ratio increases and then becomes saturated at a thickness of 3.5 mm. The offset point at a film thickness of zero is considered to be caused by the thin remaining liquid layer when the insulating plane of the calibration device makes contact with the sensor or the high impedance of the FPCB due to the complex and integrated circuit. Even if this result does not affect the measurement significantly, reducing the impedance of the circuitry is desired to improve the resolution of the sensor. A calibration curve was derived, as shown in Figure 13; this is a spline interpolation curve based on the calibration result. This analysis showed that for film thicknesses up to 1.5 mm, absolute accuracy of 0.025 mm is achieved with a relative average error of 1.6%. For film thicknesses up to 3.0 mm, the error increases to 0.25 mm and the maximum relative error is 5.7 % in a film thickness range of 0.0–3.0 mm. These values are considered as the bias error of the sensing element. This is a typical calibration result with a single sensing element, and the same procedure was repeated for all 288 sensing elements to produce calibration curves for all elements. In addition, a comparison with an ultrasonic thickness gauge was conducted with the prototype sensor to ensure of the reliability of the sensor. The maximum error between the two methods was 5.3% and the average error was 1.8% [11].

Afterwards, the characteristics of the sensing element at three different temperature conditions were investigated and the output current ratios were plotted with the film thickness (Figure 14). This graph shows that the current ratios obtained in the range of 20 °C–40 °C were in fairly good agreement with the others with an average error of 3.9%. However, if the temperature difference exceeds this value, a distorted signal results and the current ratio shows dependency on the fluid temperature, in contrast to the theoretical analytical results. The reason behind this distortion remains not fully understood. One possible reason would be the different impedance changes between the two receiver channels with respect to the water temperature. Slight distortion of the acrylic plates in the calibration device onto which the sensor was attached and how the calibration was done can also cause this unexpected result. Not only for this particular sensing element but also for most other elements, a similar magnitude of error results if the temperature difference exceeds 20 °C. This imposes a limitation on the application range of the present sensor, and this should be improved for a wider range of applications in the future.

To sum up the performance of the sensor, the three-electrode sensor has a measurement range of 0.0 mm–3.0 mm. Moreover, it has 1.6% error up to 1.5 mm and 5.7% up to 3.0 mm. With regard to temperature stability, 3.9% error was achieved in the temperature range of 20 °C–40 °C.

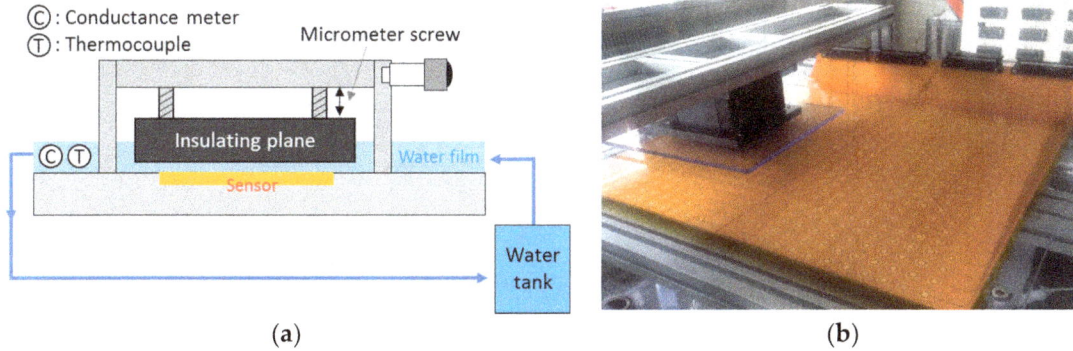

(a)   **(b)**

Figure 11. (**a**) Schematic diagram; (**b**) setup of the calibration device.

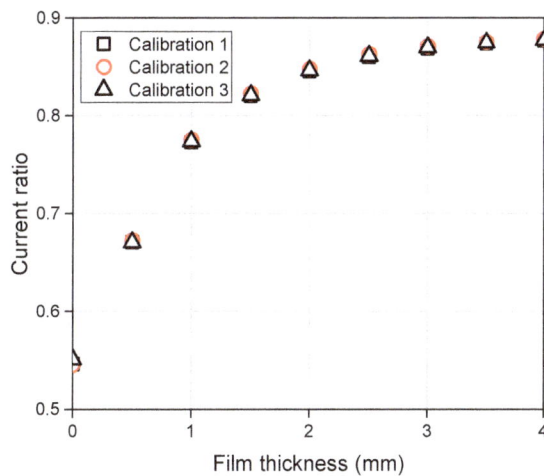

Figure 12. Calibration result (repeatability test).

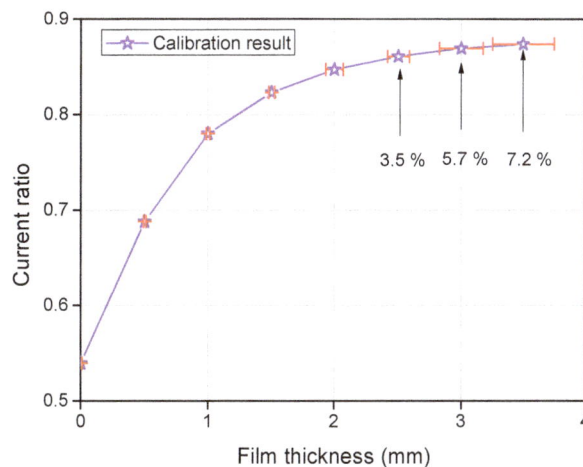

Figure 13. Calibration curve with spline interpolation.

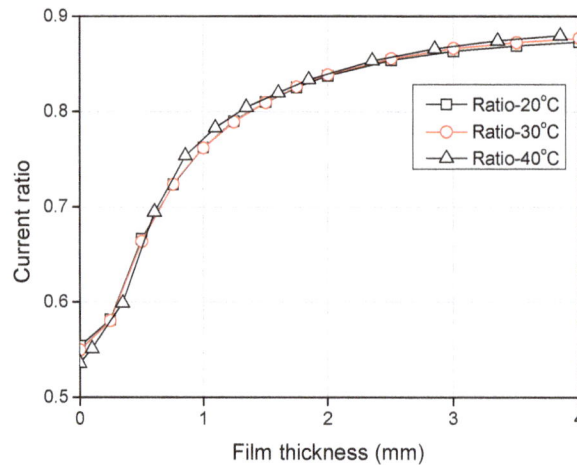

Figure 14. Current ratio with different temperature conditions.

3. Feasibility Test Result of the Liquid Film Sensor

The calibrated sensor was applied for the thickness measurement of a falling liquid film after impinging on a plate wall. The apparatus for this feasibility test is illustrated and displayed in Figure 15a,b. The water was supplied by a pump, and its flow rate and temperature were measured by an electromagnetic flow meter and a k-type thermocouple, respectively, installed downstream of the pump. The water was injected from the nozzle onto a vertical wall where the FPCB sensor was attached. The distance from the nozzle to the FPCB sensor was 25 mm and the diameter of the nozzle was 21 mm. These geometries are identical to those in Yang et al. [7]. The impinged liquid flowed down along the sensor surface, making a shaped liquid film, as shown in Figure 16. The figure also shows how the sensor was installed and applied to the test section. Subsequently, the falling water exited the test section through the drain and flowed into a storage tank. In the storage tank, a heater and a cooler were installed to control the water temperature. An electrical conductance meter was also installed in the storage tank to monitor the water conductance. The details of the measurement instrument are summarized in Table 1.

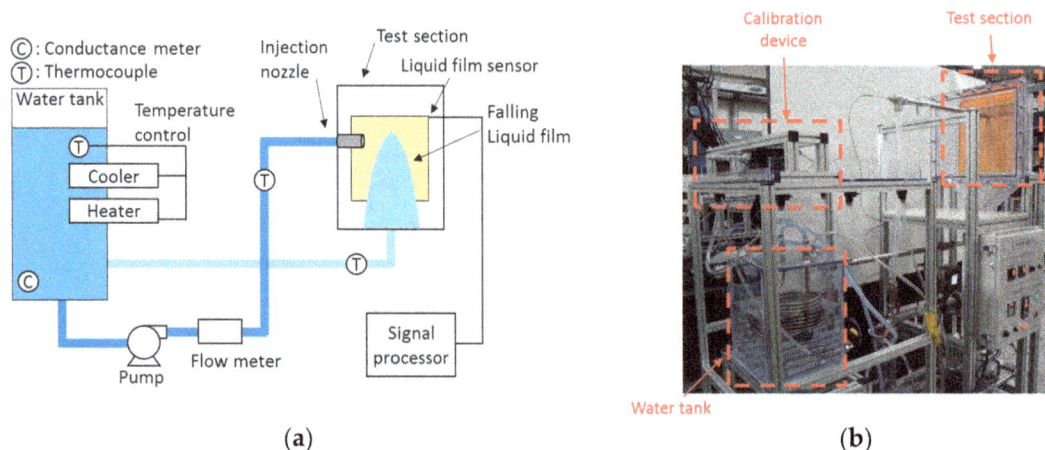

Figure 15. (**a**) Schematic diagram; and (**b**) configuration of the feasibility test.

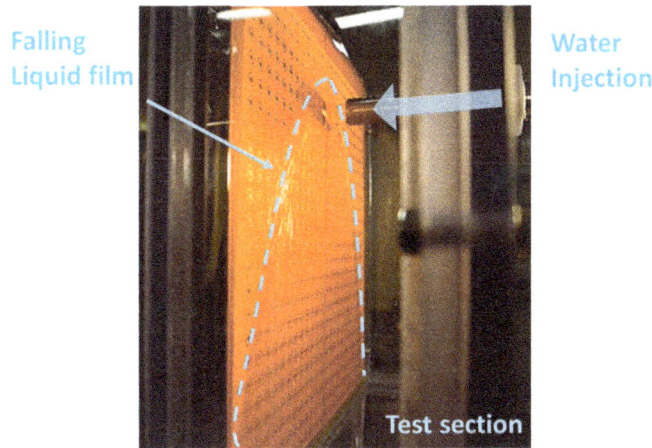

Figure 16. Edge of falling liquid film.

Table 1. Span and accuracy of the measurement instrument.

Instruments (Model)	Span	Accuracy
Thermocouple (K-type)	$-200{\sim}1250\ °C$	$\pm 2.2\ °C$
Electrical conductance meter (EUTECH Instruments CON450)	$0{\sim}200\ mS/cm$	$0.01\ \mu S/cm$
Flow meter (Toshiba GF630)	$0.0{\sim}10.0\ m/s$	$\pm 0.2\%$

Initially, the local film thickness was measured by the FPCB sensor while controlling the velocity of the water injection, with the water temperature constant at 20 °C. The water velocities at the nozzle were 0.48, 0.68, and 0.87 m/s, which cover the reference velocity condition in the experiment by Yang et al., i.e., 0.68 m/s. To determine the time-averaged thickness of the liquid film, and the measurement at each channel was continued for five seconds. Figure 17 shows the results of the local liquid film thickness measurement with three different liquid inlet velocities. The images on the left side are those of the actual liquid film observed by the camera, and dotted lines were added to indicate the edge of the film. The figures on the right depict the distributions of the local film thickness as measured by the present liquid film sensor, and the displayed contours are the time-averaged thicknesses. In this figure, the point at $(0, 0)$ is the center point of the water injection nozzle.

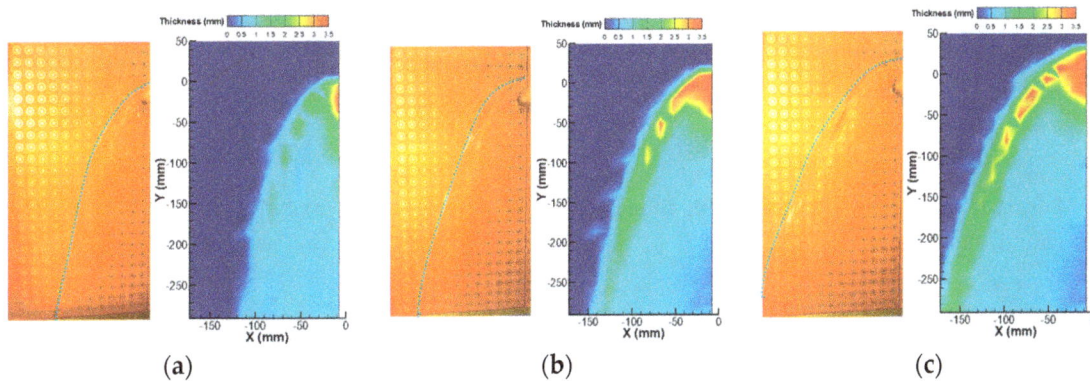

Figure 17. Local liquid film thickness of different liquid inlet velocities: (**a**) V_{in} = 0.48 m/s; (**b**) V_{in} = 0.68 m/s; and (**c**) V_{in} = 0.87 m/s.

Liquid film characteristics similar to those of the visual observation results were successfully reproduced in the film thickness measurement results. Initially, the thickest liquid film was measured near the center position where the liquid jet impingement occurs. In this region, the film thickness exceeded the measurement range of the present sensor; therefore, it had constant values of the maximum measurable thickness at 3.5 mm. The parabolic falling liquid film was reasonably reproduced by the present sensor. Near the liquid film edge, a thick film thickness was measured, consistent with the visual observation due to the surface tension force exerted on the edge of the liquid film. Oscillating edge behaviors were also observed in the visualization, and, intermittently, droplets were generated and became reattached to the wall. Owing to this oscillation and entrainment near the edge, an irregular liquid film thickness appeared near the film edge. In general, the liquid film width increased with the liquid injection velocity, and thicker film was measured with it, especially near the edge. From these measurement results and in comparison with the visual observations, it was found that the present sensor could measure liquid film thicknesses reliably under constant water temperature conditions.

A feasibility test to evaluate the working performance of the sensor under varying temperature conditions was then conducted. The liquid injection velocity was 0.63 m/s and the water temperature was changed from 20 °C to 40 °C in 10 °C steps. Figure 18 shows the measured liquid film thickness with three different water temperature conditions. Even with these temperature differences, a comparable liquid film thickness distribution was obtained. Figure 18d shows the standard deviation distributions of the liquid film thickness in these three measurements. Owing to the oscillating nature of the liquid film edge, the liquid film thickness deviation was significant at the liquid film edge compared to the other regions. In particular, the location of the liquid entrainment from the film edge varied slightly depending on the experimental conditions. Consequently, the largest deviation between the measured data was observed at the positions where droplet entrainment occurred. Except for these regions, the deviation was less than approximately 0.07 mm, and the temperature compensation capability of the three-electrode liquid film sensor was confirmed with this mild temperature variation.

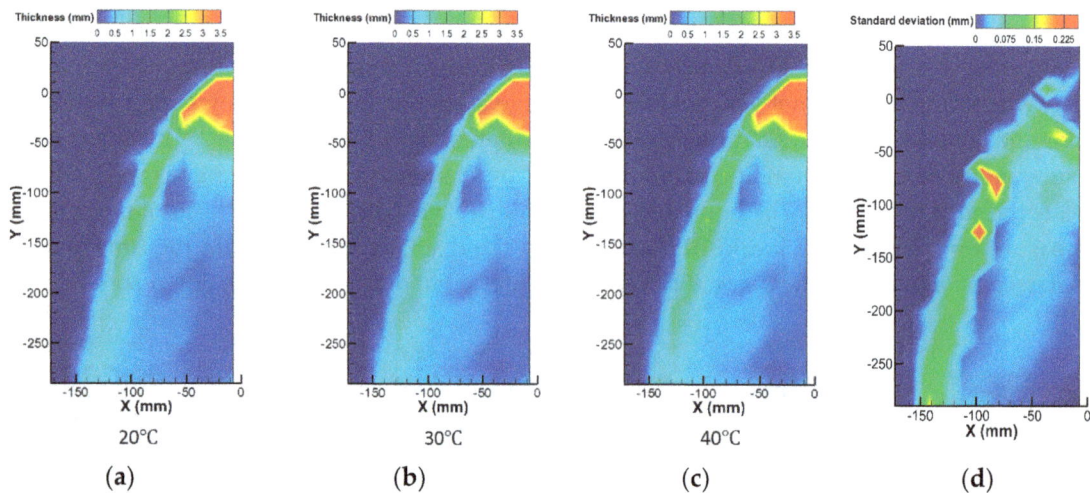

Figure 18. Liquid film thickness with different temperature conditions: (**a**) 20 °C; (**b**) 30 °C; (**c**) 40 °C; and (**d**) standard deviation.

4. Conclusions

In order to confirm the feasibility of the three-electrode conductance method for liquid film thickness measurements, which is expected to be advantageous under varying temperature conditions, two prototype liquid film sensors were fabricated on an FPCB. The designs of the prototypic sensors were determined by electrical potential analyses. The calibration tests of these sensors showed that the fabricated sensors are applicable in conditions with mild temperature variances at temperatures lower than 20 °C. Subsequently, the liquid film sensor was applied to a plate duct and to a condition with a falling liquid film thickness with impingement on a wall and was measured with a 12 × 24 sensing array. The measurement results showed reasonable consistency with the visual observations, and the feasibility of the three-electrode conductance sensor was confirmed in the temperature range of 20 °C–40 °C.

However, the possible temperature variation range is limited to 20 °C in the present design and experimental setup. In order to overcome this limitation, an improvement of the calibration devices is initially required. Additional investigations are then needed to determine the cause of the unexpected temperature dependency of the current ratio using a counterpart liquid film thickness measurement sensor. Reducing the impedance of its circuitry is required as well for more accurate resolutions by enlarging the signal-to-noise ratio.

Acknowledgments: This work was supported by the National Research Foundation (NRF) of Korea funded by the MSIP (No. NRF-2015M2B2A9028750) and also supported by the Nuclear Safety Research Program through the Korea Foundation of Nuclear Safety (KoFONS), and financial resources granted from the Nuclear Safety and Security Commission (NSSC), Republic of Korea (No. 1305011).

Author Contributions: Kyu-Byung Lee contributed by performing the simulation, designing the process of the sensor, carrying out the experiments, analyzing the results and preparing the text. Jong-Rok Kom conceived the idea and concept of the research. Goon-Cherl Park analyzed the results and devised the concept of the research. Hyoung-Kyu Cho conceived the idea, supervised the research, drew up the paper, and reviewed and revised the paper.

Conflicts of Interest: The authors declare no conflicts of interest.

References

1. Alig, I.; Oehler, H.; Lellinger, D.; Tadjbach, S. Monitoring of film formation, curing and ageing of coatings by an ultrasonic reflection method. *Prog. Org. Coat.* **2007**, *58*, 200–208. [CrossRef]

2. Takahama, H.; Kato, S. Longitudinal flow characteristics of vertically falling liquid films without concurrent gas flow. *Int. J. Multiph. Flow* **1980**, *6*, 203–215. [CrossRef]

3. Damsohn, M.; Prasser, H.-M. Experimental studies of the effect of functional spacers to annular flow in subchannels of a BWR fuel element. *Nucl. Eng. Des.* **2010**, *240*, 3126–3144. [CrossRef]

4. D'Aleo, F.P.; Papadopoulos, P.; Prasser, H.-M. Miniaturized liquid film sensor (MLFS) for two phase flow measurements in square microchannels with high spatial resolution. *Flow Meas. Instrum.* **2013**, *30*, 10–17. [CrossRef]

5. Arai, T.; Furuya, M.; Kanai, T.; Shirakawa, K. Concurrent upward liquid slug dynamics on both surfaces of annular channel acquired with liquid film sensor. *Exp. Therm. Fluid Sci.* **2015**, *60*, 337–345. [CrossRef]

6. Hayashi, M. Temperature-Electrical Conductivity Relation of Water for Environmental Monitoring and Geophysical Data Inversion. *Environ. Monit. Assess.* **2004**, *96*, 119–128. [CrossRef] [PubMed]

7. Yang, J.-H.; Cho, H.-K.; Kim, S.; Euh, D.-J.; Park, G.-C. Experimental study on two-dimensional film flow with local measurement methods. *Nucl. Eng. Des.* **2015**, *294*, 137–151. [CrossRef]

8. Coney, M.W.E. The theory and application of conductance probes for the measurement of liquid film thickness in two-phase flow. *J. Phys. E Sci. Instrum.* **1973**, *6*, 903–910. [CrossRef]

9. Kim, J.R.; Cho, H.-K.; Euh, D.-J. Experimental study of the characteristics of three flat probe conductance meter. *Trans. Korean Soc. Mech. Eng.* **2013**, *12*, 378–382.

10. COMSOL Multiphysics. *User's Guide*, Ver. 5.1; COMSOL: Burlington, MA, USA, 2015.

11. Lee, K.-B.; Kim, J.R.; Euh, D.-J.; Park, G.-C.; Cho, H.-K. Development of three-ring conductance sensor based on flexible printed circuit board for measuring liquid film thickness in two-phase flow with high resolution. *J. Sens. Sci. Technol.* **2016**, *25*, 57–64. [CrossRef]

5

A Fully Transparent Flexible Sensor for Cryogenic Temperatures Based on High Strength Metallurgical Graphene

Ryszard Pawlak [1,*], Marcin Lebioda [1], Jacek Rymaszewski [1], Witold Szymanski [2], Lukasz Kolodziejczyk [2] and Piotr Kula [2]

[1] Institute of Electrical Engineering Systems, Lodz University of Technology, 90-924 Lodz, Poland; marcin.lebioda@p.lodz.pl (M.L.); jacek.rymaszewski@p.lodz.pl (J.R.)

[2] Institute of Materials Science and Engineering, Lodz University of Technology, 90-924 Lodz, Poland; witold.szymanski@p.lodz.pl (W.S.); lukasz.kolodziejczyk@p.lodz.pl (L.K.); piotr.kula@p.lodz.pl (P.K.)

* Correspondence: ryszard.pawlak@p.lodz.pl

Academic Editors: Hyun-Joong Chung and Tae-il Kim

Abstract: Low-temperature electronics operating in below zero temperatures or even below the lower limit of the common −65 to 125 °C temperature range are essential in medical diagnostics, in space exploration and aviation, in processing and storage of food and mainly in scientific research, like superconducting materials engineering and their applications—superconducting magnets, superconducting energy storage, and magnetic levitation systems. Such electronic devices demand special approach to the materials used in passive elements and sensors. The main goal of this work was the implementation of a fully transparent, flexible cryogenic temperature sensor with graphene structures as sensing element. Electrodes were made of transparent ITO (Indium Tin Oxide) or ITO/Ag/ITO conductive layers by laser ablation and finally encapsulated in a polymer coating. A helium closed-cycle cryostat has been used in measurements of the electrical properties of these graphene-based temperature sensors under cryogenic conditions. The sensors were repeatedly cooled from room temperature to cryogenic temperature. Graphene structures were characterized using Raman spectroscopy. The observation of the resistance changes as a function of temperature indicates the potential use of graphene layers in the construction of temperature sensors. The temperature characteristics of the analyzed graphene sensors exhibit no clear anomalies or strong non-linearity in the entire studied temperature range (as compared to the typical carbon sensor).

Keywords: low-temperature; cryogenic; sensors; graphene

1. Introduction

The application of transparent conductive films and multilayer films for resistance temperature detectors in cryogenic systems was discussed for the first time in [1]. Graphene is one of the most promising materials for a diversity of modern technological applications due to its excellent electrical, optical, thermal, mechanical, electrochemical and structural characteristics [2–4]. The exceptionally high electrical conductivity of graphene combined with its transparency, flexibility and mechanical strength, make it suitable for microelectronic devices (Field Effect Transistors—FETs), photonics and optoelectronic systems, passive electronic elements and for sensing applications [3,5–7]. We propose for the first time a sensor with a graphene layer as sensitive element designed for cryogenic temperature measurement. The electrodes of the sensor are shaped using transparent conductor film on a polymer substrate using laser direct patterning, so the whole device is transparent and flexible. Encapsulation with a polymer transparent cover protects the sensor against any damage. The as-fabricated sensors have a low thermal capacity due to their construction. Changes in resistance of the sensors using

HSMG® (High Strength Metallurgical Graphene) graphene layers are measurable and significant. The sensors' sensitivity in the range of cryogenic temperatures (particularly below 50 K) is much higher than the sensitivity of comparable resistive (metallic) temperature sensors (RTD). Temperature sensors made of the graphene film transferred on the polymeric substrate have significantly better flexibility and resistance to thermal stresses in comparison to bulk graphite sensors. Application of thin conductive polymeric films (ITO—Indium Tin Oxide and AgHT^{TM}) allowed us to obtain the sensors with electrodes characterized by high optical transparency and therefore may be applicable in low-temperature optical studies (UV—Ultraviolet).

The crucial features for flexible, transparent temperature sensors are their electrical conductivity, thermal properties, and the optical transmittance of the materials used. These properties of graphene and transparent conducting oxides films are briefly discussed below.

The electrical conductivity of large-surface graphene sheets is a very complex issue, resulting from the characteristics of their carrier transport. The electronic properties of graphene are mostly considered and investigated from the viewpoint of its use as a material that could replace silicon in microelectronic devices controlled by electric fields (MOSFETs). The experimentally measured minimum conductivity in graphene ($\sigma_{min} \approx 4\,e^2/h$; resistivity $\rho \approx 6.5$ kΩ/sq) is close to the theoretically predicted quantum conductivity for Dirac fermions [2]. A rapid decrease in resistivity occurs after gate voltage switching due to the increasing carrier density. The resistivity measurements at 5 K revealed near-ballistic transport in samples of dimensions about 2 μm [8]. In longer graphene samples carrier transport has a diffusive character due to elastic and inelastic scattering [6,9]. There are different sources of elastic scattering in graphene: charged impurities, defects, absorbed atoms and those factors connected with the structural morphology of graphene sheets such as ripples and roughness on the graphene surface. Monolayer graphene in fact is not a perfectly flat 2D structure but rather, especially when placed on a substrate, it will adjust to the underlying surface roughness. The ripples of graphene can create a long-range scattering potential leading to important charge-carrier limitations (rises of resistivity) [10,11]. Discussion of the inelastic scattering from the phonons could be limited practically to the longitudinal acoustic phonons. The contribution of phonon scattering to total resistivity (on the order of some kΩ) of graphene samples is of the order of 100 Ω at 300 K [12,13]. In graphene samples of larger dimensions, e.g., in graphene nanoribbons with small widths of some tens of nanometers and lengths of some tens of micrometers, lateral confinement of the charge carriers causes the creation of an energy gap tuned by the appropriate choice of ribbon width [14,15]. There are also other causes of charge scattering, such as folding of the graphene sheets, phonons in the graphene or interfacial phonons between the graphene and the supporting substrate. Large-area graphene sheets are a 2D polycrystalline material consisting of domain and grains [7]. The grain boundary, depending of its structure, can manifest whole reflection or high transparency toward charge carrier transport [16]. In devices using as functional element large area graphene sheetd (e.g., sensors) contact/graphene interface phenomena should be taken into account. An additional charge inhomogeneity occurs in the vicinity of the contact but could extend hundreds of nanometers from the contact. Studies have proven that the contact resistance (Ti/Au-graphene) includes a component independent of the gate voltage of a value of about 800 Ω·μm, which is insensitive to temperature changes [17]. The contact resistance (700 ± 500 Ω·μm) was found to be independent of the metal work function (for Ti, Ag, Co, Cr, Fe, Ni, Pd) [18].

The thermal conductivity κ of graphene is extremely high and exceeds 5000 W/mK at room temperature [19]. These outstanding thermal properties seem to be very attractive for microelectronic and sensor applications. The predominant contribution to the high thermal conductivity of single layer graphene corresponds to acoustic phonons with a mean free path of 500–1000 nm, while the contribution of electrons is negligible [20–22]. Because of the long phonons' mean free path the thermal properties are dependent on the sample size and grain size and orientation. Defects and impurities as well as stresses in the graphene structure reduce the thermal conductivity [23,24]. The thermal properties of graphene are strongly affected by the influence of the supporting material [21]. The thermal conductivity of about

600 W/mK for single layer graphene supported on a SiO_2 membrane proved to be much less than that of suspended graphene (3000–5000 W/mK, although it should be noted that the κ value of supported graphene was still greater than that of metals (Cu, Ag—κ > 400 W/mK). The phonons' mean free path in supported graphene, predicted theoretically and estimated from measurements, is about 100 nm [21]. The interactions between graphene layers and substrate surfaces could be different for the various metals and dielectrics used as substrate materials and the method of graphene synthesis, so different scales of reduction of thermal conductivity could be observed [25].

An optical transmittance of single layer graphene equal to 97.7% was theoretically derived, assuming a constant high-frequency conductivity for Dirac fermions through the visible range of the spectrum, and proved experimentally [5,26]. This transparency is quite constant for wavelengths from 300 to 2500 nm, with a small decrease at about 270 nm [3,27]. The optical transparency for few layer graphene is reduced about 2.3% for every layer.

The above brief considerations suggest that, especially in applications that use graphene sheets of large size (of the order of millimeters), the dependence of the electrical, thermal and optical properties on methods for graphene synthesis, sample size, grain structure, purity, homogeneity and substrate type must be taken into account.

Although graphene has attracted greatest interest as a material for active microelectronic devices (graphene FETs), its exceptionally high electrical conductivity and furthermore its transparency, flexibility and mechanical strength, make it also suitable for passive electronic elements. Recently the requirements of modern optoelectronic technologies, such as photovoltaic technology, flat panel displays, OLEDs (Organic Light-Emitting Diodes) and optoelectronic devices, have caused the rapid development of new conductive transparent materials in the form of thin layers [28–32]. Of great importance for practical applications are conductive oxides transparent layers, usually indium tin oxide (ITO). The increasing price of indium and the lack of stretchability of ITO has inspired research for replacing ITO with other transparent conductive materials, such as ZnO, carbon nanotubes (CNTs) [33], conducting polymers like poly(3,4-ethylenedioxythiophene) polystyrene sulfonate—PEDOT:PSS [34], thin transparent metal films [35] and multilayer systems that consist of two outer layers of oxides (ZnO, ITO) and a thin metallic film (Ag, Cu, Au) between them, e.g., ITO/Ag/ITO multilayers. The main purpose of many of these studies was producing these multilayers on transparent electrodes for organic solar cells or OLEDs [36,37], transparent UWB (Ultra-WideBand) antennas [38,39] and also as EMI (ElectroMagnetic Interference) shielding materials [40].

Recently, several innovative temperature sensors based on graphene have been developed, thus providing an alternative to conventional rigid ceramics. Mono- or bi-layer graphene nanofabricated on a silicon substrate has been used as a thermally sensitive element, however no flexibility has been achieved in such a system. Al-Mumen et al. [41] studied the temperature sensing behavior of mono-, bi- and few layer graphene exfoliated from graphite. The resistance temperature coefficient (TCR) was determined, which was about -0.007 K^{-1} for the bi-layer graphene, about -0.003 K^{-1} for the monolayer graphene, and about -0.0015 K^{-1} for the few-layer graphene, respectively. The bilayer graphene had the highest negative TCR, measured as the temperature changed between RT and 80 °C. Kong et al. [42] fabricated a mechanically stable graphene electrode by its direct micropatterning onto a flexible polymer. The negative temperature coefficient (NTC) of the graphene electrode was similar to that of conventional NTC materials, however its response time was faster by an order of magnitude. Trung et al. [43] proposed a flexible and very sensitive sensor based on reduced graphene oxide transferred onto a transparent polymer substrate. Yang et al. [44] invented a wearable sensor by incorporation of graphene nanowalls into PDMS which value of resistance temperature coefficient exceeded by threefold the values typical for conventional sensors. Yan et al. [45] demonstrated stretchable graphene thermistors with intrinsic high stretchability that were fabricated through a lithographic filtration method based on conductive AgNW electrodes and a resistive graphene detection channel. The devices were stretched up to 50% and could maintain their functionality even in highly stretched states. Bendi et al. [46] reported a self-powered thermistor which utilizes the formation of p-i-n junctions on a graphene monolayer using ferroelectric polymer PVDF-TrFE (poly[(vinylidenefluoride-co-trifluoroethylene)]) that generates current changes when subjected to thermal stimulation (40–110 °C) under no external perturbation.

These excellent properties indicate graphene is a candidate for the preparation of the transparent, flexible electrodes, e.g., instead of ITO [47]. Many interesting nanodevices and structures using electrodes of graphene have been proposed, as organic FETs [47], flexible transparent piezoelectric energy harvesters [48] or transparent resistive memory [49]. Double layers consisting of graphene sheets and ITO layers were applied as source-drain electrodes in a thin film transistor (InGaZnO) [50] and a n GaN light emitting diode [51]. Quite recently a multilayer structure consisting of a graphene layer formed on Cu foil and doped with Au nanoparticles (anode) and with Ag-nanowires (cathode) were used to creating a through transparent quantum dot light-emitting diode [52]. Inks based on graphene obtained from the reduction of graphene oxide (GO) and suspended in liquids enabled fabrication of flexible sensors, for example, a transparent acoustic actuator [53] and a sensor for the simultaneous measurement of pressure and temperature [54]. In the construction of the organic electrochemical transistors electrodes were doped with graphene flakes prepared by ultrasonic exfoliation from graphite [55].

2. Materials and Methods

2.1. Formation of Transparent Electrodes by Laser Ablation

To create electrodes for cryogenic temperature sensors two different conductive films were used, namely commonly used ITO and ITO/Ag/ITO film (AgHT™) with surface resistance of 15 Ω/sq and 4 Ω/sq, respectively. AgHT™ is a highly conductive film on a polyester substrate which is significant in shielding against EMI/RFI (ElectroMagnetic Interference/ Radio-Frequency Interference) and also infrared heat rejection. Its good electrical conductivity, high optical transparency and flexibility determine applications of this material in membrane switches, photovoltaic structures, displays, and passive elements of flexible and transparent electronics. The ITO film on PEN (Poly(ethylene 2,6-naphthalate) substrate had a thickness of 125 nm and an ITO/Ag/ITO film thickness of 150 nm.

Laser Direct Writing (LDW) methods for nanometer films are used in the manufacture of flexible electronic circuits and sensors on the sub-millimeter scale. It is well known that laser processing of thin layers of nanometer thickness can be performed using laser beams of short wavelength and short pulse duration. Thin functional layers (ITO, AgHT, carbon nanotube layers) on transparent substrate materials (polyester, PET) have similar ablation threshold fluences, therefore damage of the substrate layers during laser processing should be avoided. Processing of most materials demands applying of UV laser beams with femto- or picosecond pulses to ensure the best effects. Patterning of ITO thin films was performed using laser beam pulses of ultraviolet to infrared wavelength of picosecond, femtosecond or nanosecond duration [56–60]. Laser direct writing was applied to electrode patterning for flat panel displays [61], fabricating a miniature transparent gas flow meter [62], electrode isolation in ITO layer on substrates used in the mobile phones [63], a pentacene thin film transistor (TFT) with source and drain electrodes patterned in ITO [64], and producing matrix array of OLEDs [65].

The main goal of our former research was to establish the possibility of using a single mode fiber laser in micromachining considering the quality of the obtained structures and their smallest dimensions while maintaining acceptable quality [66]. We have shown that the LDW method by nanosecond laser ablation ensures good conditions for prototyping structures with very high pattern fidelity. Among other uses contacts for prototyping of OLED structures [66] and samples of two types of conductive path have been prepared. The first had a meandering shape and was used in resistivity measurements and the second had the shape of a cross and was used in Hall effect measurements (Figure 1) [1].

The optical transparency was not changed, although some thermal effects were observed after laser treatment. In case of the ITO/Ag/ITO layer even improvement of the optical transmission after laser micromachining was noted. Previous studies have shown that structures patterned in AgHT conductive film can be useful for sensors in cryogenic systems and passive elements in flexible electronics [1]. A redENERGY G3 SM 20W single mode fiber laser (SPI Lasers UK Ltd., Southampton, UK), which guaranties high quality of the beam (M2 < 1.3) was used to manufacture the electrodes for cryogenic temperature sensor with graphene sensitive layer. The laser beam was scanned by a 2-Axis Scan Head (Xtreme, Nutfield Technology. Inc., Hudson, NH, USA) equipped with a 100 mm F-theta lens and was controlled by the SB-1P Waverunner software (Xtreme, Nutfield Technology. Inc., Hudson, NH, USA). The optimal parameters for creating electrodes were as follows: pulse energy—120 µJ; pulse duration—25 ns; pulse repetition frequency—72 kHz; scanning velocity—800 mm/s and for ITO film: pulse energy—145 µJ; pulse duration—25 ns; pulse repetition frequency—80 kHz; scanning velocity—1500 mm/s.

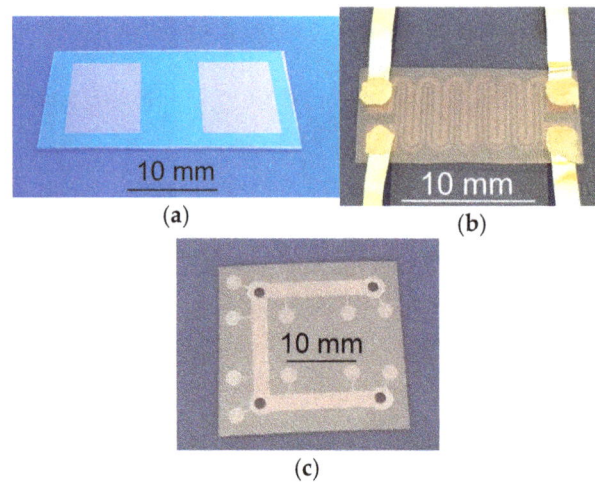

Figure 1. Examples of structures prepared by laser ablation: (**a**) electrodes for cryogenic sensors made in ITO layer on PEN substrate; (**b**) micro-heater in ITO/Ag/ITO layer with silver leads; (**c**) test structure for examination of electrical properties of ITO on PEN.

The characteristics electrode resistance changes with temperature in the range of 12–300 K are shown in Figure 2. Studies have proven that changes of substrate resistance, due to temperature changes, are continuous and repetitive in the analyzed range. This means that the conductive layer is continuous in a wide range of temperature, and the substrate is not permanently deformed or damaged.

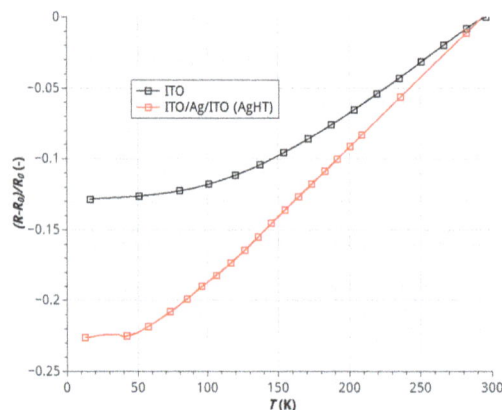

Figure 2. Temperature dependence of resistance of electrodes prepared from conductive polymers.

2.2. Synthesis and Transfer of Graphene Film

The high strength metallurgical graphene (HSMG®) sheets were synthesized in an industrially scaled thermochemical facility based on the process described in previous papers [67–69]. The copper/nickel composite substrate was heated in an argon atmosphere at a pressure of 100 kPa. After this step the chamber was evacuated to a pressure of 2 Pa and a mixture of acetylene, hydrogen and ethylene (ratio 2:1:2), at a partial pressure of 3 kPa, has been simultaneously introduced with argon for 1 min. Finally, the substrate with graphene was cooled stepwise to RT in an argon atmosphere at a pressure of 100 kPa.

The modified transfer procedure of HSMG® graphene for temperature sensors and reference samples preparations was used. The procedure is based on the commonly used method of graphene transfer from metallic substrates onto Poly(methyl methacrylate)—PMMA foil, which is described in details in our previous paper [70]. Methods utilizing a thin film of PMMA as a graphene supporting material and their variations are the most frequently used methods for transfer of 2D materials (graphene) onto any substrate. Wrinkles and cracks were observed on the graphene after the transfer process. An extensive explanation for the formation of these defects is presented in our previous work [70].

The production method of HSMG® allowed manufacturing of a single layer of graphene. The intentional and controlled modifications of the synthesis process lead to a graphene-like material (G-LM) with a slightly lower value of resistance/square.

Analysis of the graphene transferred onto reference samples was carried out using an inVia Raman spectroscope (Renishaw, New Mills, UK). All acquisitions of Raman spectra for graphene structure were performed with using an Ar+ laser at a laser excitation wavelength λ = 532 nm, exposure time was 10 s; signal was averaged from three times repeated exposition per spot. The maximum laser output power was 29.3 mW but the test was carried out at only 10% of the output power. Raman scattering was observed for the 1200–3100 cm^{-1} wavenumber range. The acquisition settings listed above did not cause any changes on the surface, like damage by local heating. Data processing was performed using the PeakFit software (Systat Software Inc., London, UK). Gauss–Lorentz curves were used for spectra deconvolution. Data obtained from the deconvolution for example: peak positions, intensities, half-widths were used for the calculation of characteristic peak ratios of graphene structures which were used in further analysis are provided in following tables (Tables 1 and 2).

Table 1. Names of Raman peaks identified in the studied graphene structures with their specific frequencies.

Peak Name	HSMG®	G-LM
	ω [cm^{-1}]	
D	1341.1	1338.3
G	1583.8	1577.6
D′	-	1615.4
2D	2678.8	2687.3

Table 2. FWHM (full width at half maximum) values and ratios of typical peaks calculated on the basis of Raman spectra deconvolution.

Peak Name	HSMG®			G-LM		
	FWHM [cm^{-1}]	I_G/I_{2D}	I_{2D}/I_G	FWHM [cm^{-1}]	I_G/I_{2D}	I_{2D}/I_G
D	36			96		
G	29			45		
D′	-	0.2	4.1	29	3.1	0.3
2D	45			136		

2.3. Encapsulation of TCO/HSMG® or G-LM Samples

The construction of the graphene cryogenic temperature sensor should eliminate or reduce the sensitivity of the sensor to stimuli other than the temperature. Particularly important is the ability to control the influence of gases on the electrical properties of graphene. This means that the graphene should be insulated from the influence of gas (vacuum inside) or the sensor should be filled with gas in a controlled way (gas inside). Moreover, the structure should provide a simple method of implementing the electrical connections. The reduction of heat exchange with the room temperature environment and heat capacity of the cooled part are just as important in cryogenic systems. The presented transfer of graphene film on a polyester substrate with highly conductive electrodes provided for a small sensor thermal capacity and reduced the heat transfer through the electrodes. The LDW method enabled patterning of electrodes in various shapes, while maintaining the transparency of the electrodes and the substrate. The transparency of the sensor is a unique feature of the presented encapsulation technology. In addition, the small cross section of the thin film electrodes effectively reduces the heat transfer to the cryogenic. Features of the applied substrate and the method of electrode patterning favored the use of a protective thin transparent polymer foil (about 10 μm). The protective film coated on graphene and partially on the electrodes isolates these elements electrically and chemically from the environment (Figure 3).

Integration of the polymer film onto the substrate is the result of a thermal activation adhesive. Proper preparation of the protective layer made it connect only to the substrate and does not affect the graphene film. The encapsulation process did not adversely affect the electrical and optical properties of the sensors. The encapsulation process may be performed either under vacuum or in a gas. This is important because thus can be used to change the parameters of the temperature sensor by doping the gas. The sensors retain the transparency and flexibility of conventional polymers after encapsulation process. The proposed encapsulation of the samples provides various options for electrical connections between the sensor and measurement systems. Two types of electrical joints were applied: 1st—a pressed contact as in butt joints, where a thin silver foil (35 μm) has been used and 2nd—an adhesive contact, where the electrically conductive silver-epoxy Elpox AX 15s (Amepox Microelectronics Ltd., Lodz, Poland) has been applied. No effects of the various types of contacts on sensor properties were observed. The proposed technology of encapsulation ensures the possibility of forming fully transparent sensors with different size and shapes (Figure 4).

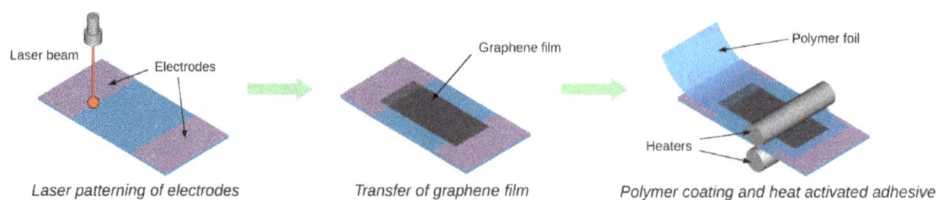

Figure 3. Method of manufacturing graphene cryogenic temperature sensor.

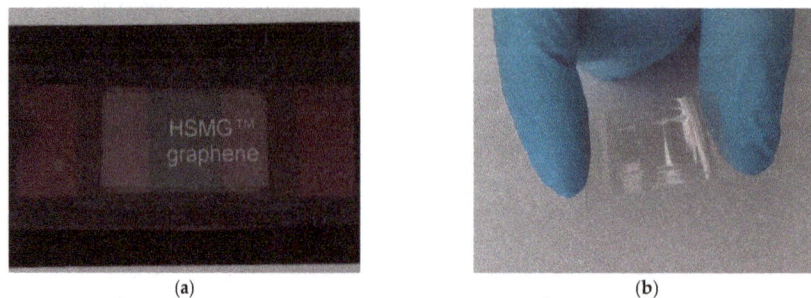

(a)   (b)

Figure 4. Examples of fully transparent temperature sensors: (**a**) transparency of cryogenic sensor—photo of the sensor on a black background with a white label; (**b**) stretchability of the sensor.

Additionally, it is possible to realize many different sensors at once using the same structure (e.g., an array temperature sensor, temperature sensor and a gas sensor). A graphene monolayer (HSMG®) and graphene-like material (G-LM) were used in the construction of sensors. The total resistance of the manufactured sensor is the sum of the resistance of electrodes R_e, the resistance of graphene-electrode interfaces R_i and the resistance of the graphene layer R_g (HSMG® or G-LM) (Figure 5).

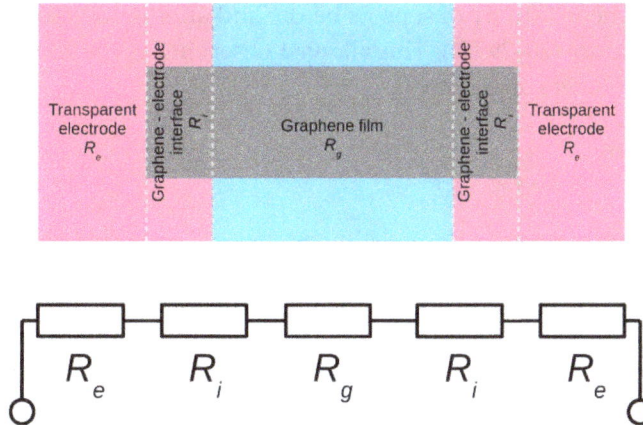

Figure 5. Equivalent circuit of graphene temperature sensor.

The temperature affects the all listed resistances to a different extent. Therefore, proper selection of materials for electrode construction is particularly important. In the proposed solution, the substrates with electrodes were made of ITO/Ag/ITO AgHT (4 Ω/sq) and ITO (15–100 Ω/sq). The resistance value of the graphene layer is relatively high (>20 kΩ/sq). A significant difference of the resistance of the graphene layer and the resistance of the substrate, resulting from the different thickness of graphene monolayer and electrodes (~120 nm), minimizes the participation of the electrode resistance and the interface between the graphene and electrodes in the total resistance of the sensor.

2.4. Instrumentation and Experimental Procedure

All tests and measurements of electrical properties of graphene cryogenic temperature sensors were performed in a helium closed-cycle DE-210 cryostat (Advanced Research Systems, Inc., Macungie, PA, USA). The sensors were cyclically cooled from room temperature (295 K) to cryogenic temperature (20 K) at a rate of about 0.1 K/s. The tested graphene sensors and reference temperature probes were placed directly on the surface of the massive copper heat exchanger (a copper block 80 mm × 80 mm × 10 mm). The encapsulation of the sensor provided the electrical and chemical isolation of the sensor from the environment and other elements of the cryogenic system. In order to eliminate the temperature gradient along the sensor the entire surface of the sensor was fixed to the heat exchanger (including electrodes). The exchanger was mounted directly on the cold finger of the cryocooler (second stage of the cryocooler) (Figure 6).

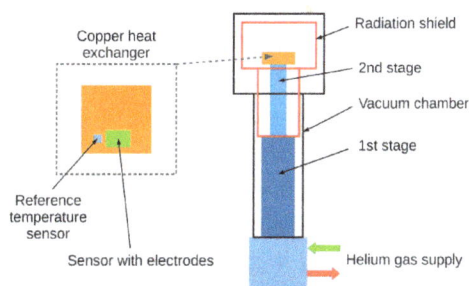

Figure 6. Scheme of cooling circuit.

The temperature of heat exchanger was measured and controlled by a precise Lakeshore 331 temperature controller and the reference sensors (silicon diodes DT-670-SD, Lakeshore Cryotronics Inc., Westerville, OH, USA). The glass-epoxy laminate G10 (MG Chemicals, Burlington, ON, Canada) was used for preparing the fastening and support elements. The observation of the temperature effect on resistivity of the graphene sensors was the main research work. Resistivity measurements were performed using a Keysight 34420A Micro-Ohm Meter (4-probe method, Keysight Technologies, Santa Rosa, CA, USA). The resistivity of sensors before and after encapsulation has been measured in wide range of temperature (20–295 K). This allowed observing any influence of the encapsulation process on electrical parameters of graphene.

3. Results and Discussion

3.1. Raman Spectroscopy

Spectra for the studied graphene structures are presented on Figure 7.

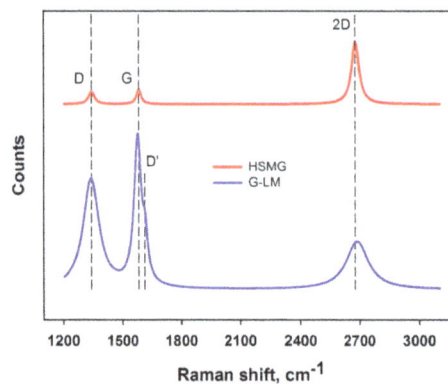

Figure 7. Raman spectra of single layer of high strength metallurgical graphene (HSMG®)—red curve and metallurgical multi-layered graphene-like material (G-LM)—blue curve.

The spectrum of HSMG® graphene consists of characteristic G and 2D peaks. The intensity ratios I_G/I_{2D} or I_{2D}/I_G are 0.2 and 4, respectively. According to the studies of Das and Ferrari it can be noted that these values correspond to a single layer of graphene [71,72]. The additional D peak which can be observed should not appear for perfect graphene (without defects), so the presence of the D peak indicates carbon atom disorders or defects such as edges, dislocations, cracks or vacancies [73]. Taking the above into consideration, we can conclude the HSMG® is defected, single-layer graphene.

The spectrum shape of G-LM graphene looks like a spectrum of graphene oxide (GO) or reduced graphene oxide (rGO) with a distinctive arrangement and shape of the G and D peaks [73–75]. In most papers on GO/rGO, the spectra have a completely different shape of the 2D peak, which intensity is low and the shape is blurred. Deconvolution of the 2D peak for GO allows one to extract an additional D + G peak [73–76]. It should be noted that the manufacturing process for HSMG® graphene and post-processing procedures preclude the formation of graphene oxide. In G-LM graphene spectra a 2D symmetrical peak can be observed. The 2D peak for G-LM graphene is broadened in comparison to the peak for HSMG® graphene. On analysis of the Raman spectrum an additional D' peak can be also observed. The D' peak, like the D peak, indicates a defect and a low ordered carbon structure [77]. Das in his paper shows the evolution of the Raman spectrum as the number of graphene layers increases from single layer graphene to the characteristic spectrum for graphite with a ratio I_G/I_{2D} equal to 3.1 [72]. The mentioned value is the same in case of G-LM but it should be noted that 2D peak shape is completely different compared to graphite. For Raman spectroscopy analysis the ratio of peak intensity is very important but the shape of the 2D peak should also be taken into account. A similar Raman spectrum to that of G-LM was described by Pimenta and defined as a nanographitic structure. However, the peak 2D is narrow compared to the G-LM 2D peak and it has an intensity almost equal to the G peak intensity [77]. Taking into account the above considerations we can say that we are dealing with a disordered, multi-layer graphene-like material structure.

3.2. Temperature Dependence of the Resistance

The results of studies on the effects of temperature on the resistance of samples made of HSMG® and G-LM showed a significant dependence of resistance on temperature in the range of 20–195 K characterized by a continuous, nearly linear decrease of the resistance (negative temperature coefficient of resistance). The process of encapsulation for samples with HSMG® and G-LM caused the relatively small, permanent increase in the resistance of the sensor (Figures 8 and 9).

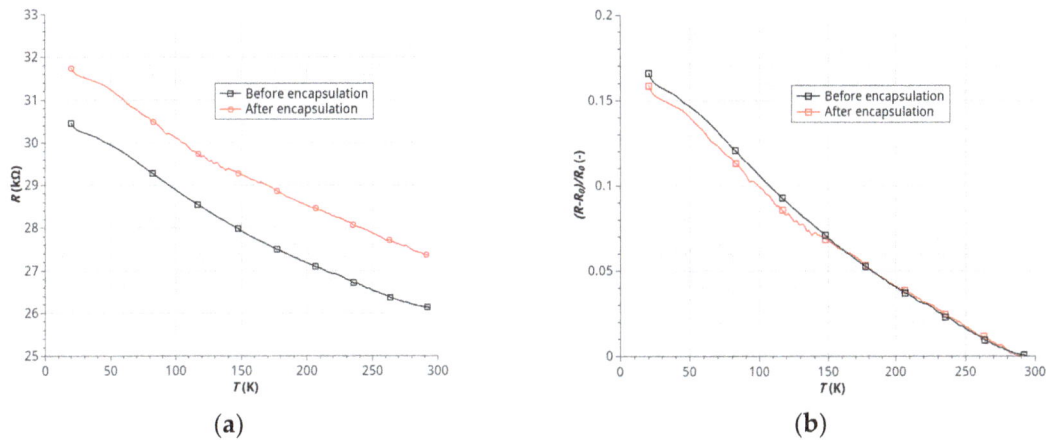

Figure 8. Temperature dependence of resistance of HSMG® sensors before and after encapsulation: (**a**) resistance R (kΩ) of sensor; (**b**) relative changes of sensor resistance (R_0—resistance value in 295 K).

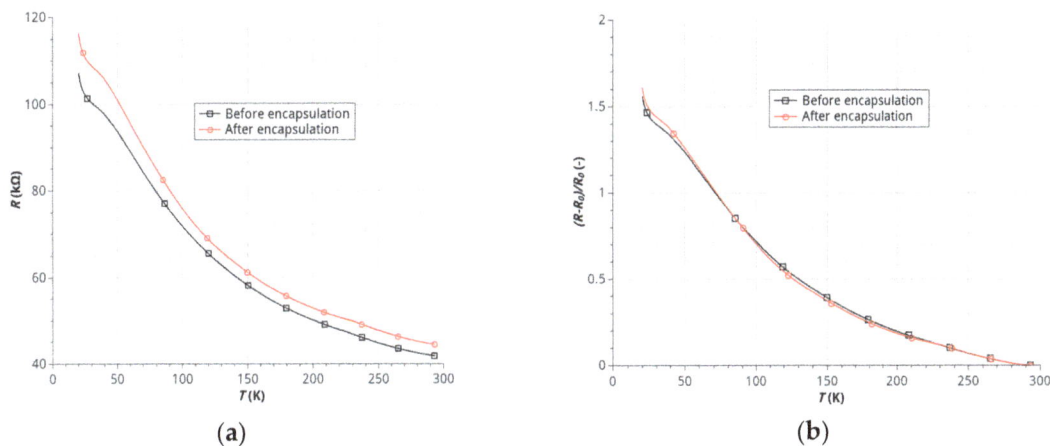

Figure 9. Temperature dependence of resistance of G-LM sensors before and after encapsulation: (**a**) resistance R (kΩ) of sensor; (**b**) relative changes of sensor resistance (R_0—resistance value in 295 K).

The increase in resistance of the graphene-electrode interface is the most likely cause of this effect. The connection resulting from, inter alia, the occurrence of weak intermolecular van der Waals forces may undergo a slight degradation. It should be noted, however, that a change in the resistance value after the encapsulation does not affect the nature of the temperature dependence of the resistance (Figures 8 and 9). This means that the process does not degrade the active layer of the sensor. In the case of the sensor made of the HSMG® layer the observed change in resistance of the sensor is on the order of 16% (Figure 8). In the case of G-LM sensors a much stronger dependence, on the order of 150%, was observed (Figure 9). The nature of the changes of resistance is less linear than for HSMG® sensors.

It should be noted that, for temperatures of less than 150 K, a growth rate of the resistance change was observed, which translates directly into increased sensitivity of the sensor in this temperature range.

4. Conclusions

The observation of the effect of temperature on the electrical characteristics of graphene indicates the potential use of graphene layers in the construction of temperature sensors. The temperature characteristics of the analyzed graphene sensors exhibit no clear anomalies or strong non-linearity in the entire studied temperature range (as compared to the typical carbon sensor). The as-fabricated sensors have a low thermal capacity due to their construction. Changes in resistance of sensors using HSMG® and G-LM layers are measurable and significant. The sensors' sensitivity in the range of cryogenic temperatures (particularly below 50 K) is much higher than the sensitivity of comparable resistive (metallic) temperature sensors (RTD). Measurements carried out at temperatures below approximately 20 K showed the increase of dynamics of the sensors' resistance changes with temperature. It should be noted that the measured changes of electrical resistance as a temperature function in the case of the sensor based on a monolayer graphene are significantly less than for a sensor prepared using a defected graphene-like material. The explanation of this phenomenon requires further study. Temperature sensors made of the graphene film transferred on a polymeric substrate have significantly better flexibility and resistance to thermal stresses in comparison to bulk graphite sensors. The application of thin conductive polymeric films (ITO and AgHT^{TM}) allowed obtaining sensors with electrodes characterized by high optical transparency. These sensors may be applicable in low-temperature optical studies (UV). In cases where transparency is not particularly required, Kapton tape with Au electrodes can be used because of its good electrical and thermal properties in cryogenic temperatures. Future studies of sensors placed in a controlled atmosphere (gas/vacuum in the capsule) are planned.

Author Contributions: R.P., M.L. and J.R. proposed idea of sensor, designed experiments, discussed and analyzed the obtained electrical data; R.P. performed laser writing of electrodes; M.L. and J.R. performed cryogenic experiments; R.P., M.L., J.R., W.S. and L.K. wrote the paper; W.S. was responsible for transfer of graphene on polymeric substrates and Raman interpretation. L.K. reviewed the paper and cooperated in Raman spectra analysis and data processing. P.K. monitored the process of graphene synthesis, discussed and analyzed the data and reviewed the paper.

Conflicts of Interest: The authors declare no conflict of interest.

References

1.   Lebioda, M.; Pawlak, R. Influence of cryogenic temperatures on electrical properties of structures patterned by a laser in ITO/Ag/ITO layers. *Phys. Status Solidi A* **2016**, *213*, 1150–1156. [CrossRef]

2.   Geim, A.K.; Novoselov, K.S. The rise of graphene. *Nat. Mater.* **2007**, *6*, 183–191. [CrossRef] [PubMed]

3.   Weiss, N.O.; Zhou, H.; Liao, L.; Liu, Y.; Jiang, S.; Huang, Y.; Duan, X. Graphene: An emerging electronic material. *Adv. Mater.* **2012**, *24*, 5782–5825. [CrossRef] [PubMed]

4.   Balandin, A.A. Thermal properties of graphene and nanostructured carbon materials. *Nat. Mater.* **2011**, *10*, 569–581. [CrossRef] [PubMed]

5.   Zhu, Y.W.; Murali, S.; Cai, W.W.; Li, X.S.; Suk, J.W.; Potts, J.R.; Ruoff, R.S. Graphene and graphene oxide: Synthesis, properties, and applications. *Adv. Mater.* **2010**, *22*, 3906–3924. [CrossRef] [PubMed]

6.   Avouris, P. Graphene: Electronic and photonic properties and devices. *Nano Lett.* **2010**, *10*, 4285–4294. [CrossRef] [PubMed]

7.   Craciun, M.F.; Russo, S.; Yamamoto, M.; Tarucha, S. Tuneable electronic properties in graphene. *Nano Today* **2011**, *6*, 42–60. [CrossRef]

8.   Bolotin, K.I.; Sikes, K.J.; Hone, J.; Stormer, H.L.; Kim, P. Temperature-dependent transport in suspended graphene. *Phys. Rev. Lett.* **2008**, *101*, 096802. [CrossRef] [PubMed]

9.   Chen, J.-H.; Jang, C.; Ishigami, M.; Xiao, S.; Cullen, W.G.; Williams, E.D.; Fuhrer, M.S. Diffusive charge transport in graphene on SiO$_2$. *Solid State Commun.* **2009**, *149*, 1080–1086. [CrossRef]

10. Katsnelson, M.I.; Geim, A.K. Electron scattering on microscopic corrugations in graphene. *Philos. Trans. R. Soc. Lond. Ser. A* **2008**, *366*, 195–204. [CrossRef] [PubMed]

11. Zwierzycki, M. Transport properties of rippled graphene. *J. Phys. Condens. Matter* **2014**, *26*, 135303. [CrossRef] [PubMed]

12. Hwang, E.H.; Das Sarma, S. Acoustic phonon scattering limited carrier mobility in two-dimensional extrinsic graphene. *Phys. Rev. B Condens. Matter* **2008**, *77*, 115449. [CrossRef]

13. Perebeinos, V.; Avouris, P. Inelastic scattering and current saturation in graphene. *Phys. Rev. B Condens. Matter* **2010**, *81*, 195442. [CrossRef]

14. Han, M.Y.; Özyilmaz, B.; Zhang, Y.; Kim, P. Energy band-gap engineering of graphene nanoribbons. *Phys. Rev. Lett.* **2007**, *98*, 206805. [CrossRef] [PubMed]

15. Moreno-Moreno, M.; Castellanos-Gomez, A.; Rubio-Bollinger, G.; Gomez-Herrero, J.; Agraït, N. Ultralong natural graphene nanoribbons and their electrical conductivity. *Small* **2009**, *5*, 924–927. [CrossRef] [PubMed]

16. Yazyev, O.V.; Louie, S.G. Electronic transport in polycrystalline graphene. *Nat. Mater.* **2010**, *9*, 806–809. [CrossRef] [PubMed]

17. Russo, S.; Craciun, M.F.; Yamamoto, M.; Morpurgo, A.F.; Tarucha, S. Contact resistance in graphene-based devices. *Physica E* **2010**, *42*, 677–679. [CrossRef]

18. Watanabe, E.; Conwill, A.; Tsuya, D.; Koide, Y. Low contact resistance metals for graphene based devices. *Diamond Relat. Mater.* **2012**, *24*, 171–174. [CrossRef]

19. Balandin, A.A.; Ghosh, S.; Bao, W.; Calizo, I.; Teweldebrhan, D.; Miao, F.; Lau, C.N. Superior thermal conductivity of single-layer graphene. *Nano Lett.* **2008**, *8*, 902–907. [CrossRef] [PubMed]

20. Ghosh, S.; Nika, D.L.; Pokatilov, E.P.; Balandin, A.A. Heat conduction in graphene: Experimental study and theoretical interpretation. *New J. Phys.* **2009**, *11*, 095012. [CrossRef]

21. Nika, D.L.; Balandin, A.A. Two-dimensional phonon transport in grapheme. *J. Phys. Condens. Matter* **2012**, *24*, 233203. [CrossRef] [PubMed]

22. Xu, Y.; Li, Z.; Duan, W. Thermal and thermoelectric properties of graphene. *Small* **2014**, *10*, 2182–2199. [CrossRef] [PubMed]

23. Lindsay, L.; Broido, D.A.; Mingo, N. Flexural phonons and thermal transport in graphene. *Phys. Rev. B* **2010**, *82*, 115427. [CrossRef]

24. Vlassiouk, I.; Smirnov, S.; Ivanov, I.; Fulvio, P.F.; Dai, S.; Meyer, H.; Chi, M.; Hensley, D.; Datskos, P.; Lavrik, N.V. Electrical and thermal conductivity of low temperature CVD graphene: The effect of disorder. *Nanotechnology* **2011**, *22*, 275716. [CrossRef] [PubMed]

25. Chen, L.; Kumar, S. Thermal transport in graphene supported on copper. *J. Appl. Phys.* **2012**, *112*, 043502. [CrossRef]

26. Nair, R.R.; Blake, P.; Grigorenko, A.N.; Novoselov, K.S.; Booth, T.J.; Stauber, T.; Peres, N.M.R.; Geim, A.K. Fine structure constant defines visual transparency of graphene. *Science* **2008**, *320*, 1308. [CrossRef] [PubMed]

27. Bonaccorso, F.; Sun, Z.; Hasan, T.; Ferrari, A.C. Graphene photonics and optoelectronics. *Nat. Photonics* **2010**, *4*, 611–622. [CrossRef]

28. Minami, T. Transparent conducting oxide semiconductors for transparent electrodes. *Semicond. Sci. Technol.* **2005**, *20*, S35–S44. [CrossRef]

29. Hosono, H. Recent progress in transparent oxide semiconductors: Materials and device application. *Thin Solid Films* **2007**, *515*, 6000–6014. [CrossRef]

30. Calnan, S.; Tiwari, A.N. High mobility transparent conducting oxides for thin film solar cells. *Thin Solid Films* **2010**, *518*, 1839–1849. [CrossRef]

31. Hecht, D.S.; Hu, L.; Irvin, G. Emerging transparent electrodes based on thin films of carbon nanotubes, graphene, and metallic nanostructures. *Adv. Mater.* **2011**, *23*, 1482–1513. [CrossRef] [PubMed]

32. Luka, G.; Krajewski, T.A.; Witkowski, B.S.; Wisz, G.; Virt, I.S.; Guziewicz, E.E.; Godlewski, M. Aluminum-doped zinc oxide films grown by atomic layer deposition for transparent electrode applications. *J. Mater. Sci. Mater. Electron.* **2011**, *22*, 1810–1815. [CrossRef]

33. Park, S.; Vosguerichian, M.; Bao, Z. A review of fabrication and applications of carbon nanotube film-based flexible electronics. *Nanoscale* **2013**, *5*, 1727–1752. [CrossRef] [PubMed]

34. Vosgueritchian, M.; Lipomi, D.J.; Bao, Z. Highly conductive and transparent PEDOT:PSS films with a fluorosurfactant for stretchable and flexible transparent electrodes. *Adv. Funct. Mater.* **2012**, *22*, 421–428. [CrossRef]

35. Ghosh, D.S.; Chen, T.L.; Pruneri, V. High figure-of-merit ultrathin metal transparent electrodes incorporating a conductive grid. *Appl. Phys. Lett.* **2010**, *96*, 041109. [CrossRef]

36. Jeong, J.A.; Kim, H.K. Low resistance and highly transparent ITO–Ag–ITO multilayer electrode using surface plasmon resonance of Ag layer for bulk-heterojunction organic solar cells. *Sol. Energy Mater. Sol. Cells* **2009**, *93*, 1801–1809. [CrossRef]

37. Lewis, J.; Grego, S.; Chalamala, B.; Vick, E.; Temple, D. Highly flexible transparent electrodes for organic light-emitting diode-based displays. *Appl. Phys. Lett.* **2004**, *85*, 3450–3452. [CrossRef]

38. Katsounaros, A.; Hao, Y.; Collings, N.; Crossland, W.A. Optically transparent ultra-wideband antenna. *Electron. Lett.* **2009**, *45*, 722–723. [CrossRef]

39. Roo-Ons, M.J.; Shynu, S.V.; Ammann, M.J.; McCormack, S.J.; Norton, B. Transparent patch antenna on a-Si thin-film glass solar module. *Electron. Lett.* **2011**, *47*, 85–86. [CrossRef]

40. EMI Shielded Conductive Film, TECKFILM™. Available online: www.chomerics.com (accessed on 10 October 2016).

41. Al-Mumen, H.; Rao, F.; Dong, L.; Li, W. Thermo-flow and temperature sensing behaviour of graphene based on surface heat convection. *Micro Nano Lett.* **2013**, *8*, 681–685. [CrossRef]

42. Kong, D.; Le, L.T.; Li, Y.; Zunino, J.L.; Lee, W. Temperature-Dependent Electrical Properties of Graphene Inkjet-Printed on Flexible Materials. *Langmuir* **2012**, *28*, 13467–13472. [CrossRef] [PubMed]

43. Trung, T.Q.; Ramasundaram, S.; Hong, S.W.; Lee, N.-E. Flexible and Transparent Nanocomposite of Reduced Graphene Oxide and P(VDF-TrFE) Copolymer for High Thermal Responsivity in a Field-Effect Transistor. *Adv. Funct. Mater.* **2014**, *24*, 3438–3445. [CrossRef]

44. Yang, J.; Wei, D.; Tang, L.; Song, X.; Luo, W.; Chu, J.; Gao, T.; Shi, H.; Dua, C. Wearable temperature sensor based on graphene nanowalls. *RSC Adv.* **2015**, *5*, 25609–25615. [CrossRef]

45. Yan, C.; Wang, J.; Lee, P.S. Stretchable graphene thermistor with tunable thermal index. *ACS Nano* **2015**, *9*, 2130–2137. [CrossRef] [PubMed]

46. Bendi, R.; Bhavanasi, V.; Parida, K.; Nguyen, V.C.; Sumboja, A.; Tsukagoshi, K.; Lee, P.S. Self-Powered Graphene Thermistor. *Nano Energy* **2016**, *26*, 586–594. [CrossRef]

47. Di, C.-A.; Wei, D.; Yu, G.; Liu, Y.; Guo, Y.; Zhu, D. Patterned graphene as source/drain electrodes for bottom-contact organic field-effect transistors. *Adv. Mater.* **2008**, *20*, 3289–3293. [CrossRef]

48. Choi, D.; Choi, M.-Y.; Choi, W.M.; Shin, H.-J.; Park, H.-K.; Seo, J.-K.; Park, J.; Yoon, S.-M.; Chae, S.; Lee, Y.H.; et al. Fully rollable transparent nanogenerators based on graphene electrodes. *Adv. Mater.* **2010**, *22*, 2187–2192. [CrossRef] [PubMed]

49. Yang, P.-K.; Chang, W.-Y.; Teng, P.-Y.; Jeng, S.-F.; Lin, S.-J.; Chiu, P.-W.; He, J.-H. Fully transparent resistive memory employing graphene electrodes for eliminating undesired surface effects. *Proc. IEEE* **2013**, *101*, 1732–1739. [CrossRef]

50. Seo, D.; Jeon, S.; Seo, S.; Song, I.; Kim, C.; Park, S.; Harris, J.S.; Chung, U.-I. Fully transparent InGaZnO thin film transistors using indium tin oxide/graphene multilayer as source/drain electrodes. *Appl. Phys. Lett.* **2010**, *97*, 172106. [CrossRef]

51. Kun, X.; Chen, X.; Jun, D.; Yanxu, Z.; Weiling, G.; Mingming, M.; Lei, Z.; Jie, S. Graphene transparent electrodes grown by rapid chemical vapor deposition with ultrathin indium tin oxide contact layers for GaN light emitting diodes. *Appl. Phys. Lett.* **2013**, *102*, 162102. [CrossRef]

52. Seo, J.-T.; Han, J.; Lim, T.; Lee, K.-H.; Hwang, J.; Yang, H.; Ju, S. Fully transparent quantum dot light-emitting diode integrated with graphene anode and cathode. *ACS Nano* **2014**, *8*, 12476–12482. [CrossRef] [PubMed]

53. Shin, K.-Y.; Hong, J.-Y.; Jang, J. Flexible and transparent graphene films as acoustic actuator electrodes using inkjet printing. *Chem. Commun.* **2011**, *47*, 8527–8529. [CrossRef] [PubMed]

54. Lee, J.S.; Shin, K.-Y.; Cheong, O.J.; Kim, J.H.; Jang, J. Highly Sensitive and Multifunctional Tactile Sensor Using Free-standing ZnO/PVDF Thin Film with Graphene Electrodes for Pressure and Temperature Monitoring. *Sci. Rep.* **2015**, *5*, 7887. [CrossRef] [PubMed]

55. Liao, C.; Zhang, M.; Niu, L.; Zheng, Z.; Yan, F. Organic electrochemical transistors with graphene-modified gate electrodes for highly sensitive and selective dopamine sensors. *J. Mater. Chem. B* **2014**, *2*, 191–200. [CrossRef]

56. Tanaka, R.; Takaoka, T.; Mizukami, H.; Arai, T.; Iwai, Y. Effects of wavelengths on processing indium tin oxide thin films using diode pumped Nd:YLF laser. *Proc. SPIE* **2003**, *4830*, 36–39.

57. Kim, K.H.; Kwon, S.J.; Mok, H.S.; Tak, T.O. Laser direct patterning of indium tin oxide layer for plasma display panel bus electrode. *Jpn. J. Appl. Phys.* **2007**, *46*, 4282–4285. [CrossRef]

58. Risch, A.; Hellmann, R. Picosecond Laser Patterning of ITO Thin Films. *Phys. Procedia* **2011**, *2*, 133–140. [CrossRef]

59. Park, M.; Chon, B.H.; Kim, H.S.; Jeoung, S.C.; Kim, D.; Lee, J.-I.; Chu, H.Y.; Kim, H.R. Ultrafast laser ablation of indium tin oxide thin films for organic light-emitting diode application. *Opt. Lasers Eng.* **2006**, *44*, 138–146. [CrossRef]

60. Račiukaitis, G.; Brikas, M.; Gedvilas, M.; Rakickas, T. Patterning of indium–tin oxide on glass with picosecond lasers. *Appl. Surf. Sci.* **2007**, *253*, 6570–6574. [CrossRef]

61. Chen, M.-F.; Chen, Y.-P.; Hsiao, W.-T.; Gu, Z.-P. Laser direct write patterning technique of indium tin oxide film. *Thin Solid Films* **2007**, *515*, 8515–8518. [CrossRef]

62. Cheng, J.-Y.; Yen, M.-H.; Hsu, W.-C.; Jhang, J.-H.; Young, T.-H. ITO patterning by a low power Q-switched green laser and its use in the fabrication of a transparent flow meter. *J. Micromech. Microeng.* **2007**, *17*, 2316–2323. [CrossRef]

63. Tseng, S.-F.; Hsiao, W.-T.; Huang, K.-C.; Chiang, D.; Chen, M.-F.; Chou, C.-P. Laser scribing of indium tin oxide (ITO) thin films deposited on various substrates for touch panels. *Appl. Surf. Sci.* **2010**, *257*, 1487–1494. [CrossRef]

64. Shin, H.; Sim, B.; Lee, M. Laser-driven high-resolution patterning of indium tin oxide thin film for electronic device. *Opt. Lasers Eng.* **2010**, *48*, 816–820. [CrossRef]

65. Noach, S.; Faraggi, E.Z.; Cohen, G.; Avny, Y.; Neumann, R.; Davidov, D.; Lewis, A. Microfabrication of an electroluminescent polymer light emitting diode pixel array. *Appl. Phys. Lett.* **1996**, *69*, 3650–3652. [CrossRef]

66. Pawlak, R.; Tomczyk, M.; Walczak, M. Ablation of selected conducting layers by fiber laser. In Proceedings of the 13th International Scientific Conference on Optical Sensors and Electronic Sensors, Lodz, Poland, 22–25 June 2014.

67. Kula, P.; Pietrasik, R.; Dybowski, K.; Atraszkiewicz, R.; Kaczmarek, L.; Szymanski, W.; Niedzielski, P.; Nowak, D.; Modrzyk, W. The growth of a polycrystalline graphene from a liquid phase. *Nanotech* **2013**, *1*, 210–212.

68. Kula, P.; Pietrasik, R.; Dybowski, K.; Atraszkiewicz, R.; Szymanski, W.; Kolodziejczyk, L.; Niedzielski, P.; Nowak, D. Single and Multilayer Growth of Graphene from the Liquid Phase. *Appl. Mech. Mater.* **2014**, *510*, 8–12. [CrossRef]

69. Kula, P.; Szymański, W.; Kolodziejczyk, Ł.; Atraszkiewicz, R.; Dybowski, K.; Grabarczyk, J.; Pietrasik, R.; Niedzielski, P.; Kaczmarek, Ł.; Cłapa, M. High strength metallurgical graphene-mechanisms of growth and properties. *Arch. Metall. Mater.* **2015**, *60*, 2535–2541.

70. Kolodziejczyk, L.; Kula, P.; Szymanski, W.; Atraszkiewicz, R.; Dybowski, K.; Pietrasik, R. Frictional behaviour of polycrystalline graphene grown on liquid metallic matrix. *Tribol. Int.* **2016**, *93*, 628–639. [CrossRef]

71. Ferrari, C.; Meyer, J.C.; Scardaci, V.; Casiraghi, C.; Lazzeri, M.; Mauri, F.; Piscanec, S.; Jiang, D.; Novoselov, K.S.; Roth, S.; Geim, A.K. Raman Spectrum of Graphene and Graphene Layers. *Phys. Rev. Lett.* **2006**, *97*, 1–4. [CrossRef] [PubMed]

72. Das, B.; Chakraborty, A.; Sood, K. Raman spectroscopy of graphene on different substrates and influence of defects. *Bull. Mater. Sci.* **2008**, *31*, 579–584. [CrossRef]

73. Sobon, G.; Sotor, J.; Jagiello, J.; Kozinski, R.; Zdrojek, M.; Holdynski, M.; Paletko, P.; Boguslawski, J.; Lipinska, L.; Abramski, K.M. Graphene oxide vs. reduced graphene oxide as saturable absorbers for Er-doped passively mode-locked fiber laser. *Opt Express.* **2012**, *20*, 19463–19473. [CrossRef] [PubMed]

74. Fu, C.; Zhao, G.; Zhang, H.; Li, S. Evaluation and characterization of reduced graphene oxide nanosheets as anode materials for lithium-ion batteries. *Int. J. Electrochem. Sci.* **2013**, *8*, 6269–6280.

75. Drewniak, S.; Muzyka, R.; Stolarczyk, A.; Pustelny, T.; Kotyczka-Moranska, M.; Setkiewicz, M. Studies of reduced graphene oxide and graphite oxide in the aspect of their possible application in gas sensors. *Sensors* **2016**, *16*, 103. [CrossRef] [PubMed]

76. Hafiz, S.M.; Ritikos, R.; Whitcher, T.J.; Razib, N.M.; Bien, D.C.S.; Chanlek, N.; Nakajima, H.; Saisopa, T.; Songsiriritthigul, P.; Huanga, N.M.; Rahman, S.A. A practical carbon dioxide gas sensor using room-temperaturehydrogen plasma reduced graphene oxide. *Sens. Actuators B Chem.* **2014**, *193*, 692–700. [CrossRef]

77. Pimenta, M.A.; Dresselhaus, G.; Dresselhaus, M.S.; Cancado, L.G.; Jorio, A.; Saito, R. Studying disorder in graphite-based systems by Raman spectroscopy. *Phys. Chem. Chem. Phys.* **2007**, *9*, 1276–1291. [CrossRef] [PubMed]

6

A Formal Methodology to Design and Deploy Dependable Wireless Sensor Networks

Alessandro Testa [1], Marcello Cinque [2], Antonio Coronato [3,*] and Juan Carlos Augusto [4]

[1]   Ministero dell'Economia e delle Finanze, Rome 00187, Italy; alessandro.testa@tesoro.it
[2]   Dipartimento di Ingegneria Elettrica e delle Tecnologie dell'Informazione,
     University of Naples "Federico II", Naples 80125, Italy; macinque@unina.it
[3]   CNR-ICAR, Naples 80131, Italy
[4]   Department of Computer Science and R.G. on Development of Intelligent Environments,
     Middlesex University of London, London NW4 2SH, UK; j.augusto@mdx.ac.uk
*   Correspondence: antonio.coronato@icar.cnr.it

Academic Editor: Leonhard M. Reindl

Abstract: Wireless Sensor Networks (WSNs) are being increasingly adopted in critical applications, where verifying the correct operation of sensor nodes is a major concern. Undesired events may undermine the mission of the WSNs. Hence, their effects need to be properly assessed before deployment, to obtain a good level of expected performance; and during the operation, in order to avoid dangerous unexpected results. In this paper, we propose a methodology that aims at assessing and improving the dependability level of WSNs by means of an event-based formal verification technique. The methodology includes a process to guide designers towards the realization of a dependable WSN and a tool ("ADVISES") to simplify its adoption. The tool is applicable to homogeneous WSNs with static routing topologies. It allows the automatic generation of formal specifications used to check correctness properties and evaluate dependability metrics at design time and at runtime for WSNs where an acceptable percentage of faults can be defined. During the runtime, we can check the behavior of the WSN accordingly to the results obtained at design time and we can detect sudden and unexpected failures, in order to trigger recovery procedures. The effectiveness of the methodology is shown in the context of two case studies, as proof-of-concept, aiming to illustrate how the tool is helpful to drive design choices and to check the correctness properties of the WSN at runtime. Although the method scales up to very large WSNs, the applicability of the methodology may be compromised by the state space explosion of the reasoning model, which must be faced by partitioning large topologies into sub-topologies.

Keywords: Wireless Sensor Networks; formal methods; dependability; metrics; modeling

1. Introduction and Motivation

Wireless Sensor Networks (WSNs) [1] are being increasingly used in critical application scenarios where the level of trust on WSNs becomes an important factor, affecting the success of industrial WSN applications. The extensive use of this kind of network stresses the need to verify their dependability—dependability is defined as the ability of a system to deliver a service that can justifiably be trusted [2]. It is an integrated concept encompassing the attributes of reliability, availability, maintainability, safety, and integrity—not only at design time to prevent wrong design choices but also at runtime in order to make a WSN more robust against failures that may occur during its operation.

Typical critical scenarios are environmental monitoring (e.g., detection of fires in forests [1]), structural monitoring of civil engineering structures [3], health monitoring (in medical scenarios) [4] and patient monitoring [5,6] by means of Ambient Intelligence (AmI [7]) systems. For example, in the case of remote patient monitoring, alerts must be raised and processed within a temporal threshold; functional incorrectness or runtime failures may result in catastrophic consequences for the patient. These types of applications are considered safety-critical and they must be designed and developed with intrinsic and stringent dependability requirements [8]. Thus, depending on the application scenarios, different dependability requirements can be defined, such as, node lifetime, network resiliency and coverage of the monitoring area.

The work presented in [9] evidenced that also in a simple deployment, a single node can be responsible for the failure of a huge piece of the network. For instance, a node that is close to the sink (i.e., the gateway of the WSN that has the role to collect all of the measures detected by the sensors) is more likely to fail due to the great demand that it is subjected to, and its failure would likely cause the isolation of a set of nodes from the sink.

Therefore, it is necessary to verify the WSNs at design time against given correctness properties, in order to increase the confidence about the robustness of the designed solution before putting it into operation. Also, in the IoT (Internet of Things) networks, the design of a dependable WSN platform has became a main focus of research. In fact, in [10], the authors propose a functional design and implementation of a complete WSN platform that can be used for a range of long-term environmental monitoring IoT applications.

Moreover, it is also important to avoid unexpected results or dangerous effects during the runtime of the WSN; this can be obtained by checking traces of events generated from the system run against the same correctness properties used at design time.

Formal methods can be used for these purposes, due to their wide adoption in the literature to verify the correctness of a system specification [11] or to perform runtime verification [12,13]. However, the practical use of formal methods for the verification of dependability properties of WSNs has received little attention, due to the distance between system engineers and formal methods experts and the need to re-adapt the formal specification to different design choices. Even if a development team were to invest on the definition of a detailed specification of WSN correctness properties, a design change (e.g., different network topology) could require one to rethink the formal specification, incurring extra undesirable costs.

To overcome these limitations, the contribution of this paper is manyfold. Specifically, we propose:

- a formal methodology to support the design and deployment of dependable WSNs both at design time (*static verification*) and at runtime (*runtime verification*);
- the definition of a unique formal specification of WSN correctness, based on the event calculus formalism, subdivided in two logical subsets: a *general correctness specification*, valid independently of the particular WSN under study, and a *structural specification* related to the properties of the target WSN (e.g., number of nodes, topology, etc.);
- the adoption of specific WSN dependability metrics, such as *connection resiliency* and *coverage*, introduced in [14], for measuring dependability degree and providing a quantitative assessment of a WSN;
- an automated verification tool, named ADVISES (AutomateD VerIfication of wSn with Event calculuS), to facilitate the adoption of the proposed approach.

The key idea of the proposed methodology is to base the verification of correctness properties on a set of specifications that can be used interchangeably at design time and at runtime. The decomposition of the specification in two sets simplifies the adoption of the approach. While general correctness specifications do not need to be adapted when changing the target WSN, structural specifications depend on the target, and are designed to be generated automatically.

The ADVISES tool facilitates the adoption of the proposed methodology by system engineers with no experience of formal methods. At design time, it can be used to perform a *robustness checking* of the target WSN, i.e., to verify the long-term robustness of the WSN (in terms of the proposed metrics) against random sequences of undesired events, useful to identify corner cases and dependability bottlenecks. At runtime, it monitors the deployed WSN. If an undesired event occurs, the tool calculates the current values of dependability metrics (e.g., raising an alarm if a given criticality level is reached) and it assesses the criticality of the network, in terms of the future hazardous scenarios that can happen, considering the new network conditions.

The proposed methodology and tool has been applied on two realistic WSNs representative of health monitoring scenarios. Whilst these scenarios are on the smaller size, they are valuable as proof-of-concept and also to explain how to use the generic problem solving approach and tool presented in this article. The case studies show how the approach is useful to deeply investigate the reasons of inefficiency and to re-target design choices. They also show how the same specification can be used at runtime to check the correct behavior of a real WSN, deployed on the field and monitored by ADVISES.

The technique presented in this paper is applicable to homogeneous WSNs with static routing topologies. In addition, for illustrative reasons, the specifications presented in the paper focus on node connectivity and isolation problems, and do not take into account the quality of produced sensor data. However, let us stress that the specifications can be easily extended to cover the aspects of WSN behavior not covered in the paper. We will provide specific examples of possible extensions in the paper.

The rest of the paper is organized as follows. A discussion of the related work is given in Section 2. Section 3 is dedicated to the description of a process underlying our the proposed methodology. Section 4 reports the definition of the correctness specifications (general and structural) of WSNs following the *event calculus* formalism. Section 5 addresses the static verification in the form of *robustness checking*. The runtime verification technique is discussed in Section 6. In Section 7 the case studies are discussed. Finally, in Section 8 the paper concludes with final remarks, a discussion of limitations, and indications for future work.

2. Related Work

Several approaches have been proposed in the literature for the dependability evaluation of WSNs properties: experimental, simulative, analytical and formal.

Experimental approaches are used to measure the dependability directly from a real distributed system during its operation, and thus they allow the analysis of dependability at runtime [15]. In the field of WSNs, Li and Liu presented in [16] a deployment of 27 Crossbow Mica2 motes that compose a WSN. They describe a Structure-Aware Self-Adaptive WSN system (SASA) designed in order to detect changes of the network due to unexpected collapses and to maintain the WSN integrity. Detection latency, system errors, network bandwidth and packet loss rate were measured; coverage and connection resiliency metrics are not considered. In the prototyping phase, it is possible to perform an accelerated testing, for example by forcing faults in sensor nodes (by means of *Fault Injection* (FI) [17,18]). In [19], a dependability model and a dependable distributed WSN framework for Structural Health Monitoring (called DependSHM) are presented. Another relevant work on dependable WSN for SHM is [20]. An approach is presented in order to repair the WSN and guarantee a specified degree of fault tolerance. The approach searches the repairing points in clusters and places a set of backup sensors at those points. Shakkira et al. [21] propose a lightweight trust decision-making scheme that has also been extended to reduce calculation overheads and improve energy efficiency by integrating an innovative protocol defined for data exchange.

Simulative approaches for assessing WSNs usually make use of behavioral simulators, i.e., tools able to reproduce the expected behavior of a system by means of a code-based description, and they are involved in the design phase. Typical simulative approaches to evaluate WSN fault/failure models are provided in [22,23]. In [22] authors address the problem of modeling and evaluating the reliability of the communication infrastructure of a WSN. The first on-line model-based testing technique [23] has been conceived to identify the sensors that have the highest probability to be faulty. Behavioral simulators, as NS-2 [24] and Avrora [25], allow the reproduction of the expected behavior of WSN nodes on the basis of the real application planned to execute on nodes. However, it is not always possible to observe non-functional properties of WSNs by means of simulative approaches, since models need to be redefined and adapted to the specific network to simulate.

The study of the performance and dependability of WSNs can be performed by means of analytical models [26–31]. In [26], authors introduce an approach for the automated generation of WSN dependability models, based on a variant of Petri nets. An analytical model to predict the battery exhaustion and the lifetime of a WSN (LEACH) is discussed in [27]. In [28] the authors present a network state model used to forecast the energy of a sensor. In [32] , a methodology to evaluate the reliability and availability of Wireless Sensor Networks in industrial environments that are subject to permanent faults on network devices is proposed. In [30], authors propose a reactive distributed scheme for detecting faulty nodes. In [29], k-connectivity in secure wireless sensor networks under the random pairwise key predistribution scheme with unreliable links is investigated. In [31], a technique to mitigate identity delegation attacks is introduced.

Formal approaches offer new opportunities for the dependability study of WSNs. Recently, different formal methods and tools have been applied for the modeling and analysis of WSNs, such as [33–35]. In [33], Kapitanova and Son apply a formal tool to WSNs. They propose a formal language to specify a generic WSN and a tool to simulate it. However, the formal specification has to be rewritten if the WSN under study changes. In [34], Man et al. propose a methodology for modeling, analysis and development of WSNs using a formal language (PAWSN) and a tool environment (TEPAWSN). They consider only power consumption as a dependability metric that is necessary but not sufficient to assess the WSN dependability (e.g., other problems of WSN, such as the isolation problem of a node have been analyzed) and also they apply only simulation. In [35], Boonma and Suzuki describe a model-driven performance engineering framework for WSNs (called Moppet). This framework uses the event calculus formalism [36] to estimate, only at design time, the performance of WSN applications in terms of power consumption and lifetime of each sensor node; other dependability metrics, such as coverage and connection resiliency are not considered. The features related to a particular WSN have to be set in the framework every time that a new experiment starts.

There are some papers ([37,38]) that considered formal methods in real-time contexts. In [37], Olveczky and Mesenguer model and study WSN algorithms using the Real-Time Maude formalism. Though authors adopt this formalism, they use an NS-2 simulator to analyze the considered scenarios, making the work very similar to simulative approaches. The work presented in [38] considers a WSN as a Reactive Multi-Agent System consisting of concurrent reactive agents. In this paper, dependability metrics are not treated and calculated and authors just describe the structure of a Reactive Decisional Agent by means of a formal language.

Finally, in [39], the authors model and evaluate the reliability and lifetime of a WSN under some typical working scenarios based on the sensor node modes (sleep and active) and the mechanism of alternating between these modes. By means of numerical examples, they illustrate the reliability and lifetime of a WSN. Wang et al. [40] implement binary decision diagrams based algorithms to facilitate the design, deployment, and maintenance of reliable WSNs for critical applications.

An open issue with formal specifications of WSNs is that they need to be adapted when changing the target WSN configuration, e.g., in terms of the number of nodes and topology. To address this problem, it is necessary to provide separated specifications and thus conceive two logical sets of specifications: a general specification for WSN correctness properties that is valid for any WSN, and a structural specification related to the topology of the target WSN, designed in order to be generated automatically. Currently, there are proposals in the literature documenting the application of formal methods to model the WSN but they only focus on some dependability metrics as lifetime and power consumption; it is necessary to provide a method of assessing the dependability in terms of other important key dependability metrics, such as coverage and connection resiliency to undesired events. Moreover, no approach has been defined using formal methods for doing static and runtime WSN verification as we propose in this paper.

3. The Process of the Methodology

Formal methods have been widely adopted in the literature to verify the correctness of a system, taking into account specifications. The verification is performed by providing proof on an abstract mathematical model of the system. Until now, there is no work that has proven how to use a unique formal approach to perform dependability assessment at design time and at runtime.

This paper defines a new, full, formal methodology to support verification of WSNs both at design time (through *static verification*) and runtime (through *runtime verification*) exploiting only one set of formal specifications divided into two subsets: general (unchangeable and valid for any WSN) and structural (variable on the basis of particular configuration of the WSN) specifications.

In this Section, we introduce our proposed methodology modeled as a process characterized in steps illustrated in Figure 1.

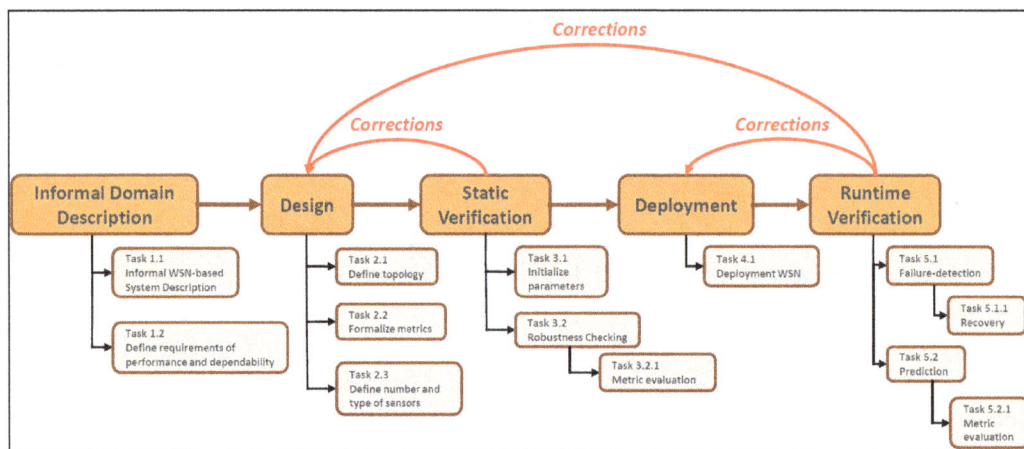

Figure 1. The process of the proposed methodology.

The process is characterized by five sequential phases: *Informal Domain Description*, *Design*, *Static Verification*, *Deployment* and *Runtime Verification*; the entire process consists of 13 tasks.

3.1. Informal Domain Description

In this phase, the application is described and its requirements of performance and dependability (e.g., coverage threshold) are defined.

3.1.1. Task 1.1: Informal WSN-Based System Description

Using natural language, the WSN-based system domain is described in a textual form that is useful to understand what is the considered scenario, what are possible events that can occur, the number of sensors of the network, what they sense, and how many redundant nodes there are.

3.1.2. Task 1.2: Define Requirements of Performance and Dependability

After having defined the application domain, in this task, the performance and dependability characteristics are described. For example, "We want to guarantee at least 65% of coverage for an area against occurred failure events", "We want the number of failure events active at the same time to be no higher than half of the total number of sensors".

The dependability metrics considered in this paper, are:

- *connection resiliency* represents the number of node failures and disconnection events that can be sustained while preserving a given number of nodes connected to the sink.
- *coverage* is the time interval in which the WSN can operate, while preserving a given number of nodes connected to the sink.

The computation of these metrics is threshold-based. The threshold expresses the fraction of failed and isolated nodes that the user can tolerate, given its design constraints. For instance, a WSN of 20 nodes and a threshold set to 50% means that, at most, 10 isolated nodes can be tolerated.

The coverage is then defined as the interval $[0, t]$, t being the timepoint of the event that caused the isolation of a number of nodes exceeding the threshold.

Clearly, the definition of the threshold depends on the specific problem and relies on expert humans who want to use the approach to make an informed estimation that is useful to improve their design (such as, how to improve the WSN topology in order to make it tolerant to at least 50% of node failures). Nevertheless, while for some applications, a fixed threshold could not be defined, the approach does not restrict the use of a single threshold. Metrics can be evaluated with varying values of thresholds, e.g., to find out the maximum number of tolerable failures to achieve the wanted resiliency level.

It is worth noting that it is easy to extend the set of metrics. Indeed, in some related work, we considered also metrics related with the correct delivery of packets and the power consumption as performance and dependability metrics.

3.2. Design

This phase includes the design of a WSN defining the topology represented by means of a tree graph with the sink node as the root. Since several WSNs are composed by sensors that are in a fixed place (i.e., in the case of structural monitoring, hospital environment monitoring, fire monitoring, home environment monitoring, etc), we focus on static routing topologies that can be represented as directed spanning trees [41] (with the sink as the root). Dependability metrics are formalized and number and type of sensors are defined.

3.2.1. Task 2.1: Define Topology

This task defines the topology; nodes and links are identified. From the structure of the WSN, we retrieve the corresponding spanning tree.

3.2.2. Task 2.2: Formalize Metrics

This task formalizes the required metrics on the basis of requirements defined in the previous step (e.g., coverage computation).

3.2.3. Task 2.3: Define Number and Type of Sensors

This task is complementary to the Task 2.1 in order to define topology and fix the number of the sensors. In particular, the properties of sensors -e.g., Receive/Transmit (RX/TX) energy consumption- are defined since they are considered in the next phase (*Static verification*).

3.3. Static Verification

The designed WSN is verified in terms of the defined dependability properties. The verification is static in the sense that the network is not still operating.

The effects of the reasoning, performed on the basis of the defined specifications (general and structural), impact on the designed WSN. For instance, if the results of the static verification do not meet the dependability requirements, it could be necessary to modify the topology of the WSN, or relax the requirements.

3.3.1. Task 3.1: Initialize Parameters

This task focuses on the choice and initialization of the parameters needed to perform static verification. The designed topology is loaded; then users can set several parameters, such as number of timepoints, sensors and tolerated failures. Moreover, dependability thresholds are set.

3.3.2. Task 3.2: Robustness Checking

In this task, the WSN design is verified by means of an event-based formal approach. Dependability metrics (formalized in the Section 3.2.2) are evaluated against random sequences of undesired events, useful to identify corner cases and dependability bottlenecks.

3.3.3. Task 3.2.1: Metric Evaluation

This task performs the computation of the dependability metrics (such as coverage and connection resiliency) analyzing the outcome produced by the robustness checking process. The metrics are calculated on the basis of their definitions and considering the output produced by the formal reasoner. Once the metric values have been obtained, this task provides the comparison among them and the threshold values set by the user.

3.4. Deployment

In this phase, the WSN, verified at design time, is actually deployed distributing wireless sensors in one or more environments.

Task 4.1: Deployment WSN

The aim of this task is to physically deploy the WSN in one or more environments.

3.5. Runtime Verification

Runtime verification is performed on the final running system. The aim is to formally check the running system (the WSN) against some correctness properties. The detection of the violation of a correctness property may be used either to trigger a recovery procedure for the running system, or to handle the incorrect status in a safe way. This feature is very useful as long as (although the correctness of the models of the system is granted by formal static verification activities) the quality of the running system may degrade to an unacceptable level, which can be identified just by runtime verification activities.

3.5.1. Task 5.1: Failure-Detection

In this task, failure events (node failures, disconnections, ...), occurring in the wireless sensors, are detected during the system running and an event is generated in a particular formalism in order to start the computing of the new dependability degree.

3.5.2. Task 5.1.1: Recovery

In this task, if the current value of metrics is lower than the desired threshold, the network characteristics (topology, position of the nodes, power of transmission, etc.) can be modified to let the WSN be able to satisfy the required dependability level. For instance, a node X can become a dependability bottleneck if it is positioned in a way that makes it the only one to be used to connect two different portions of the WSN. A failure of node X then makes a whole portion of the WSN isolated. In the recovery task, once the problem is understood, the positions of node X can be modified in order to let it share the traffic load with another node.

3.5.3. Task 5.2: Prediction

This task may allow the designing of predictive models for the running system. In particular, runtime verification is of course mainly focused on identifying the current situation of the system, which is the result of the sequence of events collected until the current moment. However, starting from the current situation and the sequence of past events, it is possible to exploit predictive models to foresee the forthcoming levels of dependability of the system.

3.5.4. Task 5.2.1: Metric Evaluation

This task operates like task 3.2.1. Of course, in this case, the target of the measure is the final running system, instead of a model of it.

4. Specifications

In this section, we describe the formal specification of WSN correctness composed by two logical sets: in the first one, we define invariant rules that are applicable to any WSN and thus written only one time; in the second set, we define variable specifications that are dependent on a given WSN structure (i.e., topology, number of nodes, sent packets, etc.). All the defined formal specification underlies the verification process described in the previous section to perform static and runtime verification:

1. *general correctness specification*—set of correctness properties' specifications, valid independently of the particular WSN under study
2. *structural specification*—a set of specifications and parameters related to the properties of the target WSN, e.g., number of nodes, network topology, quality of the wireless channel (in terms of disconnection probability), and initial charge of batteries. It has to be adapted when changing the target WSN, having thus the advantage of minimizing the effort.

4.1. Event Calculus

Since the normal and failing behavior of a WSN can be characterized in terms of an event flow (for instance, a node is turned on, a packet is sent, a packet is lost, a node stops working due to crash or battery exhaustion, or it gets isolated from the rest of the network due to the failure of other nodes, etc.), we adopt an event-based formal language. In particular, among several event-based formal languages, we choose Event Calculus, since its simplicity, its wide adoption in the sensor networks arena [35,42–44], and the possibility to formally analyze the behavior of a system as event flows, offer simple ways to evaluate the dependability metrics of our interest, even at runtime.

Event calculus was proposed for the first time in [45] and then it was extended in several ways [46]. This language belongs to the family of logical languages and it is commonly used for representing and reasoning regarding the events and their effects [47].

Fluent, event and predicate are the basic concepts of event calculus [36]. Fluents are formalized as functions and they represent a stable status of the system. For every timepoint, the value of fluents or the events that occur can be specified.

This language is also named "narrative-based": in the event calculus, there is a single time-line on which events occur and this event sequence represents the narrative. Dependability metrics can be evaluated by analyzing the narrative generated by an event calculus reasoner based on the specification of the target WSN. A narrative is useful to understand a particular behavior of a WSN.

The most important and used predicates of event calculus are:Initiates, Terminates, HoldsAt and Happens.

Supposing that e is an event, f is a fluent and t is a timepoint, we have:

- Initiates (e, f, t): it means that, if the event e is executed at time t, then the fluent f will be true after t.
- Terminates (e, f, t): it has a similar meaning, with the only difference being that when the event e is executed at time t, then the fluent f will be false after t.

- HoldsAt (f, t): it is used to tell which fluents hold at a given timepoint.
- Happens (e, t): it is used when the event e occurs at timepoint t.

Several techniques are considered to perform automated reasoning in event calculus, such as satisfiability solving, first-order logic automated theorem proving, Answer Set Programming and logic programming in Prolog.

To check the proposed correctness properties defined in event calculus, we use the Discrete Event Calculus (DEC) Reasoner [48]. The DEC Reasoner uses satisfiability (SAT) solvers [49] and by means of this we are able to perform reasoning, such as deduction, abduction, post-diction, and model finding. The DEC Reasoner is documented in detail in [50,51] in which its syntax is explained (e.g., the meaning of the symbols used in the formulas).

4.2. General Correctness Specification

The general correctness specification is described in the following. It specifies that a WSN performs correctly if no undesired events (or failures) happen. From the results of a detailed Failure Modes and Effect Analysis (FMEA) conducted on WSNs in [52], the following are examples of undesired events, ordered from the most severe one to the least severe one:

1. isolation event, i.e., a node is no longer able to reach the sink;
2. packet loss event, i.e., a packet is lost during the traversal of the network;
3. battery exhaustion event, i.e., a node stops working since it has run out of battery.

These first two types of events are actually not independent, but might be caused by simpler "basic events", such as the stop of one or more nodes (e.g., due to crash or battery exhaustion), or the temporary disconnection of a node to its neighbor(s) due to transmission errors. So, the occurrence of the third event (battery exhaustion) might cause packet losses and isolation events, if no alternative routes are present in the network. In turn, the battery exhaustion event is dependent on the power consumption of the nodes as a consequence of packet sending and receiving activities (in general assumed to be power demanding activities with respect to CPU activities [53]).

In this paper, we concentrate on the specifications related to the first event (isolation event) being the most severe one according to the FMEA in [52]; specifications of the other two events (packet loss and battery exhaustion), defined also in [52], are here omitted for a matter of space.

The isolation event happens when a node is no longer able to reach the sink of the WSN, i.e., the gateway node where data are stored or processed. For instance, considering the WSN represented in Figure 2, if node i fails, then nodes j, k, and in general all the nodes belonging to the subnet A become isolated.

More in general, if a subnet depends on a node and this node stops or becomes isolated, then all of the nodes of the subnet are isolated.

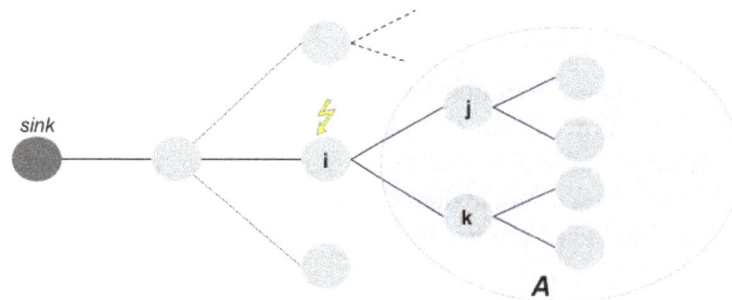

Figure 2. Isolation of a Wireless Sensor Network subnet.

In Table 1, we report the basic elements (sorts, events and fluents) used for the specification of situations like these. We distinguish basic events from generated events. These last events are generated by the reasoner on the basis of the specification and of the sequence of basic events which actually occurred.

Table 1. Basic elements of the specification for the isolation event.

Elements	Name	Description
Sorts	sensor	Reference sensor for events and fluents
	to_sensor	Sensor used in case of connection (i.e., a sensor connects to another sensor)
	from_sensor	Sensor used in case of disconnection (i.e., a sensor disconnects from another sensor)
Basic Events	Start (sensor)	Occurring event when a sensor turns on
	Stop (sensor)	Occurring event when a sensor turns off
	Connect (sensor, to_sensor)	Occurring event when a sensor connects to another sensor
	Disconnect (sensor, from_sensor)	Occurring event when a sensor disconnects from another sensor
Generated Events	Isolate (sensor)	Occurring event when a sensor is isolated from the network
Fluents	IsAlive (sensor)	True when a Start event occurs for a sensor
	IsLinked (sensor, to_sensor)	True when a Connect event occurs
	IsReachable (sensor)	True when a sensor is reachable from the sink node
	Neighbor (sensor1, sensor2)	True when sensor 1 is directly linked to sensor 2

Listing 1 shows the rules that represent the core of the specification for an isolation event. In lines 1–7, we define a rule to verify when a node becomes isolated. A sensor can be isolated if it is initially reachable, alive and, considering a link with another sensor, there is no other sensor that is alive, reachable and connected with the isolated sensor.

Also, we report (in lines 9–14) another rule which allows the checking of a Join event. In particular, a sensor that is isolated can rejoin the network and become reachable again if it still is alive, its neighbor sensor is alive and reachable from the sink node and a connection between them has been restored.

In lines 16–28, we show conditions in which Isolation and Join events cannot occur. We claim that an Isolation event cannot occur in a sensor (in lines 16–21) if at least one of its neighbor sensors is alive and reachable and is connected with the sensor. Moreover, if a sensor is not reachable, due to a previous isolation, or not alive, it cannot receive another Isolation event again. In the similar way (in lines 23–28), if a sensor is reachable or not alive or all of its neighbor sensors are not alive or not reachable, then it cannot join to the network, remaining isolated.

Listing 1: Correctness Specification for the Isolation event

```
[1]  [sensor ,from_sensor, time] Neighbor(from_sensor,sensor) & HoldsAt(
[2]       IsReachable(sensor),time) & HoldsAt(IsAlive(sensor),time) &
[3]       (!{from_sensor2} (HoldsAt(IsAlive(from_sensor2),time) &
[4]       HoldsAt(IsReachable(from_sensor2),time) & HoldsAt(
[5]       IsLinked(sensor,from_sensor2),time)) &
[6]       Neighbor(from_sensor2,sensor)) ->
[7]  Happens(Isolate(sensor),time).
[8]
[9]  [sensor ,from_sensor, time] ( !HoldsAt(IsReachable(sensor),time) &
[10]      HoldsAt(IsAlive(sensor),time) & HoldsAt(IsAlive(
[11]      from_sensor),time) & HoldsAt(IsReachable(from_sensor),time))
[12]       & HoldsAt(IsLinked(sensor,from_sensor),time) &
[13]      Neighbor(from_sensor,sensor) ->
[14] Happens(Join(sensor),time).
[15]
```

```
[16] [sensor,from_sensor, time] ((HoldsAt(IsAlive(from_sensor),time) &
[17]      HoldsAt(IsReachable(from_sensor),time) & HoldsAt(
[18]      IsLinked(sensor,from_sensor),time)) | !HoldsAt(
[19]      IsReachable(sensor),time) | !HoldsAt(IsAlive(sensor),time))
[20]      & Neighbor(from_sensor,sensor) ->
[21]!Happens(Isolate(sensor),time).
[22]
[23][sensor,from_sensor,time] ( HoldsAt(IsReachable(sensor),time) |
[24]      !HoldsAt(IsAlive(sensor),time) | !HoldsAt(
[25]      IsLinked(sensor,from_sensor),time) |
[26]      !HoldsAt(IsAlive(from_sensor),time) | !HoldsAt(
[27]      IsReachable(from_sensor),time)) & Neighbor(from_sensor,sensor)->
[28]!Happens(Join(sensor),time).
```

4.3. Structural Specification

General correctness specifications are complemented by a structural specification that comprises a set of specifications and parameters related to the properties of the target WSN, e.g., number of nodes, network topology, quality of the wireless channel (in terms of disconnection probability), and initial charge of batteries. This specification depends on a particular WSN topology and thus, differently from the specifications described in the previous sub-section, it varies on the basis of the characteristics of the target WSN.

To specify the topology, we use the predicate Neighbor (already used in the previous specifications) to indicate how nodes are linked in the topology. For instance, considering the topology in Figure 3: node i is connected with j and k and the sink (root node) is the node i.

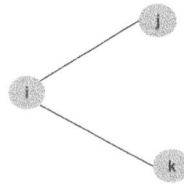

Figure 3. Example of topology of a WSN

The resulting specification is reported in Listing 2, where *sensor 1* is the parent node (i) and *sensor 2* is child nodes (j and k). Clearly, this specification can be changed easily if the topology of the WSN changes.

Listing 2: Use of the Neighbor predicate in a structural specification

```
[1] [sensor1,sensor2] Neighbor(sensor1,sensor2) <-> (
[2] (sensor1=i & sensor2=j) |
[3] (sensor1=i & sensor2=k)
[4] ).
```

The role of the Neighbor predicate is very important to understand when an axiom can be applied. Let us examine the axiom related at a possible isolation (lines 1–7 of listing reported in Listing 1) and let us apply it for Figure 3. The described implication is true when, given a couple of nodes (sensor, from_sensor), the conditions about isolation are true and there is a link between nodes (in this case, between node j and i or between node k and i). This, for instance, can never be true for the couple of nodes j and k, since there is not a physical link between them.

Regarding the parameters, their values can be used to check the correctness properties of the WSN under different conditions, i.e., under different assumptions on the initial charge of batteries (e.g., to verify a WSN in the middle of its life), or under different environmental conditions affecting the quality of the channels (impacting on the probability of having a disconnection event when checking the robustness of the WSN).

4.4. Metrics Computation

The metrics of interest can be evaluated by using the narrative produced by the reasoner starting from the specification.

Starting from the coverage, it can be calculated by considering the threshold value and by analyzing the IsReachable(sensor) and IsAlive(sensor) fluents found to be true in the event trace produced by the reasoner: if a IsReachable(x) or a IsAlive(sensor) fluent is false in the event trace, this means that node x became isolated or it stopped. For example, in the case of coverage at 50%, for a WSN with seven nodes, there is coverage when at least four nodes are not isolated (i.e., they are reachable). Hence, as soon as four different nodes are neither reachable nor alive (looking at the fluents), the network is not covered anymore. The coverage can be then evaluated as the interval $[0, t]$, t being the timepoint of the last failure or disconnection event in the narrative before the isolation (e.g., the timepoint of the event that caused the isolation of a number of nodes exceeding the threshold).

The connection resiliency can then be evaluated as the number of failure and disconnection events (namely, Stop(sensor) and Disconnect(sensor, from_sensor) events) that happen within the coverage interval, excluding the last failure/disconnection event, that is, the one that actually leads the number of isolated nodes to overcome the threshold. For example, if we have coverage in the interval $[0, 6]$, and during this period three failure/disconnection events can be counted, then the connection resiliency is 2, that is, the WSN was able to tolerate two failures or disconnections while preserving more than 50% of the nodes connected.

Let us stress that the specifications and metrics adopted in the paper are chosen for illustrative reasons to show the use of the proposed methodology for static and runtime verification in practical terms. If needed in particular application settings, the specification can be easily extended with more fluents and then the narrative used to evaluate other metrics. For instance, one can add a "battery level" fluent to evaluate the power consumption of nodes, or a "packet delivery" fluent to model the flow of packets among nodes and evaluate the probability of correct packet delivery.

4.5. Example: A Wireless Body Sensor Network

Let us consider a simple example to show the use of the specification on a WSN and how the narrative produced by the event calculus reasoner is useful to compute the metrics of interest. In particular, we consider a wireless body sensor network (WBSN) realized by Quwaider et al. [54] and illustrated in Figure 4.

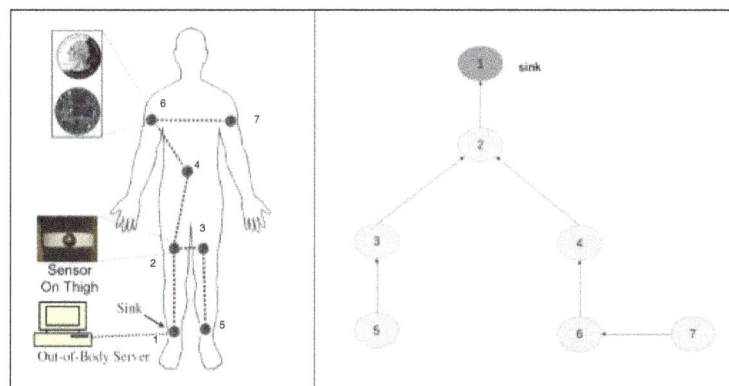

Figure 4. Quwaider wireless body sensor network (WBSN) and related topology.

This WBSN (Figure 4 on left side) is constructed by mounting seven sensor nodes attached on two ankles, two thighs, two upper-arms and one on the waist area. Each node consists of a 900 MHz Mica2Dot MOTE (running Tiny-OS operating system).

On the right side of Figure 4, we report the node tree graph corresponding to the WSN, where the arrows indicate the relationship between a couple of nodes (i.e., node 2 depends on node 1, node 3 and 4 depend on node 2, etc.).

The corresponding specification of the topology is reported in the Listing 3.

Assuming a coverage threshold of 50%, let us suppose to be interested in analyzing the behavior of the WBSN if the following events occur: Disconnect(5,3) at timepoint 1 and Stop(4) at timepoint 3.

Listing 3: Structural specification of the WBSN topology

```
[6]  [sensor1 ,sensor2] Neighbor(sensor1 ,sensor2) <-> (
[7]  (sensor1=1 & sensor2=2) |
[8]  (sensor1=2 & sensor2=3) | (sensor1=2 & sensor2=4) |
[9]  (sensor1=3 & sensor2=5) |
[10] (sensor1=4 & sensor2=6) |
[11] (sensor1=6 & sensor2=7)
[12] ).
```

If the specification is correct, we should observe a coverage interval that equals to $[0,3]$ (i.e., when node 4 stops at timepoint 3, four nodes are not reachable, namely 4, 5, 6 and 7), and a connection resiliency equals to 1 (i.e., only one event—the Disconnect (5,3) event—is tolerated). To test the desired event sequence, we add an event trace (Listing 4) to the specification, composed by a list of Happens predicates that specify nodes and timepoints in which a given event occurs. The completion statement specifies that a predicate symbol (i.e., Happens) should be subject to predicate completion.

Listing 4: Event trace

```
[14] Happens(Disconnect(5 ,  3) ,1).
[15] Happens(Stop(4) ,3).
[16]
[17] completion Happens
[18]
```

Finally, in the last part, we consider ranges of values for sensors and timepoints (Listing 5). In this case, we know that the network is composed of seven nodes and we want to observe what could happen in 10 timepoints.

Listing 5: Parameters

```
[21] range sensor 1 7
[22] range time 0 10
```

Listing 6 reports the outcome (the narrative) produced by the DEC Reasoner. The event trace confirms our expectations. We can observe that after the stop of node 4, nodes 6 and 7 become not reachable. Considering that node 5 was already not reachable, this means that a total of four nodes are isolated. The coverage is computed as the timepoint of the last failure event causing such isolation, that is 3. Consequently, the connection resiliency is computed by counting the number of failure and disconnection events in the interval $[0,3]$, excluding the last event; hence, it is equal to 1, as expected.

For this example, we consider both the coverage and connection resiliency threshold set as values chosen without particular meaning; in fact, on the basis of the desired outcome, the thresholds can be set by means of the our proposed tool.

Listing 6: Outcome of the DEC Reasoner

```
[1]  0
[2]  1
[3]  Happens(Disconnect(5, 3), 1).
[4]  2
[5]  -IsLinked(5, 3).
[6]  Happens(Isolate(5), 2).
[7]  3
[8]  -IsReachable(5).
[9]  Happens(Stop(4), 3).
[10] 4
[11] -IsAlive(4).
[12] Happens(Isolate(6), 4).
[13] 5
[14] -IsReachable(6).
[15] Happens(Isolate(7), 5).
[16] 6
[17] -IsReachable(7).
[20] 7
[21] 8
[22] 9
[23] 10
```

5. Static Verification

In the previous example, we have shown how the specifications and the reasoning performed on them can be exploited to analyze the WSN response to a given sequence of undesired events. This concept can be extended to test the WSN against a variable sequence of events, in order to verify its design (textitstatic verification) in the form of a coverage robustness checking.

Specifically, the verification consists of analyzing the robustness of the network, in terms of coverage, against a variable number of failures (stop and disconnection events), from 1 to n, where n is selected by the user, considering all combinations without repetitions. This is useful to check how many node failures the network can tolerate, while guaranteeing a given minimum level of coverage. For example, if we consider a network composed by m nodes and a threshold coverage equal to 50%, we may want to understand what are the sequences of failures causing more than m/two nodes to be isolated (i.e., coverage under the specified threshold) and how the resiliency level varies when varying the sequences of failures. This allows the evaluation of the maximum (and minimum) resiliency level reachable by a given topology and what are the critical failure sequences, i.e., the shortest ones causing a loss of coverage. These are particularly useful to pinpoint weak points in the network (so-called dependability bottlenecks).

We developed an algorithm to generate automatically the sequences of failures (stop and disconnection events specified with Happens predicates) against which to check the robustness of the WSN. The algorithm is implemented by the ADVISES tool (see Section 7.1), and it is aimed at reducing the number of failure sequences to be checked. The principles are to avoid repetitions and to end the sequence as soon as the coverage level becomes lower than the user defined threshold. For instance, we start considering all the cases when there is one failure. By means of the DEC Reasoner, we compute the coverage; if the coverage is above the threshold, the resiliency is surely greater than 1, because there is just one failure and it is tolerated in all cases. In the generic k-th step, we consider sequences of k failures. If the generic sequence $\{f_1, f_2, ..., f_k\}$ leads to a coverage below the threshold, we do not consider sequences starting with an $\{f_1, f_2, ..., f_k\}$ prefix in the $(k+1)$-th step. By considering the percentage of sequences with k failures where the coverage is above the threshold, let us say $r_k\%$, we can say that the resiliency is k in $r_k\%$ of cases.

6. Runtime Verification

The aim of this step is i) to perform a Runtime Verification (RV) [12,55] detecting failure events occurring in a real WSN, possibly to activate recovery actions and ii) to perform a prediction of the critical levels of next failure events that may occur in the WSN, in order to take countermeasures in advance. In this case, critical events, e.g., Stop and Disconnect, are not simulated anymore with Happens predicates, but they are detected from the real system, through system monitors. Details about such monitors are not in the scope of this paper, and are addressed in our previous work in [56].

An application scenario is considered to show how the runtime verification can be implemented; the aim is to describe how it is possible to catch events and observe their effects in a WSN at runtime.

We can see, in Figure 5 from the left to right, that when an event occurs in a wireless sensor node, it is detected by a system monitor that runs on a gateway. We assume that the gateway is a more powerful and stable device with respect to single WSN nodes. While this assumption is usually satisfied in practice (WSN gateways are usually dedicated computers, directly powered and stably connected to the Internet), the gateway itself can become a single point of failure for the method. This problem, not addressed in this paper, can however be mitigated by deploying more than one gateway.

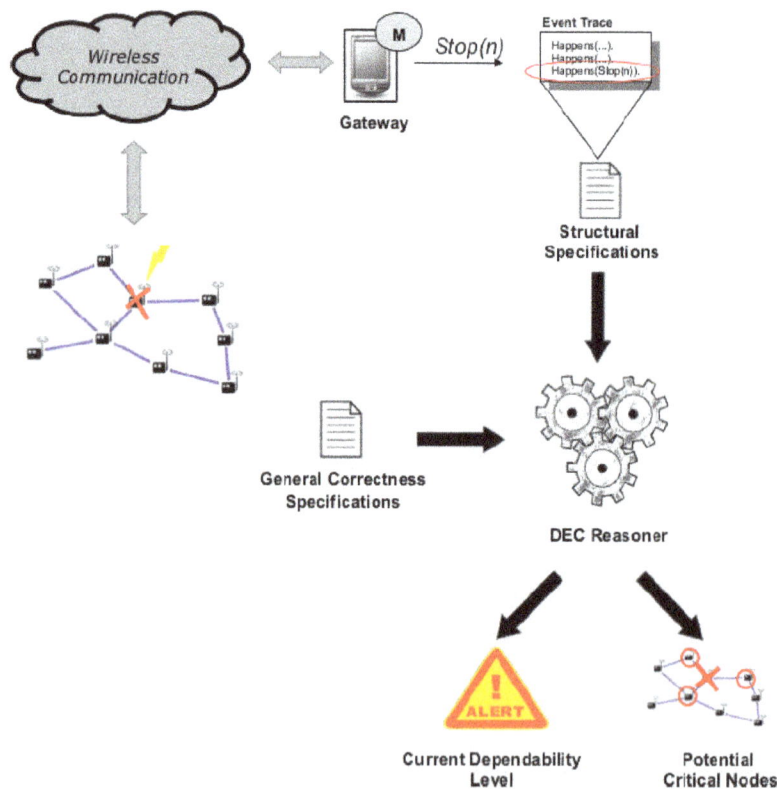

Figure 5. Application scenario in runtime context.

The failure event of a WSN node (for instance Stop(n)) is managed by the monitor deployed on the gateway and added to the current event trace to perform the reasoning. The new event trace is included in an updated structural specification; thus, the DEC Reasoner receives the structural specification with the last occurred event and, considering the general correctness specifications (initially defined), performs the reasoning, returning a couple of outcomes: (i) the Current Dependability Level of the WSN and (ii) the Potential Critical Nodes.

The first outcome reports the current WSN dependability level (i.e., the WSN now covers the X% of the monitored area, and it has been resilient to Y failures so far). The second outcome is a prediction about possible critical events that may occur after the current event (e.g., from now on, node Z represents a weak point in the WSN: it should be replicated or its batteries should be replaced). Moreover, the runtime verification is useful to further verify, at runtime, the WSN design that has been validated at design time. Even if a WSN is checked at design time, it is necessary to observe whether the implemented WSN conforms to expectations and to continuously monitor whether it is able to cope with unexpected events. If the network becomes isolated due to the failure of the only node connecting to the gateway, then the whole network will result isolated. This severe failure event could be already recognized by the method at design time, either by performing a what-if analysis or a robustness checking run, suggesting that the user reinforce the connectivity of the network.

Clearly, the quality of the information provided to the user depends on the quality of the detection. However, let us stress that the consequences of a wrong detection can be mitigated by the user. The outcome of the runtime verification is an alert to the user coupled with an indication of potential critical nodes. The alerted user can then check the actual state of the network before performing any inadequate reaction. Let us also observe that the probability of a wrong detection depends on the type of events to be detected. In the case of a fail stop behavior (such as a node crash as assumed in the paper) then the probability of wrong detection is very low since it is easy to verify for a given interval of time (through the monitors) if the node is indeed stopped. In this case, it is possible to provide reliable information to the reasoner and to the user.

7. Case Studies

In this section, by means of two case studies, we apply the methodology described in Section 3 to study and improve the robustness of two WSNs.

We have performed our experiments on a Intel P4 machine, CPU Clock 3.5 GHz, 512 MB RAM, equipped with Linux, Kernel 3.8.8. Although rather old, this hardware setting is representative of the processing power of a hypothetical gateway to be used at runtime (see Section 7.2.2).

We have focused on WSN-based healthcare systems due to their criticality related to patient monitoring scenarios.

In Table 2, we report an analysis of studied papers focused on WSN-based healthcare systems in order to select the most interesting topologies for our case studies on the basis of the number of nodes and the kind of topology (tree, grid, fully-connected, ...).

Table 2. Analysis of some WSN healthcare systems.

Work	Nodes	Topology	Sensor Platform
iNODE-based system [57]	4	tree	iNODE
BSN-based system [58]	8	fully-connected	Jennic JN5139
MEDiSN [59]	10	tree	Sentilla Tmote Mini
HM4ALL [60]	12	tree	JN5139-MOI ZigBee-based platform
Self-powered WSN [61]	13	tree	Crossbow Micaz
Multi-patient system [62]	15	grid	Tmote sky
CodeBlue [63]	16	grid	N.A.
Clinical Monitoring System [64]	18	tree	TelosB mote

After general research, we have chosen the following systems: the Self-powered WSN for remote patient monitoring (adopted in hospitals) [61] and the *MEDiSN* [59] (a WSN for automating process of patient monitoring in hospitals and disaster scenes).

The two networks have features that are useful to check the performances of our tool and thus collect interesting results. The former (Self-powered WSN) is represented by a perfect balanced tree; the latter is characterized by a sequence of three nodes in a row that could easily cause the majority of bottleneck and isolation problems; in particular, considering the MEDiSN network, we show how this network could improve by modifying the placement of the nodes.

In order to facilitate and automatize the application of the proposed methodology (comprising static and runtime verification) against a general case study, a Java-based tool, called ADVISES (available on sourgeforce) (AutomateD VerIfication of wSn with Event calculuS),has been designed and implemented.

7.1. The ADVISES Tool

The goal of the ADVISES tool [52] is to provide technical support to the methodology addressing practical aspects (e.g., setting of parameters necessary to start verification, realization of structural specification in an automatic way and merge with general specification).

This tool has been realized (i) to operate in double mode: static and runtime; (ii) to automatically generate the structural specifications given the properties of a target WSN; (iii) to perform the reasoning starting from the correctness and structural specifications; (iv) to compute dependability metrics starting from the event trace produced by the reasoner; and (v) to receive events in real-time from a WSN to start runtime verification and to evaluate current and future criticalities.

In particular, at runtime, it is like a server that is in waiting for new events coming from the WSN and that are detected by means of a system monitor.

Since the static and runtime verification do not require the same number of input parameters, we present the ADVISES tool operating in double mode: in static mode we need to select several parameters that in runtime mode they are not necessary.

7.1.1. ADVISES for Static Verification

The ADVISES tool in static verification needs several parameters, such as the number of packets, the number of failures, the battery capacity value of a node, the initial event trace, etc.

By means of the interface shown in Figure 6, a user can simply specify (i) the topology of the target WSN (using a connectivity matrix); (ii) the formal correctness specifications (e.g., for checking isolation events); (iii) the temporal window size to consider (in terms of the number of timepoints); (iv) the number of packets that each sensor can send; (v) the number of failures to be simulated (in case of robustness checking); (vi) the battery capacity of a sensor (in J) and the needed energy for RX/TX operations (in μJ); (vii) the metrics to calculate (for coverage, the threshold value is also necessary); (viii) the channel model; (ix) the initial battery level (to simulate nodes that do not start with full battery capacity; if not specified, then the battery capacity (parameter vi) is assumed as initial battery level); (x) the initial event trace (in case the user wants to perform a "what-if" analysis for a given set of events instead of a robustness check with random sequences of undesired events).

The number of failures to be simulated is a parameter that can be set by the user depending on his/her objectives. For instance, if the user wants to verify whether the WSN is resilient to at least three failures (connection resiliency equals 3) then at least three failures have to be simulated for each robustness checking random run. Clearly, the number of failures should not exceed the number of nodes or the number of paths present in the topology. However, this check is currently not performed by the tool.

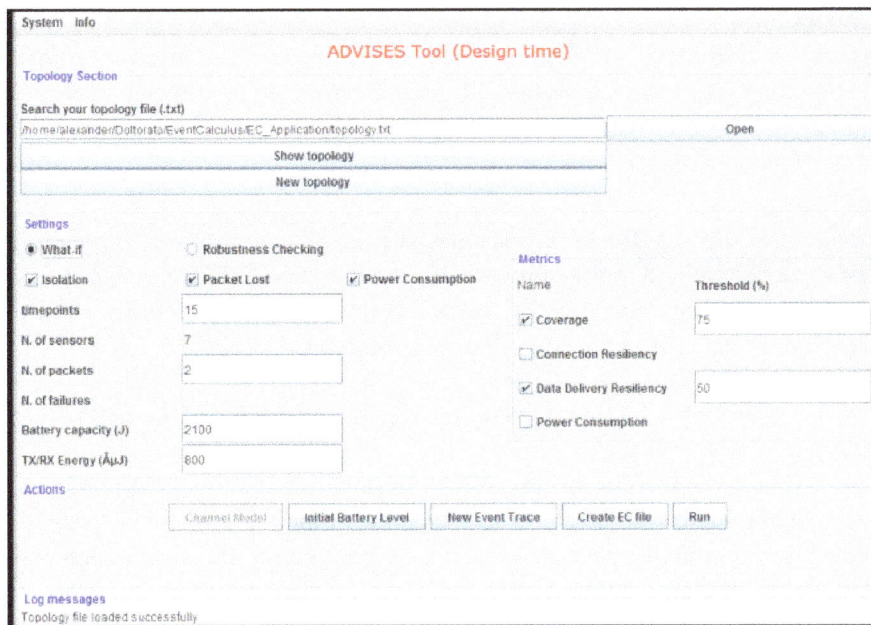

Figure 6. Outcome of ADVISES in static mode running. Parameters for What-if analysis and Robustness Checking computing are available.

The interface is subdivided by four panels:

- "Topology" section is dedicated at the loading/creating of a network topology. The user has to select a topology file that can be found by means of an "Open" button or created on-the-fly using the "New topology" button. The "Show topology" button allows the display of the selected topology in a separate frame. In this file, links between nodes are represented;

- By means of "Settings"; the user can set the several parameters (number of timepoint, of packets, of failures, battery capacity and RX/TX energy) that will be considered for the static verification; the number of sensors is automatically calculated by the ADVISES tool from the topology file. Moreover, the user can choose what general correctness specifications (e.g., the "Isolation" shown earlier) are necessary for a particular verification. Other specifications, such as Packet Lost or Power Consumption, are already available in the tool, but are not treated in the present paper.

- In the "Metrics" section, the user with a tick chooses the desired metrics and if he selects coverage or data delivery resiliency, he has to write the percentage values (default value is 50);

- In the fourth panel (*Actions*), the ADVISES tool presents the possible steps that the user can perform;

  - "Channel Model"
    Pressing this button, the users can select the disconnection probability for every link of the WSN topology choosing a percentage value (0% = never disconnection; 100% = always disconnection) in order to simulate temporary losses of connectivity between nodes. This action is disabled if the user wants to perform a what-if analysis (disconnections in this case will be directly inserted in the event trace);

  - "Initial Battery Level"
    Pressing this button, the users can set the initial battery level for every node of the WSN topology, choosing a percentage value (0% = completely discharged node; 100% = completely charged node).

  - "New Event Trace"
    Pressing this button, the users can set the initial event trace that could occur in the WSN, in the case of a what-if analysis (disabled in the case of robustness checking).

–   "Create EC File"
    Pressing this button, the ADVISES tool automatically generates an Event Calculus file in the
    same directory as the topology file. This file wraps the general correctness specifications
    chosen in the "Settings" section and the structural specifications. Once the EC file is created,
    it is displayed in a separate frame;
–   "Run"
    To obtain the output of the DEC Reasoner, the user has to press this button; analyzing the
    fluent values contained in the outcome produced by the DEC Reasoner, the ADVISES tool
    computes the desired dependability metrics. A pop-up message will appear on the screen in
    order to notify the user of the end of the computation;
–   "Log Messages"
    Panel, the ADVISES tool reports all the useful messages in order to inform the user if some
    error occurs.

From the list of parameters, it can be noted that some of them are very fine grained (e.g., energy
for transmissions), but on the other hand, some coarse grained parameters, which could also have
a significant impact on the performance and dependability, are missing (e.g., size of transmitted
messages). This depends on the fact that the Event Calculus specification adopted in the paper is used
as a proof-of-concept. Let us however point out that the tool is extensible with more specifications,
in case we want to include more aspects in the verification (e.g., size of messages or different metrics
such as power consumption). In this case, the tool also has to be modified in order to take an extended
set of parameters as input.

7.1.2. ADVISES for Runtime Verification

The ADVISES tool in runtime verification mode is simpler because it works on the basis of the
events that it receives from the WSN; there are less computations to perform. In fact, at design time,
having chosen a number of timepoints to observe, the tool has to calculate many combinations of
events; the more timepoints we consider, the more combinations of events are obtained.

The timepoints (in the use of Event Calculus) are a set of natural numbers (0, 1, 2, ...) which
are used to define a sequence of events. There is no relation between timepoints and the real time,
and the timepoints may also assume a different temporal nature (seconds, minutes, hours, ...) into the
same study.

Instead, at runtime, the ADVISES tool for each timepoint is waiting for an event to compute the
effects of this occurred event.

Figure 7 shows the ADVISES operating at runtime.

Once started, it is in server mode and waiting for events coming from the WSN; when a failure
occurs in a wireless sensor node, (for instance Stop(5)), it is managed by the monitor and sent to the
ADVISES Tool that is listening on a port ready to receive events and start reasoning. Having received
the event, it automatically updates the sequence of received events and generates the event calculus
specification file in order to perform the reasoning with the DEC Reasoner. The ADVISES tool, having
received the output from the DEC Reasoner, calculates the current values of the selected dependability
metrics to establish the current status of the WSN and performs robustness checking to predict the
future criticalities. Then, having verified the obtained metrics values with the thresholds set by the
user, the ADVISES tool sends messages to a network maintainer which may be purely informative,
or alerts in case these values are under the desired threshold. After the last step, the ADVISES tool
continues to work, waiting for the next detected events.

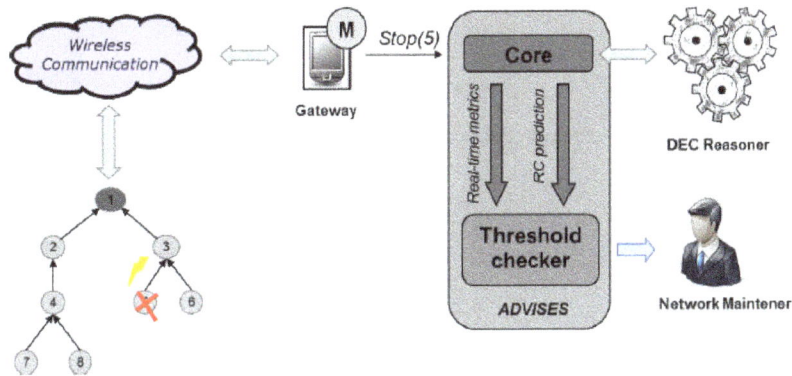

Figure 7. Outcome of ADVISES in runtime mode running. Only timepoint parameter is editable: this is the observation time in an experiment at runtime.

Figure 8 shows this interface.

By means of this interface, a user can simply specify (i) the initial topology of the target WSN, using a connectivity matrix; (ii) the temporal window size to consider, in terms of the number of timepoints to consider for the prediction; (iii) the metrics to calculate. Actually, we have to set an initial topology since a wireless sensor network could be affected by some changes due to the nature of the sensors: a wireless sensor could stop and change its neighbors. This concept is also illustrated in the Case Studies section in which we propose improvements of the initial topologies considered. The tool itself does not directly manage topology changes at runtime, but it provides indications to the user upon failures on how to manage the WSN and improve the topology in order to be more resilient in case of further undesired events. From this point, the user can run the tool with the new topology chosen after the failure.

In the "Log messages" panel, the ADVISES tool reports all the performed computations about the current state of the WSN and the risks that may affect its robustness. Also, there are useful messages in order to inform the user if some error occurs.

In Figure 9, we describe the runtime verification process (performed by the ADVISES tool) by means of a flow diagram.

Therefore, the ADVISES tool advises the network maintainer of problems related to the network and reports its critical points.

For this purpose, we have realized another interface of the ADVISES tool in order to receive events from a WSN in real-time detected through a system monitor and to start the runtime verification, evaluating both current and future criticalities.

7.2. Experimental Results

We perform two sets of experiments. For the first one, we consider the self-powered WSN [61] that is composed by 13 motes: one base station (that consists of a mote and is connected to a server), four router nodes (realized utilizing CrossBow MICAz motes) and eight sensor nodes (such as ECGs, pulse-oximeters, etc.). For the second one, we consider MEDiSN [59] that is composed of 10 motes: one base station (as gateway), four relay points (that are wireless sensors) and finally five physiological monitors (collecting patients' physiological data). For this network, we have considered a topology in which physiological monitors in the leaves are treated as if they were relay points as well, so that the topology is more complex and interesting to try our system with.

For each topology, we try to improve the networks balancing the nodes: in this way, there are more possibilities of having no isolation events and more coverage in case of some failure nodes. For our study, the ADVISES Tool helps to quickly compute the metrics of dependability by means of us being able to apply improvements.

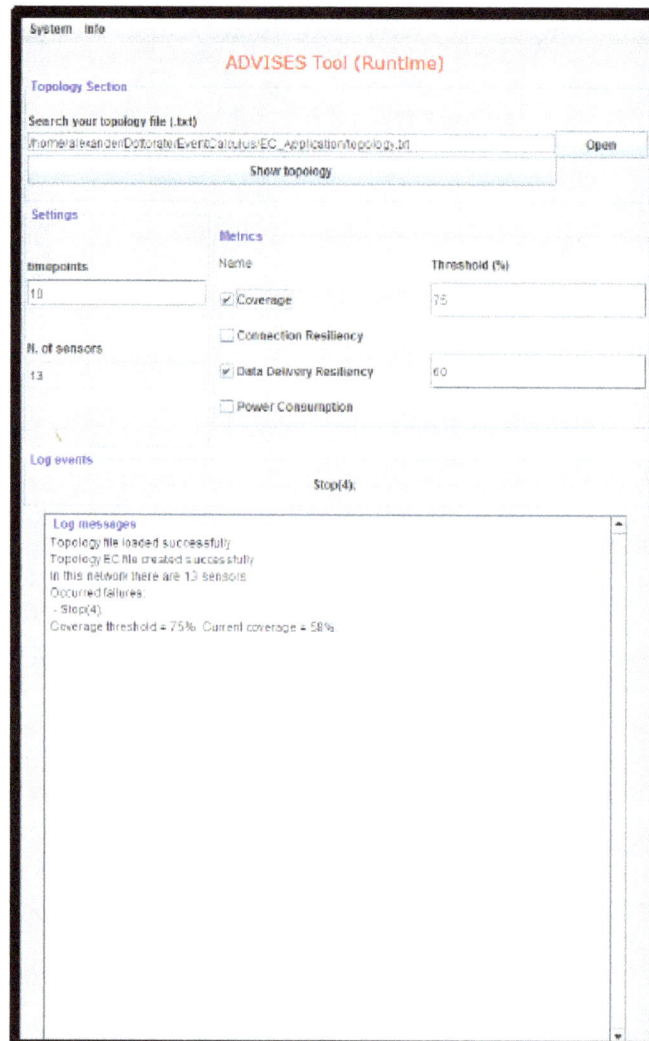

Figure 8. ADVISES in runtime mode.

7.2.1. Static Verification

For both networks, we aim to observe the percentage of cases in which there is connection resiliency to 1, 2 and 3 failures, keeping a coverage threshold at 65%, 75%, and 85%, so from a less demanding to a more demanding resiliency requirement. Moreover, starting from the original topologies of the two networks as presented in [59,61], we attempt to make them more robust, reconfiguring the connections among the nodes and observing effects.

We design the topologies, related to the two chosen networks, by means of a tree graph. The first topology is structured as a tree graph with three levels (Figure 10a); the second topology (Figure 11a) is structured as a tree graph with four levels.

Exploiting the capabilities of the ADVISES tool, we specify the characteristics of both networks and of the metrics to be evaluated; specifically, connection resiliency with different coverage thresholds. In Table 3, we present results of the robustness checking: on the columns, we identify the topologies (original and our proposed alternatives) grouped by the network (self-powered WSN and MEDiSN); on the rows, we collect the results of connection resiliency for one, two and three failures on the basis of coverage threshold (65%, 75% and 85%); the generic cell of the table represents the robustness of the network, evaluated as the percentage of cases in which the topology (identified by the column) is able to tolerate n failures (where n corresponds to the value of the connection resiliency identified by the row) guaranteeing a certain coverage value (identified by the row).

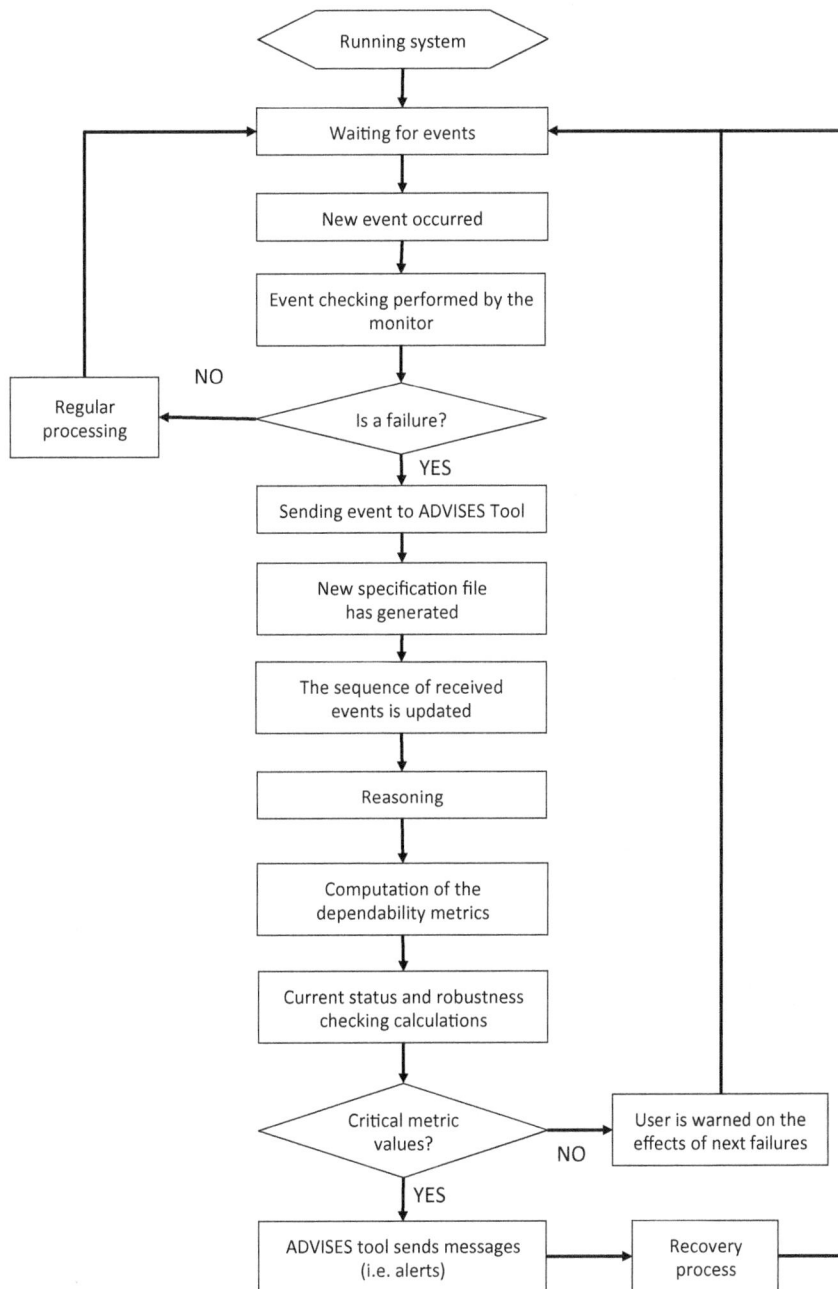

Figure 9. Runtime verification process illustrated by a flow diagram.

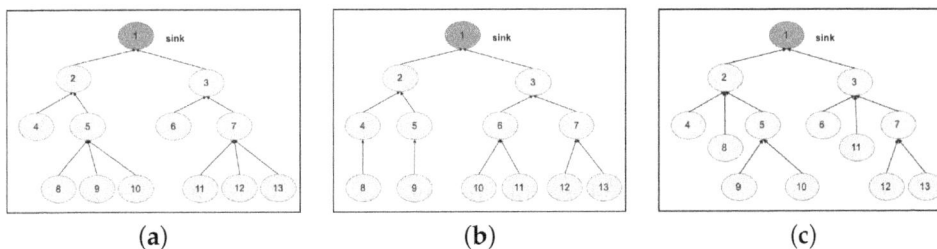

Figure 10. Topologies of the self-powered WSN. (**a**) Original; (**b**) First attempt; (**c**) Second attempt.

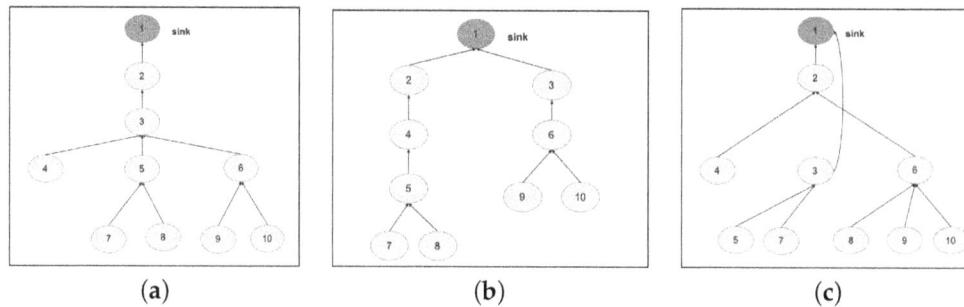

Figure 11. Topologies of MEDiSN. (**a**) Original; (**b**) First attempt; (**c**) Second attempt.

Results on column Topology 1 are achieved by performing robustness checking and dependability evaluation of original topologies. We can observe that, if we increase the requirement on connection resiliency (from 1 to 3) and on coverage (from 65% to 85%), the percentage of cases in which the WSN is able to tolerate the failures decreases, as expected. We can also note that the self-powered WSN is more robust than the MEDiSN one, due to the higher number of nodes and the lower number of levels of the tree graph.

Table 3. Percentages of connection (Conn. Resil.) resiliency for the self-powered WSN and MEDiSN. Topology 1 is the original network, Topology 2 and 3 are attempts of coverage (see "Outcome") improvement. Three thresholds of coverage (Cov. = 65%, 75% and 85%) have been selected.

Outcome		Self-Powered WSN			MEDiSN		
		Topology 1	Topology 2	Topology 3	Topology 1	Topology 2	Topology 3
	Conn. Resil. = 1	83%	83%	83%	77%	66%	77%
Cov. = 65%	Conn. Resil. = 2	52%	61%	67%	41%	30%	49%
	Conn. Resil. = 3	31%	35%	41%	18%	10%	28%
	Conn. Resil. = 1	66%	83%	83%	55%	44%	66%
Cov. = 75%	Conn. Resil. = 2	43%	48%	50%	29%	18%	43%
	Conn. Resil. = 3	27%	18%	29%	0%	0%	0%
	Conn. Resil. = 1	66%	66%	66%	55%	44%	66%
Cov. = 85%	Conn. Resil. = 2	43%	27%	43%	0%	0%	0%
	Conn. Resil. = 3	0%	0%	0%	0%	0%	0%
Examined failure sequences		114,480			46,980		

Starting from these results, we try to improve the robustness of both networks, by means of slight changes on the network topology. The aim is to show how the proposed approach and related tool are useful to drive design choices and tune the configuration of the network. Considering the self-powered WSN, we start by proposing a more uniform distribution of the nodes, balancing the tree. The resulting topology is shown in Figure 10b. What we expect to observe is a general increase of robustness, since a failure of a node causes the disconnection of a smaller number of nodes with respect to the original topology (e.g., see nodes 5 and 7). For a coverage threshold at 65%, we achieve this conclusion (see column Topology 2 in Table 3). However, opposite to what was expected, for a coverage threshold at 75% (and three failures) and a coverage threshold at 85% (and two failures), we obtain a degradation of robustness, from 27% to 18% and from 43% to 27%, respectively.

The automatic analysis performed by the reasoner on the specifications allow us to deeply investigate the exact reason of such degradation, and find better solutions. To this aim, in Table 4, we report a classification of the failure events tested by the ADVISES tool on the self-powered WSN. The events are grouped based on the obtained coverage value (reported in the Coverage column) for each of the topologies considered, and are also grouped as below or above the 75% coverage threshold. Passing from Topology 1 to Topology 2, we can observe that Topology 2 is able to reduce the overall number of events causing a coverage value below the threshold (critical events), if compared to Topology 1, improving the resiliency in several cases. However, we also note that Topology 2 introduces a new criticality, not present in Topology 1, related to node 3: if this node fails or gets disconnected, the coverage suddenly drops down to 47%. Hence, node 3 represents a dependability bottleneck for Topology 2. Guided by this result, we propose another slight modification to the topology, obtaining a new version (Topology 3) shown in Figure 10c, reducing both the number of child nodes of node 3 and the path length of nodes 8 and 11. In this case, looking at the Topology 3 column in Table 4, we can observe that the criticality on node 3 disappears, while reducing the overall number of critical events compared to Topology 1. The benefits of this new topology are also visible in Table 3; specifically, topology 3 is able to improve the robustness in all cases with respect to Topologies 1 and 2.

In a similar way, we exploit our approach to find better topologies for the MEDiSN case study. In this case, we have observed that if we keep the same number of tree levels but reduce the number of leaf nodes (see the topology in Figure 11b), the robustness level decreases, as shown in the Topology 2 column in Table 3. Looking at Table 5 (reporting the classification of critical events achieved for the MEDiSN WSN), we can note that, even if Topology 2 is able to reduce the criticality of nodes 2 and 3, which are clearly two dependability bottlenecks for the original MEDiSN topology, it introduces new critical events (e.g., the failure of node 4). Moved by these results, we increase the number of leaf nodes and decrease the tree levels, obtaining a third topology (Topology 3 in Figure 11c) able to reduce the number of critical events (see Topology 3 column in Table 5) and to improve the overall resiliency level in all cases (see the Topology 3 column in Table 3).

Table 4. Classification of failure events for a self-powered WSN (Topology 1, 2 and 3).

	Failure Event			Coverage
	Topology 1	Topology 2	Topology 3	
Coverage < 75% (critical events)	no event failure	Stop(3) OR Disconnect(3,1)	no event failure	47%
	Stop(2) OR Stop(3) OR Disconnect(2,1) OR Disconnect(3,1)	no event failure	Stop(2) OR Stop(3) OR Disconnect(2,1) OR Disconnect(3,1)	54%
	no event failure	Stop(2) OR Disconnect(2,1)	no event failure	62%
	Stop(5) OR Stop(7) OR Disconnect(5,2) OR Disconnect(7,3)	no event failure	no event failure	70%
Coverage ≥ 75%	no failure event	Stop(6) OR Stop(7) OR Disconnect(6,3) OR Disconnect(7,3)	Stop(5) OR Stop(7) OR Disconnect(5,2) OR Disconnect(7,3)	77%
	no failure event	Stop(4) OR Stop(5) OR Disconnect(4,2) OR Disconnect(5,2)	no failure event	85%
	Stop(4) OR Stop(6) OR Stop(8) OR Stop(9) OR Stop(10) OR Stop(11) OR Stop(12) OR Stop(13) OR Disconnect(4,2) OR Disconnect(6,3) OR Disconnect(8,5) OR Disconnect(9,5) OR Disconnect(10,5) OR Disconnect(11,7) OR Disconnect(12,7) OR Disconnect(13,7)	Stop(8) OR Stop(9) OR Stop(10) OR Stop(11) OR Stop(12) OR Stop(13) OR Disconnect(8,5) OR Disconnect(9,5) OR Disconnect(10,5) OR Disconnect(11,7) OR Disconnect(12,7) OR Disconnect(13,7)	Stop(4) OR Stop(6) OR Stop(8) OR Stop(9) OR Stop(10) OR Stop(11) OR Stop(12) OR Stop(13) OR Disconnect(4,2) OR Disconnect(6,3) OR Disconnect(8,5) OR Disconnect(9,5) OR Disconnect(10,5) OR Disconnect(11,7) OR Disconnect(12,7) OR Disconnect(13,7)	93%

Table 5. Classification of failure events for MEDiSN (Topology 1, 2 and 3).

	Failure Event			Coverage
	Topology 1	**Topology 2**	**Topology 3**	
Coverage < 75% (critical events)	Stop(2) OR Disconnect(2,1)	no event failure	no event failure	10%
	Stop(3) OR Disconnect(3,2)	no event failure	no event failure	20%
	no event failure	no event failure	Stop(2) OR Disconnect(2,1)	40%
	no event failure	Stop(2) OR Disconnect(2,1)	no event failure	50%
	no event failure	Stop(3) OR Stop(4) OR Disconnect(3,1) OR Disconnect(4,2)	Stop(6) OR Disconnect(6,2)	60%
	Stop(5) OR Stop(6) OR Disconnect(5,3) OR Disconnect(6,3)	Stop(5) OR Stop(6) OR Disconnect(5,4) OR Disconnect(6,3)	Stop(3) OR Disconnect(3,1)	70%
Coverage ≥ 75%	Stop(4) OR Stop(7) OR Stop(8) OR Stop(9) OR Stop(10) OR Disconnect(4,3) OR Disconnect(7,5) OR Disconnect(8,5) OR Disconnect(9,6) OR Disconnect(10,6)	Stop(7) OR Stop(8) OR Stop(9) OR Stop(10) OR Disconnect(7,5) OR Disconnect(8,5) OR Disconnect(9,6) OR Disconnect(10,6)	Stop(4) OR Stop(5) OR Stop(7) OR Stop(8) OR Stop(9) OR Stop(10) OR Disconnect(4,2) OR Disconnect(5,3) OR Disconnect(7,3) OR Disconnect(8,6) OR Disconnect(9,6) OR Disconnect(10,6)	90%

In both the cases, we have seen how the proposed approach is useful to precisely spot dependability bottlenecks and define more robust configurations. For both experiments, the ADVISES tool has performed a total of 161,460 reasonings on the specifications (114,480 for the self-powered WSN and 46,980 for MEDiSN). In particular, Table 6 shows the number of failure sequences and the related time spent by the ADVISES Tool to perform the reasoning in the self-powered WSN and in MEDiSN networks considering a coverage threshold value equal to 75%. In the worst case (12,144 different failure sequences tested for the self-powered WSN) the tool takes about 9 h to perform the evaluation on our commodity hardware. While this time could still be acceptable at design time (and it does not affect the conceptual validity of the approach), it clearly represents a practical limit, especially for large WSNs. This issue, and related solutions that we are currently investigating, are further discussed in Section 8.

Table 6. Reasoning time of the ADVISES Tool for a self-powered WSN and MEDiSN considering a threshold value equal to 75%.

Connection Resiliency	Self-Powered WSN		MEDiSN	
	Failure Sequences	**Elapsed Time (s)**	**Failure Sequences**	**Elapsed Time (s)**
1	24	1500	18	600
2	552	9720	306	2160
3	12,144	32,580	4896	2820

7.2.2. Runtime Verification

After the static analysis, we physically deploy the best topologies (topology 3 for both WSNs) at our labs to perform runtime verification.

For this purpose, we have designed and implemented a system monitor with the aim of detecting failure events from the real-world WSN. The monitor runs on a machine and listens for packets coming from all sensors trough the sink node of the WSN. The detection of events (such as the stop of a node) is performed assuming that each sensor sends packets periodically, with a known rate, which is common to several WSN applications. Hence, for every node, the monitor sets a timeout, which is reset each time the monitor receives a packet from the given node. If the timeout expires for a node X, the monitor sends a Stop (X) event to ADVISES (running in runtime mode). The use of time-out may also detect temporary disconnections or delays. In this case, when packets from a node X are received again after a stop, the monitor sends a Start (X) event to ADVISES. Clearly, different failure detection approaches could be used as well, however this is not relevant for our experiment and out of the scope of the paper.

The monitor has been implemented as a java application running on a server (a Pentium 4 machine in our case) and connected via USB to a MIB520 Base station by Crossbow. As sensor nodes, we have adopted Iris Motes by Crossbow equipped with a ZigBee RF Transceiver and a TinyOS 2.0 operating system, running the BlinkToRadio application, just to perform a periodic sensing and sending of packets of all nodes to the sink.

Considering the deployment of third self-powered WSN topology, we start the system monitor that is in communication with the ADVISES Tool.

As a first test, we stop node 5. The failure event is detected by the system monitor at runtime (elapsed time ≈ 1 s) and it is sent to ADVISES. Once the failure event has been received, ADVISES starts the reasoning and shows the result in the "Log message" panel as described in Figure 12a (metric computation time ≈ 1 min and 30 s): coverage is equal to 77%. ADVISES also identifies potential critical nodes (prediction phase) that could compromise the robustness of the network; in fact, from Figure 12a, we can see that the tool "advises" the user (prediction time ≈ 4 m) that if the next failure event that occurs is, for example, Stop (7), then the network is not robust anymore (the coverage goes below the desired threshold, 65% in this test). To validate this suggestion, we also stop sensor node 7. From the log posted by the ADVISES tool (Figure 12b), we observe that effectively the coverage decreases (54%) and goes under the desired threshold.

From the experiment, we can note that the adoption of the approach to monitor a real WSN at runtime is straightforward. Using the same formal specifications used at design time, the tool is able to tell, at runtime, what is the current dependability level of the WSN (e.g., coverage at 77%) and to pinpoint critical nodes (e.g., nodes 2, 3 and 7 in the example) that need to be maintained (e.g., by replacing batteries) or strengthened (e.g., by replicating them with other nodes) to avoid the whole mission of the WSN being compromised.

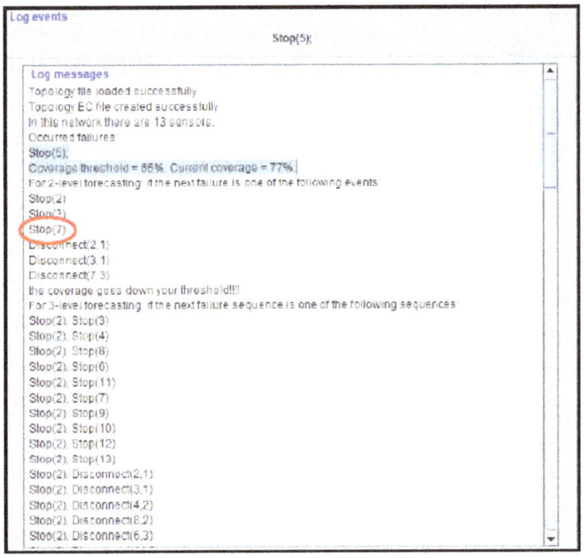

(a)

Figure 12. *Cont.*

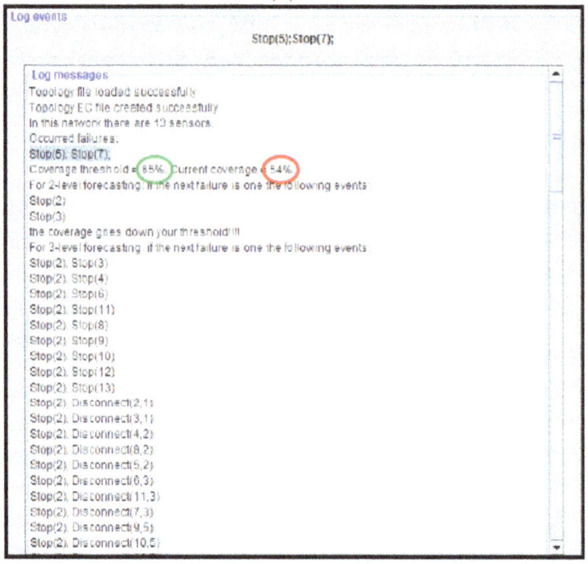

(b)

Figure 12. Outcomes of an example of the ADVISES Tool when it is set in Runtime Verification mode. (**a**) Current coverage value is 77% (against the coverage threshold value 65%) in the case of node 5 failure and warnings for next Stop events; (**b**) Current coverage value is 54% (against the coverage threshold value 65%) in the case of both node 5 and 7 failures; in this case coverage is under the threshold value set.

8. Conclusions and Future Work

This paper addressed the problem of the dependability assessment of WSNs with formal methods. Assessing the dependability of WSNs is a crucial task since these kinds of networks are more and more used in critical application scenarios where the level of trust becomes an important factor; depending on application scenarios, different dependability requirements can be defined, such as, node lifetime, network resiliency, and coverage of the monitoring area. From a preliminary analysis, the need emerged for verification of a WSN at design time in order to increase the confidence about the robustness of designed solutions before putting it into operation; and the need for monitoring a WSN during operation in order to avoid unexpected results or dangerous effects and thus to perform what in the literature is defined as continuous monitoring.

The research activity dealt with the definition of formal specifications used for the behavioral checking of WSN-based systems at design and runtime phases; a set of correctness specifications applied to a generic WSN has been defined using event calculus as formal language since the behavior of a WSN can be characterized in terms of an event flow (e.g., a node turns on, a packet is sent, a node stops due to failure, etc.) and the event calculus formalism allowed the easy specification of the system in terms of events.

This paper demonstrated through proof-of-concept scenarios that it is possible to assess the dependability of WSNs by means of formal methods, in particular event calculus formalism, using the narrative it generates. By means of two case studies, we have shown how the adoption of a formal specification is helpful to deeply investigate the reasons for inefficiencies, in terms of the degree of dissatisfaction of given dependability requirements, and to suggest viable improvements to the design that have positive effects on the global dependability level of the system. The implementation of a tool, namely ADVISES, has also shown how the approach can be easily adopted by technicians with no experience of formal methods, the structural specification being generated automatically and completely hidden to users. Finally, the paper described how the same specification can be adopted to perform continuous monitoring at runtime, once the designed WSN is implemented and deployed on the field.

The implementation and actual use of the approach also helped us to highlight practical limits to be solved. We have considered topologies of WSNs adopted in typical critical scenarios, such as healthcare (with 7/15 nodes). The simplicity of such topologies, although taken from real examples, may lead to doubts regarding the general usefulness of the approach (the same reasoning could be performed intuitively with no need of a tool). However, we selected the examples with the aim of pointing out how ADVISES allows the quantitative justification of design choices against precise requirements and metrics, helping technicians to have the instruments to make informed design decisions, e.g., when he/she is forced to choose between a few topologies to be considered, due to restrictions. The selection has been done also with the objective of implementing the topology on the field, with real WSN nodes, in order to practically show how the same approach and tool can be used both at design time and at runtime for continuous monitoring.

Concerning the time complexity of the method, it is strictly dependent on that of the EC, which has been estimated (assuming n the number of events) as $O(n^{-3})$ in case of absolute time and total ordering of the events [65], $O(2^{n-3})$ in the case of relative times and partial ordering [66]. The proposed methodology relies on two kinds of analysis (runtime verification and static verification) based on relative times and partial ordering of events. Therefore, the time complexity is upper bounded to $O(2^{n-3})$. For runtime verification, the reasoner is invoked just one time at the occurrence of the adverse event in the network. For static verification, the reasoner is invoked a number of times, corresponding to the combinations without repetitions of the emulated failures. Generally, in both cases, the number of events k generated for the analysis in EC is the square of the number of nodes N of the network.

During the experimental phase, we have also tested the correct functioning of the specification with topologies with more nodes (about 100) but we experienced a remarkable deterioration of the performance on our commodity hardware (about 1 h to test all the sequences with only one failure), due to the state space explosion problem. While this issue does not affect the conceptual validity of the approach and its use on typical critical WSNs, it undermines its practical adoption for very large WSNs. The issue can be partially solved at design time by using more powerful hardware and by considering that the user, at design time, is more favourable to waiting for the reasoning result if this can help his/her design decisions. However, the issue remains for the runtime monitoring, where the output of the reasoning must be provided with due timing constraints and with no assumptions on the processing power of gateways. To face this scalability problem, we are currently conceiving a method to divide a large topology in several sub-topologies (equivalent to WSN clusters). During the runtime, the different sub-topologies can be monitored by different gateways (acting as cluster heads), performing the reasoning on each sub-topology in parallel and then finally joining the results taking into account the dependences between the sub-topologies.

Although the methodology here presented may not always identify an optimal topology from all possibly desirable features, it still improves the support to analyze similarly complex topologies where the difference may not be obvious to the designer. The generic strategy and tool presented in this article facilitates the experimentation and comparison of different heuristics to address different problems. Two obvious areas of experimentation and improvement are the relaxation of restrictions and the improvement of efficiency. From this perspective, the computational platform used for experimentation in this article is anecdotal and the main contribution is in problem solving strategies.

We base our approach on sensors that are fixed (such as beacons) and with an established data routing reducing a topology, such as a spanning tree that is valid for a WSN. As future work, we plan to extend the use of the specification also for mobile scenarios, specifying further events that notify wireless sensor movements within clusters and from cluster to cluster.

Acknowledgments: The research work reported in this paper has been partially supported by the eAsy inteLligent service Platform for Healthy Ageing (ALPHA) Project.

Author Contributions: Alessandro Testa, Marcello Cinque, Antonio Coronato and Juan Carlos Augusto conceived the methodology, designed the experiments and analyzed the results. Alessandro Testa developed the software system and conducted the experiments.

Conflicts of Interest: The authors declare no conflict of interest.

References

1. Stankovic, J.A. Wireless sensor networks. *Computer* **2008**, *41*, 92–95.
2. Avizienis, A.; Laprie, J.C.; Randell, B.; Landwehr, C. Basic concepts and taxonomy of dependable and secure computing. *IEEE Trans. Dependable Secure Comput.* **2004**, *1*, 11–33.
3. Yick, J.; Mukherjee, B.; Ghosal, D. *Wireless Sensor Network Survey*; Elsevier: Amsterdam, The Netherlands, 2008; pp. 2292–2330.
4. Alemdar, H.; Ersoy, C. Wireless sensor networks for healthcare: A survey. *Comput. Netw.* **2010**, *54*, 2688–2710.
5. Coronato, A.; de Pietro, G. Formal design of ambient intelligence applications. *Computer* **2010**, *43*, 60–68.
6. Augusto, J.C.; Zheng, H.; Mulvenna, M.D.; Wang, H.; Carswell, W.; Jeffers, W.P. Design and modelling of the nocturnal AAL care system. In Proceedings of the 2nd International Symposium on Ambient Intelligence (ISAmI 2011), Salamanca, Spain, 6–8 April 2011; pp. 109–116.
7. Aarts, E.; Wichert, R. *Ambient Intelligence*; TechnologyGuide: Newton, MA, USA, 2009; pp. 244–249.
8. Coronato, A.; De Pietro, G. Formal specification and verification of ubiquitous and pervasive systems. *ACM Trans. Auton. Adapt. Syst.* **2011**, *6*, 1–6.
9. Cinque, M.; Cotroneo, D.; Di Martinio, C.; Russo, S. Modeling and assessing the dependability of wireless sensor networks. In Proceedings of the 26th IEEE International Symposium on Reliable Distributed Systems (SRDS'07), Beijing, China, 10–12 October 2007; pp. 33–44.
10. Lazarescu, M.T. Design of a WSN platform for long-term environmental monitoring for IoT applications. *IEEE J. Emerg. Sel. Top. Circuits Syst.* **2013**, *3*, 45–54.
11. Woodcock, J.; Larsen, P.G.; Bicarregui, J.; Fitzgerald, J. Formal methods: Practice and experience. *ACM* **2009**, *41*, 19.
12. Leucker, M.; Schallhart, C. A brief account of runtime verification. *J. Log. Algebraic Program.* **2009**, *78*, 293–303.
13. Bakhouya, M.; Campbell, R.; Coronato, A.; Pietro, G.D.; Ranganathan, A. Introduction to special section on formal methods in pervasive computing. *ACM Trans. Auton. Adapt. Syst.* **2012**, *7*, 6.
14. Testa, A.; Coronato, A.; Cinque, M.; Augusto, J.C. Static verification of wireless sensor networks with formal methods. In Proceedings of the IEEE 2012 Eighth International Conference on Signal Image Technology and Internet Based Systems (SITIS), Naples, Italy, 25–29 November 2012; pp. 587–594.
15. Bondavalli, A.; Ceccarelli, A.; Falai, L.; Vadursi, M. A New Approach and a Related Tool for Dependability Measurements on Distributed Systems. *IEEE Trans. Instrum. Meas.* **2010**, *59*, 820–831.
16. Li, M.; Liu, Y. Underground coal mine monitoring with wireless sensor networks. *ACM Trans. Sens. Netw.* **2009**, *5*, 1–29.
17. Cinque, M.; Cotroneo, D.; Martino, C.D.; Russo, S.; Testa, A. AVR-INJECT: A tool for injecting faults in Wireless Sensor Nodes. In Proceedings of the IEEE International Symposium on Parallel & Distributed Processing (IPDPS), Rome, Italy, 23–29 May 2009; pp. 1–8.
18. Cinque, M.; Cotroneo, D.; di Martino, C.; Testa, A. An effective approach for injecting faults in wireless sensor network operating systems. In Proceedings of the 2010 IEEE Symposium on Computers and Communications (ISCC), Riccione, Italy, 22–25 June 2010; pp. 567–569.
19. Bhuiyan, M.; Wang, G.; Wu, J.; Cao, J.; Liu, X.; Wang, T. Dependable structural health monitoring using wireless sensor networks. *IEEE Trans. Dependable Secure Comput.* **2015**, *PP*, 1.
20. Bhuiyan, M.Z.A.; Wang, G.; Cao, J.; Wu, J. Deploying wireless sensor networks with fault-tolerance for structural health monitoring. *IEEE Trans. Comput.* **2015**, *64*, 382–395.
21. Shakkira, K.; Mohamed, M.T. Advanced Lightweight, Dependable and secure Trust System for Clustered wireless sensor networks. In Proceedings of the 2015 International Conference on Innovations in Information, Embedded and Communication Systems (ICIIECS), Coimbatore, India, 19–20 March 2015; pp. 1–4.
22. Shrestha, A.; Xing, L.; Liu, H. Infrastructure communication reliability of wireless sensor networks. In Proceedings of the 2nd IEEE International Symposium on Dependable, Autonomic and Secure Computing, Indianapolis, India, 29 September–1 October 2006; pp. 250–257.
23. Koushanfar, F.; Potkonjak, M.; Sangiovanni-Vincentelli, A. On-line fault detection of sensor measurements. In Proceedings of the 2003 IEEE Sensors, Toronto, ON, Canada, 22–24 October 2003; pp. 974–979.

24. Zhang, J.; Li, W.; Cui, D.; Zhao, X.; Yin, Z. The NS2-based simulation and research on wireless sensor network route protocol. In Proceedings of the 5th International Conference on Wireless Communications, Networking and Mobile Computing (WiCom'09), Beijing, China, 24–26 September 2009; pp. 1–4.

25. Titzer, B.L.; Lee, D.K.; Palsberg, J. Avrora: Scalable sensor network simulation with precise timing. In Proceedings of the 4th International Symposium on Information Processing in Sensor Networks (IPSN'05), Los Angeles, CA, USA, 24–27 April 2005.

26. Di Martino, C.; Cinque, M.; Cotroneo, D. Automated generation of performance and dependability models for the assessment of wireless sensor networks. *IEEE Trans. Comput.* **2012**, *61*, 870–884.

27. Heinzelman, W.; Chandrakasan, A.; Balakrishnan, H. Energy-efficient communication protocol for wireless microsensor networks. In Proceedings of the 33rd Annual Hawaii International Conference on System Sciences, Maui, HI, USA, 4–7 January 2000; p. 10.

28. Mini, A.F.; Nath, B.; Loureiro, A.A.F. A probabilistic approach to predict the energy consumption in wireless sensor networks. In Proceedings of the IV Workshop de Comunicao sem Fio e Computao Mvel, Sao Paulo, Brazil, 23–25 October 2002; pp. 23–25.

29. Yavuz, F.; Zhao, J.; Yağan, O.; Gligor, V. Designing secure and reliable wireless sensor networks under a pairwise key predistribution scheme. In Proceedings of the 2015 IEEE International Conference on Communications (ICC), London, UK, 8–12 June 2015; pp. 6277–6283.

30. Sharma, K.P.; Sharma, T.P. rDFD: Reactive distributed fault detection in wireless sensor networks. *Wirel. Netw.* **2016**, *22*, 1–16.

31. Khalil, I.M.; Khreishah, A.; Ahmed, F.; Shuaib, K. Dependable wireless sensor networks for reliable and secure humanitarian relief applications. *Ad Hoc Netw.* **2014**, *13*, 94–106.

32. Silva, I.; Guedes, L.A.; Portugal, P.; Vasques, F. Reliability and availability evaluation of wireless sensor networks for industrial applications. *Sensors* **2012**, *12*, 806–838.

33. Kapitanova, K.; Son, S. MEDAL: A coMpact event description and analysis language for wireless sensor networks. In Proceedings of the 2009 Sixth International Conference on Networked Sensing Systems (INSS), Pittsburgh, PA, USA, 17–19 June 2009; pp. 1–4.

34. Man, K.L.; Vallee, T.; Leung, H.; Mercaldi, M.; van der Wulp, J.; Donno, M.; Pastrnak, M. TEPAWSN—A tool environment for wireless sensor networks. In Proceedings of the 2009 4th IEEE Conference on Industrial Electronics and Applications (ICIEA 2009), Xi'an, China, 25–27 May 2009; pp. 730–733.

35. Boonma, P.; Suzuki, J. Moppet: A model-driven performance engineering framework for wireless sensor networks. *Comput. J.* **2010**, *53*, 1674–1690.

36. Shanahan, M. The event calculus explained. *Lect. Notes Comput. Sci.* **1999**, *1600*, 409–430.

37. Ölveczky, P.; Meseguer, J. Specification and analysis of real-time systems using Real-Time Maude. In *Fundamental Approaches to Software Engineering*; Springer: Berlin, Germany, 2004; pp. 354–358.

38. Romadi, R.; Berbia, H. Wireless sensor network a specification method based on reactive decisional agents. In Proceedings of the 3rd International Conference on Information and Communication Technologies: From Theory to Applications (ICTTA 2008), Damascus, Syria, 7–11 April 2008; pp. 1–5.

39. Wang, C.; Xing, L.; Vokkarane, V.M.; Sun, Y.L. Reliability and lifetime modeling of wireless sensor nodes. *Microelectron. Reliab.* **2014**, *54*, 160–166.

40. Wang, C.; Xing, L.; Vokkarane, V.M.; Sun, Y.L. Infrastructure communication sensitivity analysis of wireless sensor networks. *Qual. Reliab. Eng. Int.* **2016**, *32*, 581–594.

41. Liang, Y.; Liu, R. Routing topology inference for wireless sensor networks. *ACM SIGCOMM Comput. Commun. Rev.* **2013**, *43*, 21–28.

42. Zoumboulakis, M.; Roussos, G. *Complex Event Detection in Extremely Resource-Constrained Wireless Sensor Networks*; Kluwer Academic Publishers: Hingham, MA, USA, 2011; Volume 16, pp. 194–213.

43. Bromuri, S.; Stathis, K. Distributed agent environments in the Ambient Event Calculus. In Proceedings of the Third ACM International Conference on Distributed Event-Based Systems (DEBS'09), Nashville, TN, USA, 6–9 July 2009; pp. 1–12.

44. Blum, J.; Magill, E. Telecare service challenge: Conflict detection. In Proceedings of the 2011 5th International Conference on Pervasive Computing Technologies for Healthcare (PervasiveHealth), Dublin, Ireland, 23–26 May 2011; pp. 502–507.

45. Kowalski, R.; Sergot, M. A logic-based calculus of events. *New Gener. Comput.* **1986**, *4*, 67–95.

46. Mueller, E.T. Automating commonsense reasoning using the event calculus. *Commun. ACM* **2009**, *52*, 113–117.

47. Van Harmelen, F.; Lifschitz, V.; Porter, B. *Handbook of Knowledge Representation*; Foundations of Artificial Intelligence; Elsevier: Amsterdam, The Netherlands, 2008.

48. Kim, T.W.; Lee, J.; Palla, R. Circumscriptive event calculus as answer set programming. In Proceedings of the 21st International Jiont Conference on Artifical Intelligence (IJCAI'09), Pasadena, CA, USA, 11–17 July 2009; pp. 823–829.

49. Hamadi, Y.; Jabbour, S.; Sais, L. ManySAT: A parallel SAT solver. *J. Satisfiability Boolean Model. Comput.* **2009**, *6*, 245–262.

50. Mueller, E.T. DECReasoner. 2005. Available online: http://decreasoner.sourceforge.net (accessed on 12 March 2014).

51. Muller, E.T. Discrete Event Calculus Reasoner Documentation. 2008. Available online: http://decreasoner.sourceforge.net/csr/decreasoner.pdf (accessed on 12 March 2014).

52. Testa, A. Dependability Assessment of Wireless Sensor Networks with Formal Methods. Ph.D. Thesis, University of Naples Federico II, Napoli, Italy, 2013.

53. Gungor, V.C.; Hancke, G.P. Industrial wireless sensor networks: Challenges, design principles, and technical approaches. *IEEE Trans. Ind. Electron.* **2009**, *56*, 4258–4265.

54. Quwaider, M.; Biswas, S. DTN routing in body sensor networks with dynamic postural partitioning. *Ad Hoc Netw.* **2010**, *8*, 824–841.

55. Coronato, A.; de Pietro, G. Tools for the rapid prototyping of provably correct ambient intelligence applications. *IEEE Trans. Softw. Eng.* **2012**, *38*, 975–991.

56. Cinque, M.; Coronato, A.; Testa, A. Dependable Services for Mobile Health Monitoring Systems. *Int. J. Ambient Comput. Intell.* **2012**, *4*, 1–15.

57. Ying, H.; Schlosser, M.; Schnitzer, A.; Schafer, T.; Schlafke, M.E.; Leonhardt, S.; Schiek, M. Distributed intelligent sensor network for the rehabilitation of Parkinson's patients. *Trans. Inform. Technol. Biomed.* **2011**, *15*, 268–276.

58. Wu, C.H.; Tseng, Y.C. Data compression by temporal and spatial correlations in a body-area sensor network: A case study in pilates motion recognition. *IEEE Trans. Mob. Comput.* **2011**, *10*, 1459–1472.

59. Ko, J.; Lim, J.H.; Chen, Y.; Musvaloiu-E, R.; Terzis, A.; Masson, G.M.; Gao, T.; Destler, W.; Selavo, L.; Dutton, R.P. MEDiSN: Medical emergency detection in sensor networks. *ACM Trans. Embed. Comput. Syst.* **2010**, *10*, 1–29.

60. Fernández-López, H.; Afonso, J.A.; Correia, J.; Simões, R. HM4All: A vital signs monitoring system based in spatially distributed zigBee networks. In Proceedings of the IEEE 2010 4th International Conference on Pervasive Computing Technologies for Healthcare (PervasiveHealth), Munich, Germany, 22–25 March 2010; pp. 1–4.

61. Hande, A.; Polk, T.; Walker, W.; Bhatia, D. Self-powered wireless sensor networks for remote patient monitoring in hospitals. *Sensors* **2006**, *6*, 1102–1117.

62. Fariborzi, H.; Moghavvemi, M. Architecture of a wireless sensor network for vital signs transmission in hospital setting. In Proceedings of the 2007 International Conference on Convergence Information Technology (ICCIT'07), Gyeongju, Korea, 21–23 November 2007; pp. 745–749.

63. Qiu, Y.; Zhou, J.; Baek, J.; Lopez, J. Authentication and key establishment in dynamic wireless sensor networks. *Sensors* **2010**, *10*, 3718–3731.

64. Chipara, O.; Lu, C.; Bailey, T.C.; Roman, G.C. Reliable clinical monitoring using wireless sensor networks: Experiences in a step-down hospital unit. In Proceedings of the 8th ACM Conference on Embedded Networked Sensor Systems (SenSys'10), Zurich, Switzerland, 3–5 November 2010; pp. 155–168.

65. Chittaro, L.; Montanari1, A. Efficient temporal reasoning in the cached event calculus. *Comput. Int.* **1996**, *12*, 359–382.

66. Chittaro, L.; Montanari, A. Speeding up temporal reasoning by exploiting the notion of kernel of an ordering relation. In Proceedings of the Second International Workshop on Temporal Representation and Reasoning (TIME'95), Melbourne Beach, FL, USA, 26 April 1995; pp. 73–80.

7

Target Detection over the Diurnal Cycle Using a Multispectral Infrared Sensor

Huijie Zhao, Zheng Ji, Na Li *, Jianrong Gu and Yansong Li

School of Instrumentation Science & Opto-Electronics Engineering, Beihang University, 37 Xueyuan Road, Haidian District, Beijing 100191, China; hjzhao@buaa.edu.cn (H.Z.); jizhengss1988@buaa.edu.cn (Z.J.); karon@buaa.edu.cn (J.G.); lysbuaa@buaa.edu.cn (Y.L.)
* Correspondence: lina_17@buaa.edu.cn

Academic Editor: Gonzalo Pajares Martinsanz

Abstract: When detecting a target over the diurnal cycle, a conventional infrared thermal sensor might lose the target due to the thermal crossover, which could happen at any time throughout the day when the infrared image contrast between target and background in a scene is indistinguishable due to the temperature variation. In this paper, the benefits of using a multispectral-based infrared sensor over the diurnal cycle have been shown. Firstly, a brief theoretical analysis on how the thermal crossover influences a conventional thermal sensor, within the conditions where the thermal crossover would happen and why the mid-infrared (3~5 μm) multispectral technology is effective, is presented. Furthermore, the effectiveness of this technology is also described and we describe how the prototype design and multispectral technology is employed to help solve the thermal crossover detection problem. Thirdly, several targets are set up outside and imaged in the field experiment over a 24-h period. The experimental results show that the multispectral infrared imaging system can enhance the contrast of the detected images and effectively solve the failure of the conventional infrared sensor during the diurnal cycle, which is of great significance for infrared surveillance applications.

Keywords: infrared sensor; multispectral; diurnal cycle; thermal crossover

1. Introduction

Infrared imaging detection systems are becoming more prevalent in numerous fields, including remote sensing [1], medical monitoring [2], military surveillance [3], and scientific research [4,5]. These systems offer major advantages over visual detection systems, such as their continuous day and night imaging capabilities, especially for target detection and acquisition [6].

When targets are aimed to be detected over the diurnal cycle using a conventional mid-infrared (3~5 μm) sensor, the results are generally affected by thermal crossover, where the infrared image contrast from the target and the background is difficult to discriminate from each other as the target would have integrated with the background and the radiation difference between the target and background was too low to be sensed by the infrared thermal sensor. Moreover, this could cause the targets to be blended into the background, lowering the detection accuracy, and even make the thermal sensor lose the target. In addition, the thermal crossover may also occur at any point in the day, because of solar loading, clouds, rain and fog. Therefore, it is critical to solve this problem for the conventional mid-infrared thermal sensor, especially for the infrared surveillance system.

In the last few decades, research has focused on how to solve the problem of infrared detection during thermal crossover periods and the thermal polarization technique, which is proposed as a method to enhance conventional thermal imaging, has been employed. Felton et al. [7–9] compared the crossover periods for mid-and long-wave infrared polarimetric and conventional thermal imagery. The mid-infrared (3~5 μm) imaging polarimeter they used was based on a division-of-aperture (DoA) lens technology developed by Polaris Sensor Technologies, which employed a 2×2 array of mini-lenses followed by four linear polarizers at different orientations, forming four identical images of the scene on four quadrants of the sensor focal plane array. The long-wave infrared (8~12 μm) polarimeter they used was a microbolometer-based rotating retarder imaging polarimeter developed by Polaris Sensor Technologies, which could capture up to 12 images sequentially in time with each image at a different orientation. Their experimental results showed that the polarimetric technology could be used as a method to enhance the conventional infrared image contrast between the targets and background during thermal crossover periods. However, their infrared image contrast improvement was not direct but resulted from the calculation of Stokes vector formula, which might not be suitable for the systems that requires high-performance of real-time processing. Still, as their work mainly focused on polarimetric detection experiments, what the pictures are when the target integrated with the background during the diurnal cycle and the theoretical analysis on how the thermal crossover influences the conventional thermal sensor and why the polarimetric technology could be used to solve the thermal crossover detection problem was also not mentioned. Based on Felton's research, Wilson et al. [10,11] used a single pixel scanning passive millimeter-wave polarimetric sensor, operating at a frequency of 77 GHz with a noise equivalent temperature difference (NETD) of 0.5 K, to measure the infrared image contrast during thermal crossover periods. As the passive millimeter-wave sensor is designed with capabilities to measure two linear polarization states simultaneously, it breaks the limitation that many of millimeter wave (mmW) sensors are only able to detect a single linear polarization state and improve the detection accuracy. Additionally, Retief et al. [12] studied the prediction method of thermal crossover based on imaging measurements under different weather conditions over the diurnal cycle. They used a series of infrared background objects images as the basis to establish the heat balance model and, on this basis, to predict when the thermal crossover may occur. In addition to the thermal polarization technique, the infrared multispectral technology is also considered as an important approach to solve the thermal crossover detection problem. The prior studies [13–17] on infrared multispectral technology mainly showed the potential benefits of infrared multispectral processing for clutter-limited ground target detection. However, due to constraints on the spectral resolution, band coverage, and radiometric sensitivity of existing sensors at that time, accurate measurement data and the real experimental image data were not available. Despite this, these studies firstly made the infrared multispectral technology a potential method for target/background identification. Furthermore, Schwartz and Eismann et al. [18–20] conducted a series of multispectral field measurements at Redstone Arsenal using a Bomem-developed high-sensitivity infrared Fourier Transform Spectrometer, which operates in the IR region (3–12 μm) with 8 cm^{-1} spectral resolution and noise equivalent spectral radiance (NESR, in nW/cm^{2}sr·cm^{-1} units) 7.5@3.8 μm, to enhance the capabilities of passive infrared surveillance. With the instrument, the data of several test panels, military vehicles and vegetated backgrounds at different times and under various environmental condition were obtained, their analysis of the experimental results statistically showed that the thermal sensor could detect the target hidden in vegetated and desert backgrounds with the use of multispectral techniques. As their work mainly focused on post-collection data analyses of infrared hyperspectral measurements and multispectral target detection algorithms, the design of the instrument, the real experimental image data and how the multispectral technology could be employed as an effective supplementary method for the conventional mid-infrared broadband thermal detection over the diurnal cycle was not mentioned. Nevertheless, their research results showed the potential and capacity of multispectral processing to detect low-contrast ground targets by providing valid estimates of targets to the background spectral contrast.

Overall, from the abovementioned research results, although the polarization technique was an effective solution to thermal crossover detection, there were still some disadvantages. Firstly, the improvement of the infrared image contrast resulted from the calculation of Stokes vector formula, which means that the contrast enhancement is not direct. Secondly, the time division imaging or simultaneous imaging technique are usually used in polarization detection, which would increase image processing time or the system size and weight. In addition, the environmental factors could affect the polarimetric contrast. Potential sources include vehicles, buildings, trees, clouds, water vapor, etc., which are not necessarily visible within the scene but still illuminate the objects in the field of view of the detector could be a reduction in the magnitude of polarimetric signature of a target. Compared with the polarization technique, as the target's infrared spectrum signature only differs with materials, one or some characteristic wavelengths could be enough to reflect the difference between the target and background without any redundant calculation. Thus, it would be faster and more direct to distinguish the target from the background in a complex environment with the multispectral technology if the characteristic wavelengths were acquired in advance according to prior knowledge. In this paper, our goal is to discuss how the multispectral technology could be employed to solve the problem of thermal crossover, design a fast, compact and light infrared multispectral prototype with the known characteristic wavelengths according to the prior knowledge and conclude that multispectral technology is capable of enhancing conventional thermal imaging.

Overview of Thermal Detection over the Diurnal Cycle

Thermal crossover is defined as a natural phenomenon that normally occurs twice daily, but may occur at any time throughout the day when temperature conditions are such that there is a loss of contrast between two adjacent objects on the infrared sensor. Figure 1 pictorially shows a schematic of an infrared system measuring the target radiance L_t and the background radiance L_{bg}. The infrared system can be any conventional infrared sensor or camera and located at any arbitrary orientation. The target can be any typical common objects, such as vehicles, and the background can be any natural or artificial objects, such as grass, tree, or road. To simplify, without considering the scattering, the total received radiance at the infrared system can be expressed by two components:

$$\begin{cases} L_{bg}(\lambda, \theta_v, \theta_s, \varphi) = L_{bg}^r(\lambda, \theta_v, \theta_s, \varphi) + L_{bg}^e(\lambda, \theta_v, \theta_s, \varphi, T) \\ L_t(\lambda, \theta_v, \theta_s, \varphi) = L_t^r(\lambda, \theta_v, \theta_s, \varphi) + L_t^e(\lambda, \theta_v, \theta_s, \varphi, T) \end{cases} \tag{1}$$

where L_{bg}^e and L_t^e are the emissive radiance (the radiant flux emitted by a surface, per unit solid angle, per unit projected area, per wavelength) of the background and target, L_{bg}^r and L_t^r are the reflection of the solar irradiance on the background and the target, λ is the wavelength of light, θ_v is the viewing zenith angle of the detection system, θ_s is the solar zenith angle, and φ is the azimuth angle between θ_v and θ_s, T is the temperature. As $DN = a \cdot L + b$, the DN difference between the targets and background objects (represented by C) can be expressed as [21]:

$$C = \left| DN_t - DN_{bg} \right| = \left| a(L_t - L_{bg}) \right| = a \left| (L_t^e - L_{bg}^e) + (L_t^r - L_{bg}^r) \right| \tag{2}$$

Furthermore, assuming that the reflectivity of the target and the background objects are ρ_t and ρ_{bg}, respectively, if ignoring the scattering and transmittance, the target and background objects' absorptivity would be $\alpha_t = 1 - \rho_t$ and $\alpha_{bg} = 1 - \rho_{bg}$. As the vast majority of objects in nature produce diffuse reflection, the reflectivity ρ_t and ρ_{bg} should be replaced by the Bidirectional Reflectance Distribution Function (BRDF, a function which defines the spectral and spatial reflection characteristic of a surface and is the ratio of reflected radiance to incident irradiance at a particular wavelength [22]) to represent the anisotropic properties of solar radiation effects on the reflectivity of objects. Therefore, Equation (2) can be rewritten as:

$$C = a \cdot \left| \begin{array}{l} \int_{\lambda_1}^{\lambda_2} \left[BRDF_t(\lambda, \theta_s, \theta_v, \varphi) - BRDF_{bg}(\lambda, \theta_s, \theta_v, \varphi) \right] L_s(\lambda) d\lambda \\ + \int_{\lambda_1}^{\lambda_2} \left([1 - BRDF_t(\lambda, \theta_s, \theta_v, \varphi)] L_t^e(\lambda, T) - [1 - BRDF_{bg}(\lambda, \theta_s, \theta_v, \varphi)] L_{bg}^e(\lambda, T) \right) d\lambda \end{array} \right| \tag{3}$$

where $L_s(\lambda)$ is the solar radiation and $\lambda_1 \sim \lambda_2$ is the working wavelength range of the infrared thermal sensor. In the case that θ_v, θ_s and λ are constant, BRDF only differs with the object's material.

(a)

(b)

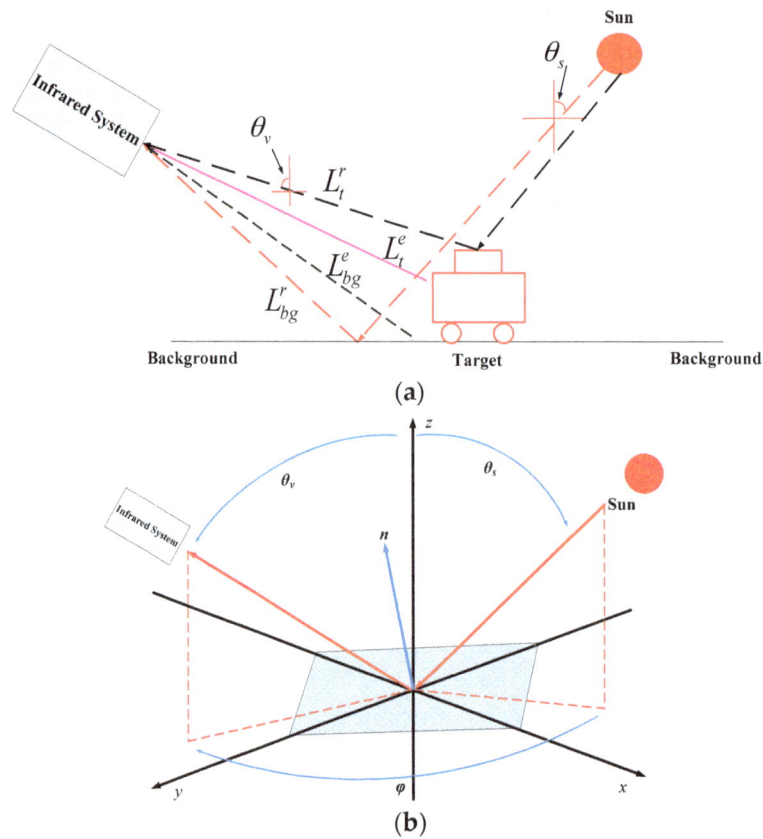

Figure 1. (**a**) Schematic of the infrared system measuring the radiance L_t and the background radiance L_{bg}. (**b**) Description of θ_v, θ_s and φ.

As can be seen from Equation (3), in general, the thermal crossover over the diurnal cycle would occur when C between the target and the background is zero or below the threshold value required to execute a specific task by the conventional infrared thermal sensor. Specifically, we divide one day, 24 h, into five time zones, as shown in Figure 2.

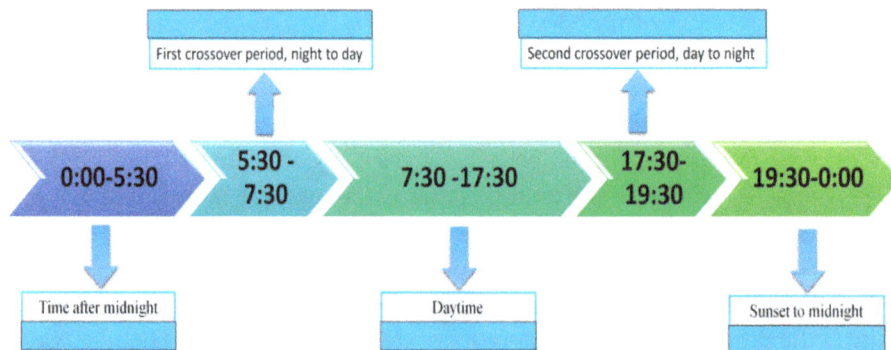

Figure 2. Five time zones from over one day.

Time after midnight:

$$C = a \left| \int_{\lambda_1}^{\lambda_2} \left([1 - BRDF_t(\lambda, \theta_s, \theta_v, \varphi)] L_t^e(\lambda, T) - \left[1 - BRDF_{bg}(\lambda, \theta_s, \theta_v, \varphi)\right] L_{bg}^e(\lambda, T) \right) d\lambda \right| \qquad (4)$$

provided that the target and background have different material; in this case, the thermal crossover would not occur as C would not be zero.

First crossover period: in case of a sunny day, for the target with lower thermal inertia, such as metal, thermal crossover would occur. In this case, the multispectral exploration technique can be used to find the emissivity difference between the target and the background in $\Delta\lambda$ to enhance the contrast C. For the target with higher thermal inertia, such as water, thermal crossover might not occur.

Daytime: The circumstance is more complicated as $L_s(\lambda)$ would have an effect on thermal crossover. No matter whether the target has lower or higher thermal inertia, thermal crossover may occur at any time, depending both on temperature differences and environmental factors, such as rain, and fog. In this case, the multispectral exploration technology can still be used to find the emissivity difference in $\Delta\lambda$ to enhance the contrast C if thermal crossover occurs.

Second crossover period: similar to the "first crossover period", for the target with lower thermal inertia the thermal crossover would occur and the multispectral exploration technology can be used to solve the problem of infrared detection during thermal crossover periods.

Sunset to midnight: similar to the "time after midnight", provided that the target and background have different materials C would not be zero and thermal crossover would not occur.

2. Materials and Methods

2.1. Why the Infrared Multispectral Technology Works

From the abovementioned discussion and Equation (3), it can be seen that it is the combined impact of temperature difference, emissivity difference between the targets and background objects, and reflected solar radiation that leads to the occurrence of thermal crossover. To simplify the problem analysis, the single factor analysis of temperature and emissivity was specified in the following two cases.

In the first case, we assume that the targets and background objects have the same emissivity and use $RRD(\lambda, T)$ to represent the relative thermal radiation differences between the targets and background objects, which is shown as Equation (5).

$$RRD(\lambda, T) = \frac{\frac{1}{\pi}\int_{\lambda_1}^{\lambda_2} \left(\alpha_t L_t^e(\lambda, T_1) - \alpha_{bg} L_{bg}^e(\lambda, T_2) \right) d\lambda}{\frac{1}{\pi}\int_{\lambda_1}^{\lambda_2} \alpha_{bg} L_{bg}^e(\lambda, T_2) d\lambda} \tag{5}$$

Figure 3 shows the graphed outputs of Equation (5), provided that the ambient temperature was 300 K and the temperature difference between the targets and the background objects changes within ± 5 K. As can be seen from Figure 3, between the 3.7 μm–4.8 μm region, which is also the typical working wavelength range for a commercial infrared detector, $RRD(\lambda, T)$ changes within -20%–25% In addition, with the decrease of wavelength, the curve $RRD(\lambda, T)$ becomes steeper and would be more sensitive to the changes in temperature. Particularly, the calculation of Equation (5) in the whole 3.7 μm–4.8 μm region was also made (not shown in Figure 3) and $RRD(\lambda, T)$ changes within a smaller region, -15%–15%, which points out that, to a certain degree, for the traditional infrared broadband thermal sensor, compared to the one with several narrow wavebands, the thermal crossover would be more likely to happen and affect thermal detection for a longer time under the same conditions.

In the second case, we assume that the targets are grey plate and steel plate, and background objects are road and sand, respectively, both of them have the same temperature, 300 K. With the emissivity data obtained from the IR module using the software Sensors, the calculation results of Equation (4) is shown as Figure 4. As can be seen from Figure 4, in the 3.7 μm–4.8 μm region, the $RRD(\lambda, T)$ curve changes from 65% to 900%. Compared with $RRD(\lambda, T)$ in the first case, obviously, the change of $RRD(\lambda, T)$ caused by emissivity presents a greater volatility and wider range than that caused by temperature in Figure 3, which, in other words, indicates that the emissivity difference under characterized bands between the targets and background objects could be utilized to solve the detection problem during the thermal crossover periods.

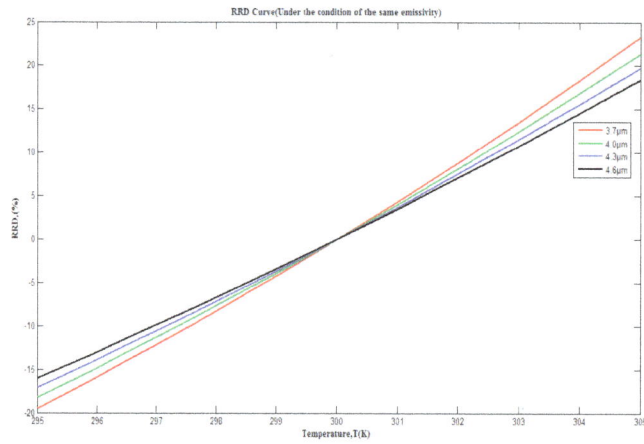

Figure 3. Relative thermal radiation differences Curve under the condition of the same emissivity in 3.7 μm–4.8 μm region.

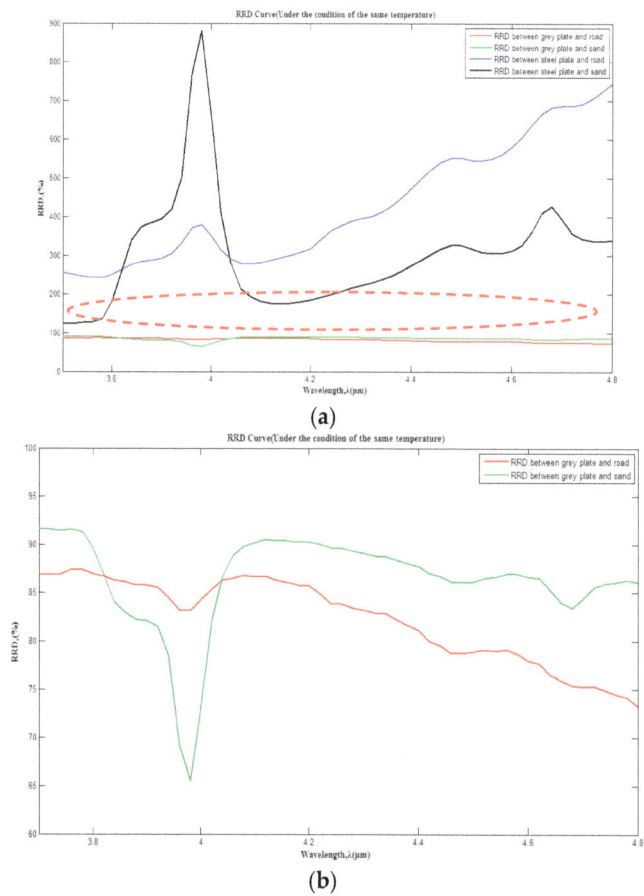

(a)

(b)

Figure 4. RRD curve under the condition of the same temperature in the 3.7 μm–4.8 μm region. (a) RRD curve among the targets grey plate, steel plate, and the background objects road and sand; (b) RRD curve among the target grey plate and the background objects road and sand.

2.2. Design of the Multispectral Infrared Imaging System

In order to verify the effectiveness of the multispectral infrared technology in solving the thermal crossover detection problem. A multispectral infrared imaging system prototype was designed and employed to conduct field experiment. The prototype consists of the infrared optical system, which is composed of the front infrared optical system and rear infrared optical system, a filter wheel with five band-pass filters, and a mid-infrared detector, as shown in Figure 5. The infrared camera lens has a focal length 100 mm. The mid-infrared detector is a France Sofradir Ltd. Model Mars 320×256 detector operating in region of 3.7–4.8 μm with a $5.5° \times 4.4°$ field of view and up to 100 fps; this detector has a geometrical resolution of 0.3 mrad and a minimum detectable temperature difference between pixels of 0.03 °C and NETD of 9 mK. The five band-pass filters are produced by Sweden Spectrogon Ltd. with central wavelengths of 3700 nm, 3800 nm, 4120 nm, 4420 nm, and 4720 nm, respectively, and mounted on the filter wheel, which is driven by a stepper motor. In addition, the filter wheel reserves a hole without any filters so that the image comparison between the traditional broadband infrared image and narrowband infrared multispectral images can be conducted. Additionally, the cold reflection impact on the image has been considered and reduced to the minimum. The mid-infrared detector, filter wheel, and data acquisition and storage are controlled by a PC. The laboratory prototype is shown in Figure 6.

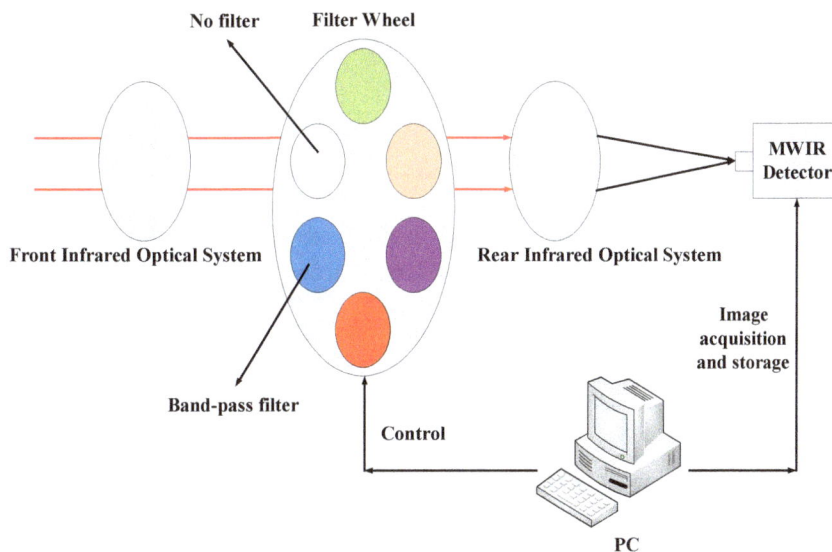

Figure 5. Schematic of multispectral infrared imaging system.

Figure 6. Prototype of multispectral infrared imaging system.

For the polarization technique used by Felton in [7–9], the MWIR imaging polarimeter employed division-of-aperture lens technology and the infrared image contrast improvement resulted from the calculation of Stokes vector formula. Thus, its image processing time, the system size and the system weight were longer and larger, the three parameters were 87 fps, the system size was $420L \times 90W \times 210H$ and weight was 4.99 kg. However, for our designed multispectral imaging system prototype, one or some characteristic wavelengths would be enough to reflect the difference between target and background without any redundant sensors and calculation if the characteristic wavelengths were acquired in advance according to the prior knowledge. Thus the prototype is faster, more compact, lighter and less costly. The image processing time of our prototype was up to 100 fps, the system size was $200L \times 180W \times 135H$ and was 3 kg, making it more suitable for practical applications. The specific specifications of the two sensors are listed in Table 1.

Table 1. Specifications of imaging polarimeter and multispectral infrared thermal sensor.

Parameter	Value of Felton's MWIR Imaging Polarimeter	Value of Our MWIR Imaging Spectrometer
Technological type	Polarization technique with four polarizers	Multispectral technique with five filters
FOV (°)	5.5	5.5×4.4
Focal Length (mm)	100	100
F/#	2.3	2
Total FPA pixels	Four FPA arrays 640×512 (single 220×220)	Single FPA 320×256
Pixel size (µm)	24×24	30×30
Max Frame Rate (fps)	87	up to 100
Sensor Dimensions (mm)	$420L \times 90W \times 210H$	$200L \times 180W \times 135H$
Sensor weight (kg)	4.99	3
Central wavelength (nm)	-	3700 nm, 3800 nm, 4120 nm, 4420 nm, 4720 nm
FWHM (nm)	-	80~100 nm
Sensitivity	10^{-7} W/cm^{2}sr	9 mK

Prior to the field experiment, the infrared multispectral imaging system is radiometrically calibrated in the field laboratory through a calibration procedure developed by EOI Ltd. so that the imaging capabilities of each wavelength can be assessed. The imaging capabilities measurement results are summarized in Table 2.

Table 2. Noise equivalent temperature difference (NETD) Measurement Results.

Wavelength	NETD (Background 298 K, f/2, Integration Time 6 ms)
3700 nm	476.6 mK
3800 nm	556.2 mK
4120 nm	533.9 mK
4420 nm	513.6 mK
4720 nm	490.1 mK

When conducting reconnaissance or surveillance tasks with the infrared multispectral imaging system, the blank hole without any filters is initially rotated to the optical axis of the system and the imaging system is just a conventional broadband infrared sensor under the initial state. With the change of time and weather conditions around the observed area, thermal crossover may occur. Once the observed target is hidden in the background caused by thermal crossover, the PC will rotate the filter wheel and control the detector to acquire the images under different multispectral wavebands to find the emissivity difference in the narrow wavebands between the targets and background objects and solve the thermal crossover problem. Afterwards, in order to highlight the multispectral information of the image and conform to the human eye's visual acquity at the same time, the narrow-band multispectral images and infrared broadband thermal image are blended to enhance the target recognition and solve the thermal crossover problem by using an HSV fusion algorithm [23]. HSV is one of the color systems that is used to pick a color from the color palette (H is hue, S is saturation and V is Value) and it is closer to people's experience and perception of color, compared with RGB.

2.3. Consideration of Experiment Design

To test the validity of the prototype, the field experiment was performed at the New Main Building at Beihang University. The infrared multispectral sensor was situated on the eighth floor of Tower B of the New Main Building (approximately 40 m) looking out of the window in the direction towards the target site, which was at approximately 100 m in distance. This open area was selected for the purpose of long-period image acquisition. The targets consisted of three different plates, galvanized sheet, steel sheet, and a wooden plate, and the natural backgrounds included grass, trees, and concrete road, which are shown in Figure 7a,b.

(a)

(b)

Figure 7. (a) Visible image of the target site consisting of three different plates and natural background; (b) The visible image of the test scene obtained on 15 March 2016.

The wooden plate was selected to make comparative experiments in order to prove the thermal radiation difference between the object with high inertia and one with low inertia over the diurnal cycle. The galvanized sheet and steel sheet were selected to demonstrate the thermal radiation difference under their characteristic wavebands and confirm the effectiveness of the multispectral technology. As some environmental parameters, like ambient temperature, relative humidity, and solar irradiance, may affect the experiment result, the experiment date was selected in advance according to the weather forecast, and the environmental parameters with a large ambient temperature difference, fewer clouds during the testing period, and relatively stable humidity, are advantageous to the experiment. The environmental data was collected on 15 March 2016 and the image data and environmental parameters were acquired continuously between 00:00 on 15 March 2016 and 23:59 on 15 March 2016 with a speed of half a minute per image and half a minute per measurement, respectively. The sunrise and sunset on 15 March 2016 occurred at roughly 06:25 and 18:21, respectively.

3. Results and Discussion

According to Figure 2, in which the 24 h day was divided into five time zones, the experimental results were demonstrated similarly. The contrast ratio between the DN values of the target and the background $C' = DN_{target}/DN_{bg}$ can be employed to reflect the contrast change of the regions of interest (ROI). At the time, after midnight, due to the different materials and temperature among the target galvanized sheet, steel sheet, and wooden plate, and the background road, C' did not approach 1 and thermal crossover did not occur, as shown in Figure 8a. At the first crossover period, for the wooden plate with higher thermal inertia, C' was 0.837 and thermal crossover did not occur, while for the galvanized sheet and steel sheet with lower thermal inertia, C' was approximately 1 (the exact number was 0.962 and 1.025, respectively) and thermal crossover did occur, as shown in Figure 8b. During the daytime, as $L_s(\lambda)$ had an effect on thermal crossover, the temperature difference among the background and the galvanized sheet and steel sheet increased gradually, C' was significantly greater than 1 and thermal crossover did not occur unless there was a rapid change in the weather conditions, as shown in Figure 8c. At the second crossover period, for the galvanized sheet and steel sheet with lower thermal inertia, C' was 1.011 and 1.045, respectively, and thermal crossover did occur again while, for the wooden plate, C' was 0.924 and the thermal crossover still did not occur, as shown in Figure 8d. From sunset to midnight, for each target C' was far less than 1 and thermal crossover did not occur, as shown in Figure 8e. Additionally, the average grey value of the image was larger than Figure 8a at the time after midnight because of the higher temperature. The contrast values among the three targets and the background in Figure 8 are listed in Table 3.

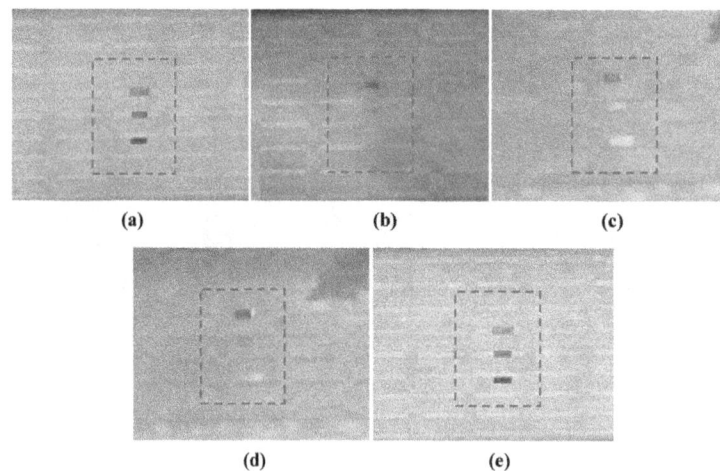

Figure 8. Infrared image obtained in each time zone. (**a**) Infrared image obtained at 03:00; (**b**) infrared image obtained at 06:50; (**c**) infrared image obtained at 12:30; (**d**) infrared image obtained at 18:05; and (**e**) infrared image obtained at 21:00.

Table 3. Contrast among the three targets and background in Figure 8.

Figure Number	Wooden Plate	Galvanized Sheet	Steel Sheet
a	0.822	0.784	0.632
b	0.837	0.962	1.025
c	0.825	1.098	1.151
d	0.810	1.011	1.045
e	0.807	0.786	0.612

In order to clarify the effectiveness of the multispectral technology to solve the thermal crossover problem, the multispectral images under central wavelengths of 4120 nm, 4420 nm, and 4720 nm at the first crossover period were obtained, and the results are presented in the form of image contrast plots, calculated using $C' = DN_{target}/DN_{bg}$. Included with each of these plots are the corresponding environmental data, as shown in Figure 9. Specifically, Figure 9b,c clearly show that the contrast curve of the galvanized sheet and steel sheet varied more significantly than the wooden plate during the 24-h test period and the contrast of the galvanized sheet and steel sheet was close to 1 during two diurnal cycles, while the contrast of the wooden plate fluctuated between 1.0 and 1.16 throughout the experiment time, proving the existence of thermal crossover for the objects with lower thermal inertia once again. Figure 9d showed the multispectral images, which were obtained at the same period with Figure 9b. As can be seen from Figure 9d, the multispectral technology was used to find the emissivity difference between the target and the background at 3700 nm, 3800 nm, 4120 nm, 4420 nm, and 4720 nm to enhance the contrast among the galvanized sheet, steel sheet, and road. Among the five wavebands the best contrast improvement was at 4720 nm, with 4420 nm following, which presented a consistent trend in accordance with Figure 4a and indicated the effectiveness of the multispectral technology in solving the thermal crossover problem. The contrast values among the three targets and the background under different wavebands in Figure 9d are listed in Table 4. Compared with the Figure 8, the image contrast enhancement in the target area is direct after employing the narrow band-pass filters.

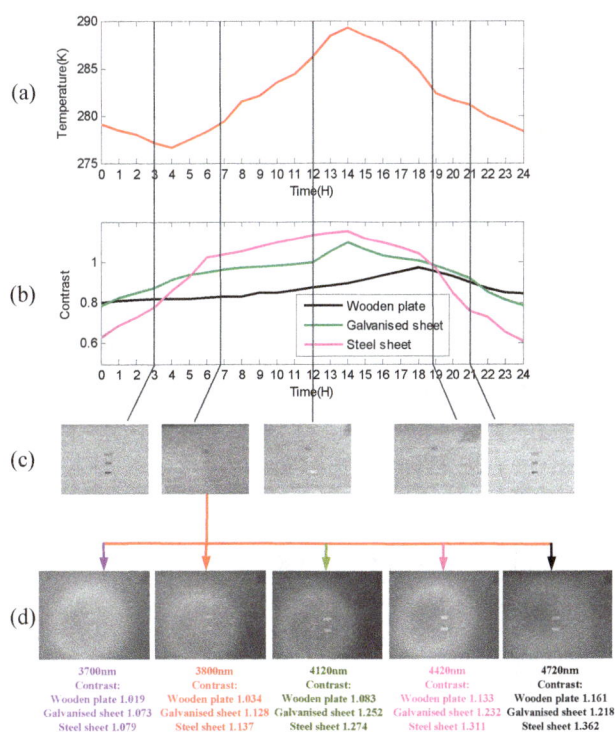

Figure 9. (a) Ambient temperature during the 24-h test; (b) contrast curve among the three targets and the background road; (c) images of the targets in the five time zones; and (d) images of the targets at 3700 nm, 3800 nm, 4120 nm, 4420 nm, and 4720 nm at the first diurnal cycle.

Figure 10 showed the pseudo-color image obtained by running the HSV image fusion algorithm described in Section 3 with Figures 8b and 9d. Combining the infrared broadband image with the infrared images under characteristic wavebands, the three targets were marked with different colors and presented clearly, by which the multispectral technology employed an effective supplementary method for the conventional mid-infrared broadband thermal sensor to solve the thermal crossover detection problem.

Further, it can be noted that the magnitude of contrast improvement is not as large as the calculation results in Figure 4a because of the solar radiation effect, stray radiation caused by band-pass filters, and the difference between the actual emissivity value and the real emissivity value. However, this does not influence our experimental conclusions that multispectral technology can be employed to solve the thermal crossover problem.

In order to further show the advantage of multispectral technology in solving the thermal crossover problem, the same field experiment with polarization technique by using $0°$, $45°$, $90°$, $135°$ four linear polarizers was also conducted and the four polarization state polarization images obtained at 07:00 were shown in Figure 11.

Table 4. Contrast among the three targets and the background in Figure 9.

Wavebands	Wooden Plate	Galvanized Sheet	Steel Sheet
3700 nm	1.019	1.073	1.079
3800 nm	1.034	1.128	1.137
4120 nm	1.083	1.252	1.274
4420 nm	1.133	1.232	1.311
4720 nm	1.161	1.218	1.362

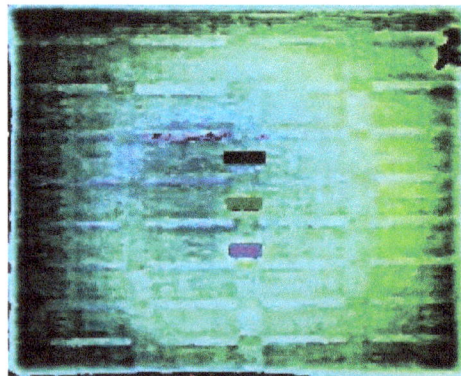

Figure 10. Pseudo-color image fused by Figures 7b and 8d.

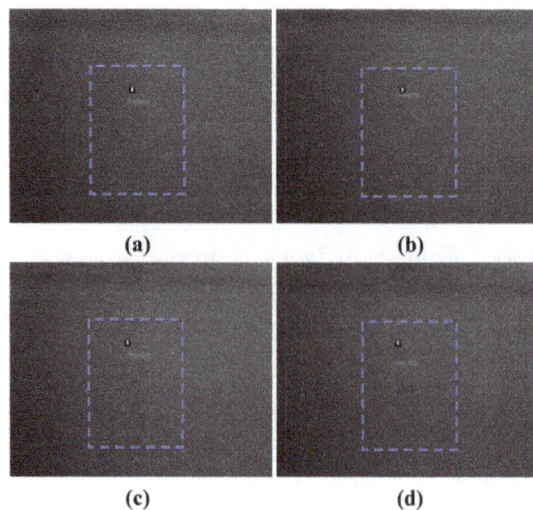

Figure 11. Infrared polarization images with four polarization states. (**a**) $0°$; (**b**) $45°$; (**c**) $90°$; (**d**) $135°$.

As can be seen from Figure 11, for the wooden plate, thermal crossover still did not occur, while for the galvanized sheet and steel sheet, the thermal crossover did occur, which was similar to the results with multispectral technology. The contrast values between the galvanized sheet, steel sheet and the background in Figure 11 are listed in Table 5.

Table 5. Contrast between galvanized sheet, steel sheet targets and background in Figure 11.

Figure Number	Galvanized Sheet	Steel Sheet
a	0.909	0.942
b	0.977	0.964
c	0.979	0.981
d	0.965	0.976

From Table 5, it can be found that, compared with the infrared multispectral images, without further image processing, the image contrast enhancement in the target area in the infrared polarization images with four polarization states were not obvious. Thus, the Stokes vectors, which completely characterized the polarization states of targets from the scene need to be calculated. The data products used in this experiment included S_0 and S_1 Stokes parameter images where S_0 is the horizontal (0°) plus the vertical (90°) components of polarization and the S_1 Stokes parameter is the horizontal minus the vertical components of polarization. The S_0 and S_1 Stokes parameter images are shown as Figure 12a,b respectively.

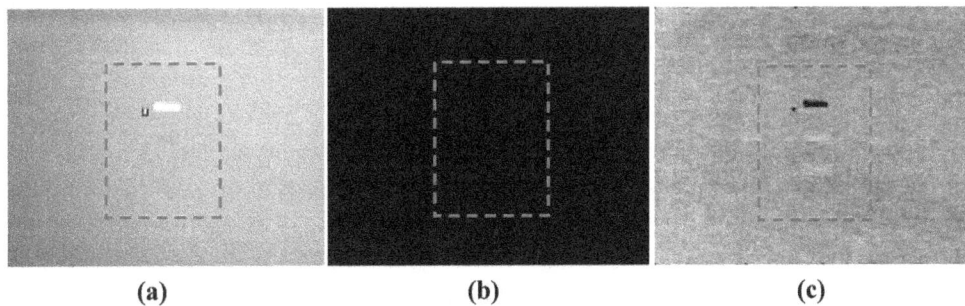

| (a) | (b) | (c) |

Figure 12. S_0 and S_1 Stokes parameter images (**a**) S_0 image; (**b**) S_1 image; (**c**) S_1 after contrast stretching.

The image contrast of the galvanized sheet and steel sheet in Figure 12a was 0.921 and 1.073, respectively, which showed some extent of improvement compared with Figure 11. However the DN difference between the galvanized sheet, steel sheet targets and background in Figure 12a were only 16 and 7. In Figure 12b, although the calculated image contrast, according to $C' = DN_{target}/DN_{bg}$, was improved, it was meaningless as the DN of targets and background were too low to be sensed by eyes. In fact, the DN difference between the galvanized sheet, steel sheet targets and background in Figure 12b were only 3 and 2, respectively. In order to show the targets in Figure 12b relatively clearly, the images were further processed with the contrast stretching algorithm, as shown in Figure 12c. Through the data processing procedure, it could be found that even with the Stokes parameter calculation, the difference between target and background had still not been improved significantly so that further image processing procedures were required. The main reason for the polarization detection experiment result was that the abundant geometry information contained in the background weakened the polarization characteristics differences. Because the polarization technology achieves distinction between target and background through the perception of their polarization characteristics differences, the background information might have an influence on the target detection. However, for the multispectral technology, as stated previously, compared with the polarization technique, as the target's infrared spectrum signature only differs with materials, it would be faster and more direct to distinguish the target from the background in the complex environment only if the characteristic wavelengths of the targets and backgrounds were acquired in advance.

4. Conclusions

As the thermal crossover has great influence on the infrared sensors working in a single wide range, it is significant to solve this problem for the conventional mid-infrared thermal sensor, especially for the infrared surveillance system. In this study, we analyze theoretically how the thermal crossover disables the conventional thermal sensor and under what conditions the thermal crossover would happen. Furthermore, based on the analysis, a fast, compact and light optical prototype based on infrared multispectral technology is designed with the known characteristic wavelengths according to the prior knowledge. Then the experimental process has been optimized and more image data is provided, especially regarding what the pictures are when the target integrated with the background during the diurnal cycle. Then, the whole process of employing the multispectral technology to solve the thermal crossover detection problem is clearly shown. In addition, a comparison experiment with polarization technique is also conducted to further show the advantage of multispectral technology.

The field experiment with multispectral technology was conducted over a 24-h period with the targets of galvanized sheet, steel sheet, and wooden plate, and the background road on a sunny day. The results showed that, for the galvanized sheet and steel sheet targets, the thermal crossover could affect a contrast for up to four hours at two diurnal cycles, jeopardizing the success of surveillance missions. For the wooden plate target, although the image contrast reduced over the diurnal cycle, it could still distinguish the targets from the background objects, which means that thermal crossover might not always occur, or even possibly not exist at all over the diurnal cycle for the objects with higher thermal inertia. Through employing the infrared narrow band-pass filters, thermal crossover in the first diurnal cycle was relieved as the contrast was upgraded to the levels such that the metal targets could be distinguished from the background objects. Furthermore, the experimental results provided us with the information about what the characterized bands between the targets and background objects were, which would be useful for system design in the future. In addition, the pseudo-colored image produced by multi-spectral image fusion method showed the effectiveness of the multispectral technology for contrast promotion of each target. Then, as a comparison, the same field experiment with polarization technique by using $0°$, $45°$, $90°$, $135°$ four linear polarizers was also conducted and the S_0 and S_1 Stokes parameter images showed that the image contrast showed some extent of improvement but no obvious improvement, as the background weakened the polarization characteristics differences.

While promising, the field experiment should just be considered as very preliminary practical application and the experimental results should also just be viewed as a proof-of-principle. Nevertheless, the conclusion that the multispectral technology can be employed to solve the thermal crossover problem is unambiguous. In future, it might be possible to further extend the range of applications for the conventional thermal infrared broadband sensor into the thermal crossover periods by exploiting the emissivity of infrared spectral signatures and fusing multispectral images from the perspective of mid-infrared thermal detecting system design. Research focusing on the characterized bands between different common targets and background objects and how the weather conditions influence the thermal crossover will be undertaken.

Acknowledgments: This work was supported by the National Natural Science Foundation of China (No. 61571029), the CAST Innovation Foundation of the China Academy of Space Technology, and the Changjiang Scholars and Innovative Research Team in University (No. IRT1203).

Author Contributions: Zheng Ji wrote the manuscript and was responsible for the research design, data collection, and analysis. Huijie Zhao, Na Li, Jianrong Gu and Yansong Li assisted in the methodology development and research design and participated in the writing of the manuscript and its revision.

Conflicts of Interest: The authors declare no conflict of interest.

Abbreviations

The following abbreviations are used in this manuscript:

mmW — Millimeter Wave
NESR — Noise Equivalent Spectral Radiance
NETD — Noise Equivalent Temperature Difference
DN — Digital Number
BRDF — Bidirectional Reflectance Distribution Function
RRD — Relative thermal radiation differences
IR — Infrared
PC — Personal Computer

References

1. Sidran, M. Broadband reflectance and emissivity of specular and rough water surfaces. *Appl. Opt.* **1981**, *20*, 3176–3183. [CrossRef] [PubMed]
2. Elsner, A.E.; Weber, A.; Cheney, M.C.; VanNasdale, D.A.; Miura, M. Imaging polarimetry in patients with neovascular age-related macular degeneration. *J. Opt. Soc. Am. A* **2007**, *24*, 1468–1480. [CrossRef]
3. Cooper, A.W.; Lentz, W.J.; Walker, P.L.; Chan, P.M. Infrared polarization measurements of ship signatures and background contrast. *Proc. SPIE* **1994**, *2223*, 300–309.
4. Jin, L.; Hamada, T.; Otani, Y.; Umeda, N. Measurement of characteristics of magnetic fluid by the Mueller matrix imaging polarimeter. *Opt. Eng.* **2004**, *43*, 181–185. [CrossRef]
5. Wijngaarden, R.J.; Heeck, K.; Welling, M.; Limburg, R.; Pannetier, M.; van Zetten, K.; Roorda, V.L.G.; Voorwinden, A.R. Fast imaging polarimeter for magneto-optical investigations. *Rev. Sci. Instrum.* **2001**, *72*, 2661–2664. [CrossRef]
6. Shaw, J.A. Degree of linear polarization in spectral radiances from water-viewing infrared radiometers. *Appl. Opt.* **1999**, *38*, 3157–3165. [CrossRef] [PubMed]
7. Felton, M.; Gurton, K.P.; Pezzaniti, J.L.; lt, D.B.C.; Roth, L.E. Measured comparison of the crossover periods for mid- and long-wave IR (MWIR and LWIR) polarimetric and conventional thermal imagery. *Opt. Express* **2010**, *18*, 15704–15713. [CrossRef] [PubMed]
8. Felton, M.; Gurton, K.P.; Pezzaniti, J.L.; Chenault, D.B.; Roth, L.E. *Comparison of the Inversion Periods for Mid-wave IR (MidIR) and Long-Wave IR (LWIR) Polarimetric and Conventional Thermal Imagery*; Army Research Laboratory: Adelphi, MD, USA, 2010; pp. 20783–21197.
9. Felton, M.; Gurton, K.P.; Roth, L.E.; Pezzaniti, J.L.; Chenault, D.B. Measured comparison of the inversion periods for polarimetric and conventional thermal long-wave IR (LWIR) imagery. *Proc. SPIE* **2009**, *7461*. [CrossRef]
10. Wilson, J.P.; Schuetz, C.A.; Harrity, C.E.; Kozacik, S.; Eng, D.L.K.; Prather, D.W. Measured comparison of contrast and crossover periods for passive millimeter-wave polarimetric imagery. *Opt. Express* **2013**, *21*, 12899–12907. [CrossRef] [PubMed]
11. Wilson, J.P.; Murakowski, M.; Schuetz, C.A.; Prather, D.W. Simulations of polarization dependent contrast during the diurnal heating cycle for passive millimeter-wave imagery. *Proc. SPIE* **2013**, *8873*. [CrossRef]
12. Retief, S.J.P.; Willers, C.J.; Wheeler, M.S. Prediction of thermal crossover based on imaging measurements over the diurnal cycle. *Proc. SPIE* **2003**, *5097*, 58–69.
13. Stotts, L.B.; Winter, E.M.; Hoff, L.E.; Reed, I.S. Clutter Rejection Using Multi-Spectral Processing. *Proc. SPIE* **1990**, *1305*. [CrossRef]
14. Winter, E.M. *Infrared Spectral Analysis*; Technical Research Associates Report No. TRA-90D-109; Air Force Research Laboratory: Dayton, OH, USA, 1990.
15. Stocker, A.D.; Reed, I.S.; Yu, X. Multi-Dimensional Signal Processing for Electro-Optical Target Detection. *Proc. SPIE* **1990**, *1305*, 218–231.
16. Stocker, A.D.; Yu, X.; Winter, E.M.; Hoff, L.E. Adaptive Detection of Sub-Pixel Targets Using Multi-Band Frame Sequences. *Proc. SPIE* **1991**, *1481*. [CrossRef]
17. Cederquist, J.N.; Johnson, R.O.; Reed, I.S. *Infrared Multispectral Imagery Program. Phase I: Model-Based Performance Predictions*; ERIM Final Report No. 232300-41-F to AF Wl/AARI-4, Contract No. F33615-90-C-1441; Environmental Research Institute of Michigan: Ann Arbor, MI, USA, 1993.

18. Eismann, M.T. Infrared Multispectral Target/Background Field Measurements. *Proc. SPIE* **1994**, *2235*, 135–147.

19. Stocker, A.D.; Seldin, A.O.H.; Cederquist, J.N.; Schwartz, C.R. Analysis of Infrared Multi-Spectral Target/Background Field Measurements. *Proc. SPIE* **1994**, *2235*, 148–161.

20. Schwartz, C.R.; Eismann, M.T.; Cederquist, J.N. Thermal multispectral detection of military vehicles in vegetated and desert backgrounds. *Proc. SPIE* **1996**, *2742*, 286–297.

21. Zhao, H.; Ji, Z.; Zhang, Y.; Sun, X.; Song, P.; Li, Y. Mid-infrared imaging system based on polarizers for detecting marine targets covered in sun glint. *Opt. Express* **2016**, *24*, 16396–16409. [CrossRef] [PubMed]

22. Nicodemus, F.E. Directional reflectance and emissivity of an opaque surface. *Appl. Opt.* **1965**, *4*, 767–775. [CrossRef]

23. Chen, W.; Wang, X.; Jin, W.; Li, F.; Cao, Y. Experiment of target detection based on medium infrared polarization imaging. *Infrared Laser Eng.* **2011**, *40*, 7–11.

8

TPLE: A Reliable Data Delivery Scheme for On-Road WSN Traffic Monitoring

Rui Wang [1,2,*], **Fei Chang** [1] and **Suli Ren** [1]

[1] School of Computer and Communication Engineering, University of Science and Technology Beijing, Beijing 100083, China; changfeifei@outlook.com (F.C.); rensuli@126.com (S.R.)

[2] Beijing Key Laboratory of Knowledge Engineering for Materials Science, Beijing 100083, China

* Correspondence: wangrui@ustb.edu.cn

Academic Editor: Paolo Bellavista

Abstract: In an on-road environment, motor-engines severely disturb the wireless link of a sensor node, leading to high package loss rate, high delivery delay, and poor radio communication quality. The existing data delivery mechanisms, such as the ACK-based retransmission mechanism and window-based link quality estimation mechanism, could not handle these challenges well. To solve this challenge, we propose a Target-Prediction-based Link quality Estimation scheme (TPLE) to realize high quality data delivery in an on-road environment. To perform on-road link quality estimation, TPLE dynamically calculates the track of a nearby vehicle target and estimates target impact on wireless link. Based on the local estimation of link quality, TPLE schedules radio communication tasks effectively. Simulations indicate that our proposed TPLE scheme produces a 94% data delivery rate, its average retransmission number is around 0.8. Our conducted on-road data delivery experiments also indicated a similar result as the computer simulation.

Keywords: intelligent transportation systems; reliable data delivery; wireless sensor networks

1. Introduction

Traffic monitoring is an important part of modern city administration. Real-time road traffic data helps improve transportation efficiency [1], such as road information broadcast and real time weather publication.

Due to features such as low cost, flexible deployment, and easy maintenance, Wireless Sensor Networks are very suitable for establishment of extensive Intelligent Transportation Systems [2]. Some studies [2–5] have been conducted to collect on-road data by WSNs.

To realize large scale data collection by WSN upon on-road monitoring, there are still many challenges. One challenge is the reliability of data transmission from a sensor node to its destination in an on-road transportation environment. In an on-road traffic environment, the engine and other noises are much stronger as compared to other environments, especially the noise caused by engines. So the reliable wireless data delivery in on-road environments should counter for these obstacles.

We conducted an experiment to point out the aforementioned communication complexity. In that experiment, we used two sensor nodes equipped with CC1000 [6], one as a sender and one as a receiver. We set the distance between two sensor nodes to be 6 m and set the radio power to be 0 dB/m. We conducted the same experiment in three different environments:

(1)   Two nodes placed an indoor environment.

(2)   Two sensor nodes placed an outdoor non-traffic environment.

(3)   Two sensor nodes placed along the road sides (where heavy vehicles run on the road).

Each experiment lasted 10 min duration. As shown in Figure 1, the meaning of y axis is package loss rate in three different environments. We define 10 percentage points of its value as a span. You can see the package loss rate in environment (**a**) is the minimum, and the maximum packet loss rate achieved was 64.15% where heavy vehicles run on the road. Figure 1 shows the overall package loss rate. In an on-road context, the radio communication quality in non-traffic outdoor environment is better than in the traffic road. The only difference between the two scenarios is the busy vehicle targets running in the road.

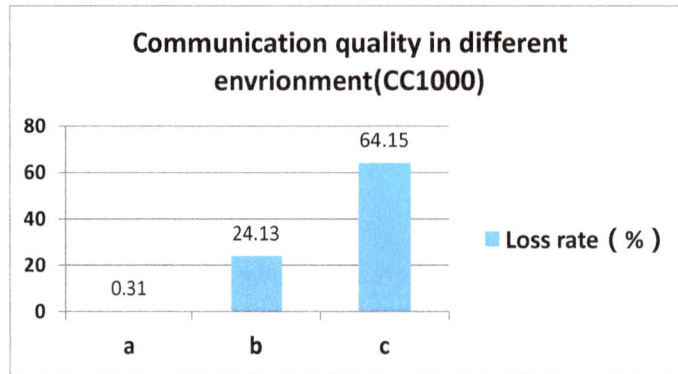

Figure 1. Radio loss rate in different settings.

From Figure 1 we made the following observations:

(1)   The wireless communication quality in on-road environments is relatively poor compared to other environments. The main reason for the worst radio communication quality is engine noises nearby the wireless vicinity.

(2)   The running and passing vehicles caused huge distortion to wireless link quality. The unstable and unpredictable nature of link quality gave rise to poor performance of data delivery, such as low delivery rate, high time delay, etc.

Traditionally, in wireless communication, link quality estimation is adopted to support the realization of reliable and low-delay data transmission. We argue that in an on-road traffic environment, the link quality cannot be described by existing link models. Apart from the normal factors that influence link quality, such as background noises, etc., when an on-road traffic environment is referred, the running vehicles cause disruption to wireless links, especially the radio link used in wireless sensor networks. From Figure 1, by comparing scenarios (**b**) and (**c**), intuitively, almost 40% of sent Packets are lost due to the noise of motor engines.

In this paper, we focus on the reliability of on-road data delivery, we have proposed and implemented a reliable data transmission scheme, called TPLE, which can conduct node data delivery according to real time traffic position around the sensor node. Our proposed scheme can effectively estimate the link quality and provide good communication quality. Our contributions are as follows:

1.   We proposed the TPLE scheme which can effectively deal with strong noises caused by on-road motor engines targets.

2.   With this scheme, we obtain relatively high radio communication quality with the acceptable cost of time delay with respect to data delivery.

3.   We developed a real-time application to verify our proposed scheme; our system can be applied in pervasive on-road data collection.

2. Related Work

In this part, we briefly introduce the research of applying WSN in on-road monitoring and the research of reliable data delivery mechanisms.

Coleri et al. [2] proposed the idea of adopting WSN in on-road monitoring. The author proposed a single-magnetic-sensor-based solution for on-road vehicle identification. Also, a two-sensor-based on-the-spot vehicle speed estimation scheme was proposed. In [3], the author proposed a data collection scheme to detect vehicle on a grass. Wireless radio delivers both raw measurement and in-network control information. Radio communication mainly occurs around those sensor nodes near the moving vehicle.

In the existing research of WSN 802.11 networking, the challenges to building reliable data delivery are packet losses and errors. To handle these two challenges, solutions can be classified into two categories:

2.1. Frame-Based Solutions

In [4], the author proposed a solution for traffic lights management. Communications occurred among on-vehicle sensors and roadside relays and the solution is that they needed continuous RSSI (Received Signal Strength Indication) sampling, which is energy inefficient.

In [5], the author proposed a detection and classification solution for motor vehicles. The vehicle classification is accomplished by processing data from multiple sensor nodes. A dynamical central node receives radio messages from correlated neighbors and performs the collaborative detection or classification tasks.

All of the aforementioned research work by default assumed a reliable wireless data delivery. However, as we have mentioned earlier, the reliability of data delivery in terms of an on-road environment cannot be omitted to build a practical data collection system. The on-road environment presents new challenges to wireless link estimation research.

2.2. Physical-Based Solutions

COLLIE [7] detects collisions from weak signal for 802.11 networks by identifying error patterns within a physical-layer symbol. Moreover, COLLIE adopts a feedback of error packets from the receiver side. COLLIE is not applicable in the WSN field because this mechanism cost is too high with respect to measurement overheads and energy consumption.

SoftRate [8] uses SoftPHY hints to estimate sudden changes in the BER, so that collisions and packet losses can be detected. SoftRate needs access to a physical layer which is difficult to realize in WSN.

Present research on the chip error patterns on IEEE 802.15.4 standards shows that the packet losses can be estimated [9]. Access to a physical layer is also essential here and makes this solution impractical in WSN.

Carried out research about packet loss estimation prediction in WSN with a joint RSSI-LQI based classifier [10], the received packet can be classified into four categories: lost, successfully received, error by collision, and error by weak signal. Then corresponding actions can be adopted to improve the reliability of the data delivery. The disadvantage of this scheme is solutions, [11–13] use RTS/CTS to assure a reliable data communication in 802.11 networks. The RTS/CTS overhead is relatively high and it is always disabled in WSN design.

In summary, the existing research in 802.11 networks on how to predict packet loss to realize a reliable data delivery cannot be applied to WSN. There are some studies, such as [10], trying to realize reliable data delivery. However, as we have demonstrated in Figure 1, the noise and disturbance in an on-road context is much higher than typical WSN deployment sites, such as in indoor [14], open field [15], and outdoor [16,17] contexts. In this paper, we present our research on how to tackle the unprecedented scale of disruption in an on-road WSN communication, and we also present our implementation of Target-Prediction-based on a Link quality Estimation system which neither needed any overheads in the physical layer nor used RTS/CTS.

3. TPLE: Target-Prediction-Based Link Quality Estimation Mechanism

In this section we propose a mechanism, namely TPLE for on-road data collection systems to reliably deliver data. We firstly describe the ideas of TPLE, and then we describe the TPLE data collection procedures for data collection regarding on-road traffic. The framework of TPLE is shown in Figure 2 as follows.

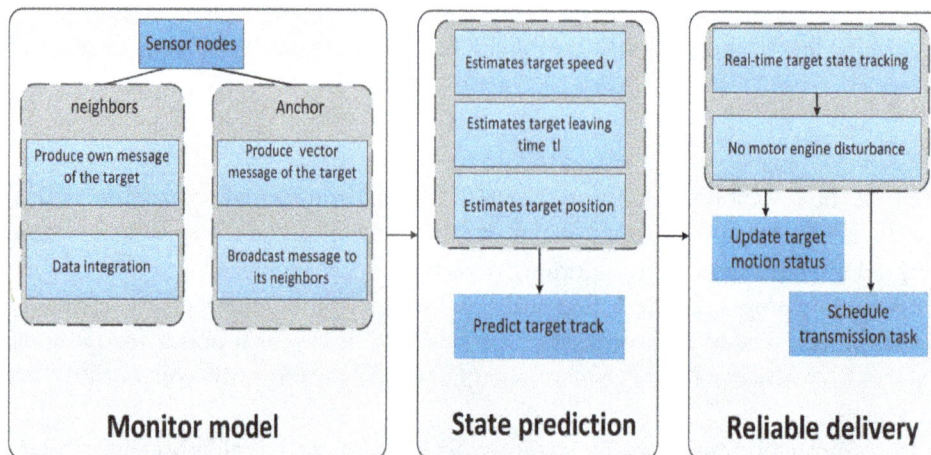

Figure 2. The framework of TPLE.

As shown in Figure 2, TPLE includes three parts: monitor model, state prediction, and reliable delivery. The target can be sensed by sensor nodes when it comes, and the vector message about the incoming target which is produced by Anchor and its neighbors will be integrated to predict the target state. According to real time motion state tracking, the state of the motor engine which causes disturbance to data delivery can be calculated dynamically locally. When the engine is out of the influential scope of a sensor, the sensor will update the target motion status and schedule the transmission task effectively. Thus, a reliable data delivery mechanism is built.

In detail, when the target is sensed by the Anchor, it can produce a message of feature vector I, which includes the message of the time instant that the target passes the Anchor and the signal strength of monitoring target when the target reaches the Anchor. Then, the Anchor will broadcast the feature vector I to its next neighbor node. When the next neighbor received the I, in a short time, it can also produce a feature vector about the target, and will broadcast to its next neighbor node. We can estimate the target speed and calculate the leaving time when the target is out of the influential scope of each sensor by using such a form for message transmission of the feature vector. So, the sensor can reschedule the transmission tasks according to the state of the target. It perfectly avoids the motor engine disturbance of the vehicle target when the sensors are in the process of delivering data and the radio communication quality can be greatly enhanced.

3.1. TPLE Algorithm for Link Quality Estimation

As we mentioned before, the extra complexity of on-road traffic context lies in the unpredictable and sudden appearance of a vehicle target. Vehicle engines cause great disturbance to wireless links when they are close to them.

According to our previous work in the on-road environment, the radio communication quality of a sensor node will decline when vehicle engines appear around a sensor. So an intelligent on-road data delivery scheme should be established for a radio communication task by reducing the disturbances of traffic motor targets.

Assume that a motor engine's magnetic field is like a disk having radius of R. The motion model of a nearby vehicle can be described as follows:

$$x_{k+1} = F_k x_k + G_k u_k + w_k \qquad (1)$$

The measurement model of a sensor node is as follows:

$$z_k = H_k x_k + v_k \qquad (2)$$

In these two equations, x_k is the state vector at time instant k, F_k is the targets matrix of state transition, G_k is the matrix of input noise, u_k is the dynamic noise, w_k is the process noise, Z_k is the measurement vector of sensor node, H_k is the measurement matrix, v_k is the measurement noise.

By adopting target tracking methods [18–20], such as Kalman filter [21], the target states such as speed and position can be estimated. According to the radius R of the influential scope, the time instant when a target leaves the influential scope can be predicted. When a target's track is estimated by a sensor node, the sensor node can reschedule the transmission task and improve its quality.

Based on this idea, we propose a target-prediction-based link quality estimation mechanism, (TPLE). In TPLE mechanism, the downstream nodes can be informed appearance of an incoming vehicle target and dynamically calculate the link quality at local. By this mechanism, the radio communication quality can be greatly enhanced.

3.2. Reliable Data Delivery Scheme

In this part, we describe the TPLE-based reliable data delivery scheme. First, we introduce some scenarios that are very common in an on-road traffic monitoring environment. Figure 2 shows typical on-road monitoring settings. Several sensor nodes are deployed along a road, the sensor nodes are fixed in position, the distance between two neighbor nodes is d(For the simplicity consideration, here we suppose the distance between neighbor nodes are equal in one system) [22]. Also, we suppose that the vehicle target is in a state of uniform rectilinear motion. We suppose the sensor nodes are synchronized with their local time in the system.

Here we introduce the concept of an Anchor. We define an Anchor to be the first node in a monitoring system that notices the incoming target. According to [23], a sensor node can accurately report the time instant of a passing vehicle. When the vehicle target is monitored by a sensor node N_m, this sensor node can produce a message of feature vector I_m, where $I_m = \{t_m, s_m\}$. Here, t_m is the time instant that the target passing a sensor node N_m, s_m is the signal strength of the monitoring target when the target reaches a sensor node N_m. In Figure 3, N_1 is supposed to be the Anchor. When a vehicle target is identified by an Anchor, a message of feature vector I_1 will be broadcast to its relevant neighbors N_2 in this figure.

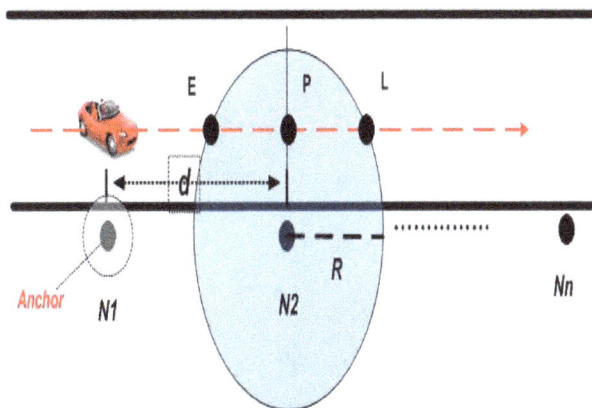

Figure 3. Monitoring settings.

As the data transmission process may possibly fail to accomplish the task, in a real system, a simple ACK-based retransmission mechanism, for such acknowledgement provides an Active Message in TinyOS [24] that can be added to an Anchor to ensure that this packet transmission process has been successfully performed.

As to the relevant neighbor, N_2 in Figure 3, it will be informed the feature vector of an incoming target from upstream N_1. Meanwhile, in a short time, N_2 produces its own feature vector about that target I_2, $I_2 = \{t_2, s_2\}$. With the combination of I_1 and I_2, N_2 can obtain an estimation of target speed locally:

$$\overline{v} = \frac{d}{|t_2 - t_1|} \tag{3}$$

where d is the distance between sensor N_1 and N_2; t_1, t_2 is the time instant when the target reaches the sensor N_1, N_2 respectively, the value of them is derived from feature vector I_1 and I_2. With the estimated target speed $\overline{v}$, N_2 can predict the target track. The leaving time of the coming target can be estimated as:

$$\overline{t_l} = t_2 + \frac{d_{PL}}{\overline{v}} \tag{4}$$

where d_{PL} is the distance between point P and point L. Because it is technically difficult to calculate the value of d_{PL}, we replace d_{PL} with R in (1), and get $\overline{t'_l}$. As R > d_{PL}, so $\overline{t'_l} \geq \overline{t_l}$. Besides simplicity, another advantage of this replacement is that the increased $\overline{t_l}$ assures the package transmission which will be scheduled upon vehicle departure. The tradeoff of this replacement is the tiny increment in package delay. After this calculation of t_l, t_l can schedule its radio transmission tasks locally.

As we define the Anchor to be the first node that discovers the incoming target. In some cases, the physically first node which is supposed to discover the vehicle target may miss the target and fail to report to relevant nodes. In such a case, the first discovery of that vehicle serves as the Anchor and finishes the report task.

As all the sensor nodes in our system are homogenous, it is not difficult to implement the aforementioned mechanisms in a sensor node. Figure 4 indicates a diagram chart of our proposed TPLE algorithm.

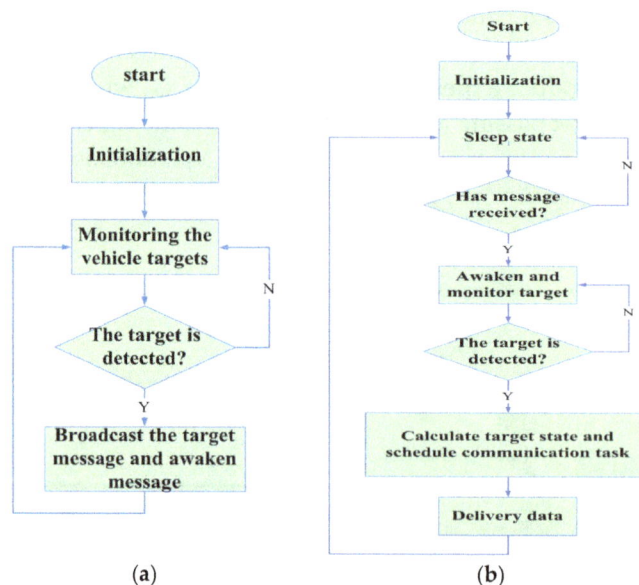

Figure 4. The diagram chart of TPLE, where (**a**) represents the flow diagram of the Anchor; (**b**) represents the flow diagram of sensor nodes.

4. Verification and Experiment

In traditional wireless communication a periodic sleep-wake mechanism for sensor nodes has been designed and employed in order to save energy better, in which every node in the sleep mode periodically wakes up to communicate with its neighbor nodes [25]. Once the communication is over the node goes in to sleep mode again until the next frame begins. However, in traffic monitoring circumstances, the sensor nodes need to monitor targets frequently and communicate with their neighbors during active time. Besides, a node cannot receive messages during sleep mode, so, the messages targeted for such a node may be lost. In such a situation, the traditional duty cycle mechanism can neither save energy effectively nor obviously improve transmission quality of the WSN. The TPLE algorithm we proposed can predict the target status and then the active opportunity of sensor nodes can be estimated. Our algorithm does not need to monitor targets frequently, especially if the data delivery is in an environment without motor engine disturbance. Therefore, our TPLE algorithm can save energy of the WSN effectively and better improve the quality of data delivery. In this part, we conducted both simulations and on-road experiments to verify our proposed scheme.

4.1. Simulations

We conduct computer simulations to verify our TPLE data delivery scheme. In our simulations, a sensor node is supposed to report its measurement when a vehicle target passes from its side.

(1) Baseline algorithms and simulation settings

We use two algorithms as comparison baselines, a simple acknowledgement-based single hop delivery provided by CC1000 radio stack (CRS) of Active Message in TinyOS [24] and EWMA algorithm adopted in [26]. The maximal retransmission number in the MRS algorithm is set to be 3. The length of temporal statistical window in EWMA is set to be 10 s. The EWMA algorithm sends a probing message every 0.5 s. In EWMA, a link is supposed to be good if the statistical package delivery rate is higher than 0.7.

The data delivery rate, average package delay, energy consumption, and expected package retransmission lead to performance evaluations.

As vehicles are the largest disruption to link quality, we adopt different traffic densities (low, medium, high) to evaluate the performances. In low traffic density, we generate 5 vehicle appearances every minute; in the medium density, 10 vehicles are generated every minute; and in the high density, 20 vehicles are generated every minute. For each traffic density, we conducted 20 groups of simulation.

The simulation settings are listed in Table 1. In Table 1, according to a large amount of on-road collected data, we set the radius of sensor nodes radio influence to 4 m, which means a vehicle in this range will reduce the sensor nodes' wireless communication quality. Also, we assigned 0.80, which is a relatively ideal radio communication quality in outdoor non-vehicle-appearance environment. The minimal distance between two vehicles is set to be 5 m, which is a normal setting in the case of no traffic jam. Moreover, our proposed TPLE mechanism is applicable in this assumption. The Baud rate of a sensor node is assigned to be 19.2 kbps, which is a common rate in CC1000 radio communication modules. The packet size has been fixed to 28 bytes, which is the standard packet size in TinyOS. We use the Baud Rate and Packet size to calculate packet delay between sender and receiver; we ignore the other factors like transmission delay and software delay.

Table 1. Simulation Settings.

Parameter	Value	Description
D_{n2n}-D_{n2n}	6 m	Distance between neighbor nodes
V	5 m/s–15 m/s	Vehicle speed
$M_{ax}D_{v2v}$ $m_{ax}D_{v2v}$	5 m	Minimal distance between two vehicles
R	4 m	Radius of a nodes magnetic field
F1-F1	0.8	Packet delivery rate without vehicle in influential scope
Pf-Pf	0.25	Packet delivery rate with vehicle in influential scope
Baud rate	19.2 kbps	Radio rate
Packet size	28 bytes	Radio packet size 0

(2) Data Delivery Rate

The data delivery rate is an important criterion for link quality. Figure 5 indicates the data delivery results in the simulation. According to this figure, the data delivery rate of our proposed TPLE scheme is higher than 94% in different traffic density, which is very ideal. The delivery rate of CRS algorithm in all three traffic scenarios remains consistent to 70% almost. The reason is that the CRS algorithm does not take any link quality into consideration, the ACK mechanism is the only solution for high data delivery rates. As to the EWMA algorithm, although it takes the link state into consideration, its link state estimation accurately reflects the real link status. In the high traffic density scenario, the performance of EWMA is mostly observed as being unstable. The reason for this phenomenon is that dense traffic introduces extra-high unpredictability factors for the link state estimation parts of EWMA.

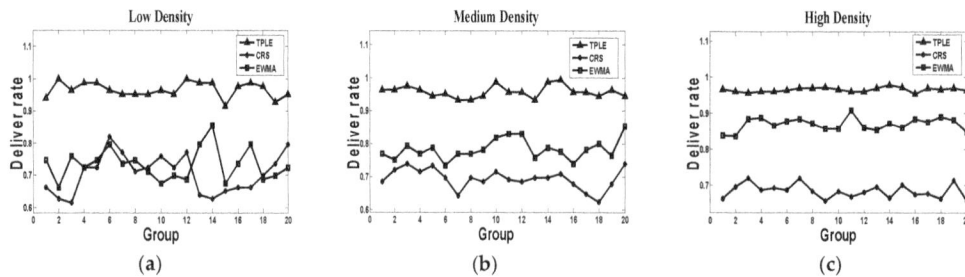

Figure 5. Data delivery rate simulation, where (**a**) represents the data delivery rate in low traffic density; (**b**) represents the data delivery rate in medium traffic density; (**c**) represents the data delivery in high traffic density. The representation of horizontal coordinates is the groups of simulation, the representation of vertical coordinates is the data delivery rate.

To summarize, our proposed TPLE scheme provides a stable and high data delivery rate for on-road sensors, it deals well with the different traffic scenarios.

(3) Average packet delivery delay

Figure 6 indicates the packet delay in simulation. EWMA algorithm has the largest average packet delay; the reason is that the link estimation in the EWMA algorithm enables wireless transmission when the statistical link quality is better than the threshold. The target-related information transmission experiences an unavoidable delay because of the statistical link quality estimation mechanism. The CRS algorithm has the smallest delivery delay because it starts packet transmission once the target is detected. The cost of low packet delay is a relatively high retransmission number and low packet delivery rate. We will demonstrate it later. Our proposed TPLE has a relatively high delivery delay, because it schedules transmission task when the target leaves, not like the immediate transmission in the CRS algorithm.

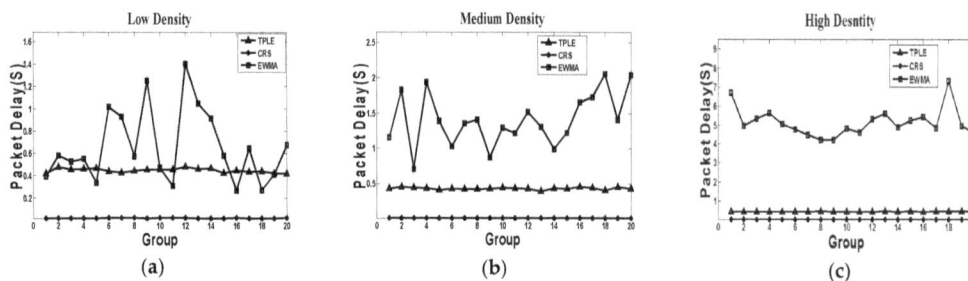

Figure 6. Packet delay simulation, where (**a**) represents the packet delay in low traffic density; (**b**) represents the packet delay in medium traffic density; (**c**) represents the packet delay in high traffic density. The representation of horizontal coordinates is the groups of simulation, the representation of vertical coordinates is the packet delay.

(4)   Average packet retransmission number

Figure 7 indicates the average packet retransmission number in the simulation. From this figure, we can see along with the change in traffic density, the packet retransmission number of CRS and TPLE algorithms keep stable. TPLE has the lowest packet retransmission number while the CRS algorithm has the largest retransmission number. The average retransmission number of the EWMA algorithm decreases as the traffic density increases. The reason for this phenomenon is that more traffic results in the link quality estimation of EWMA becoming more accurate and the transmission quality is improved. Compared with two baseline algorithms, TPLE has the least retransmission because it schedules the radio communication tasks when it has a relatively good link quality.

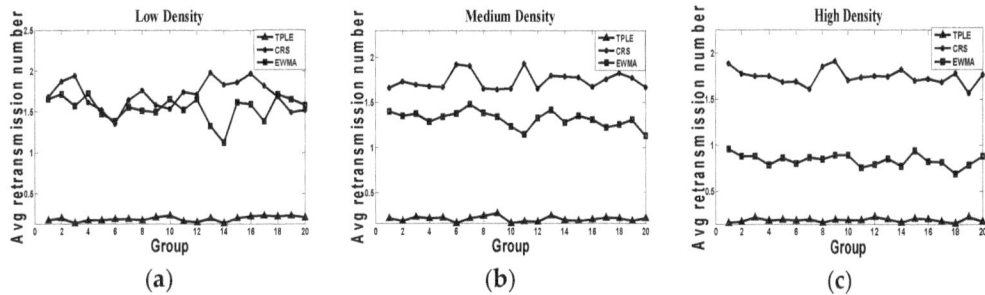

Figure 7. Average retransmission number, where (**a**) represents the packet retransmission number in low traffic density; (**b**) represents the packet retransmission number in medium traffic density; (**c**) represents the packet retransmission number in high traffic density. The representation of horizontal coordinates is the groups of simulation, the representation of vertical coordinates is the average packet retransmission number.

(5)   Energy consumption

As radio communication accounts for the greatest energy cost of a sensor node, we use total packet number to evaluate the energy consumption. In our algorithm, we use the CC1000 as a communication module which with very low current consumption, its unit energy consumption is the change of energy in unit time at a certain power. In the three algorithms (EWMA, CRS, and TPLE), we define the total packet number, the transmission power, and the transmission time as the reference standard for energy consumption. The working power of the CC1000 can be set before the experiment. So, we can evaluate the energy consumption according to the total packet number of the sensors. From Figure 8, we can see the EWMA algorithm definitely has the largest energy consumption because of the periodical link probing messages. As TPLE adopts the most effective link estimation mechanism, it consumes the least energy.

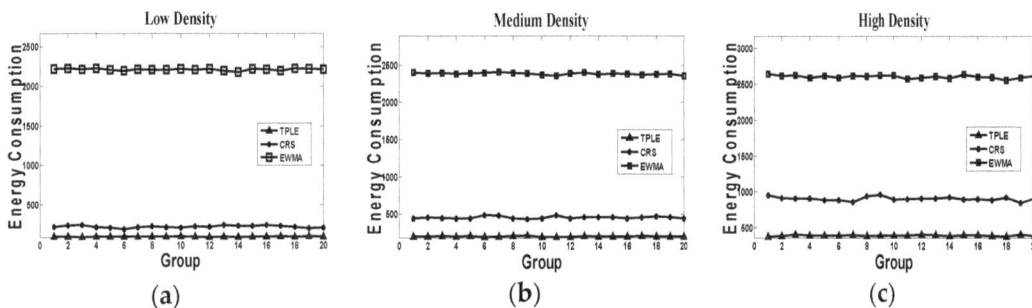

Figure 8. Energy cost simulation, where (**a**) represents the energy consumption in low traffic density; (**b**) represents the energy consumption in medium traffic density; (**c**) represents the energy consumption in high traffic density. The representation of horizontal coordinates is the groups of simulation, the representation of vertical coordinates is the energy consumption.

4.2. On-Road Verification Experiments

We also conducted real experiments to verify our proposed data delivery scheme. The experiments can be seen in Figure 9. Based on the platform EasiTia [23], we implemented the TPLE algorithms in sensor nodes.

Then the sensor nodes were deployed alongside a road to detect and report vehicle messages. Figure 8 indicates our on-road experiment settings. We used four nodes along the road-side to perform data collection tasks, two nodes on each side of the road. A sensor node is supposed to send its measurement about a vehicle to a sink node.

We conducted three groups of experiments, each experiment lasted 10 min, we use packet delay, packet delivery rate, and retransmission numbers to evaluate our experiment result. Table 2 indicates results of our on-road experiments. We compared the three algorithms (EWMA, CRS, and TPLE algorithms) in delivery rate, packet delay, and retransmission number. In Table 2, Group 1, Group 2, and Group 3 represent the EWMA algorithm, the CRS algorithm, and the TPLE algorithm respectively. Each experiment of three algorithms is the same, and the result values of delivery rate, packet delay, and retransmission number are the average values of all the experiments.

From this table we can see that, the on-road data delivery rate of TPLE is not as good as the results in simulations. The reason of this phenomenon may result from the inaccurate definition of the influential scope of the sensors' radio link. However, we may see that the packet delivery delay and average retransmission number is ideal as indicated by our simulation.

On the basis of both simulations and on-road experiments, we verified our proposed TPLE data delivery scheme. We demonstrated that our proposed scheme performed relatively better data delivery with less cost for a small packet delivery. The delay is acceptable in an on-road monitoring environment.

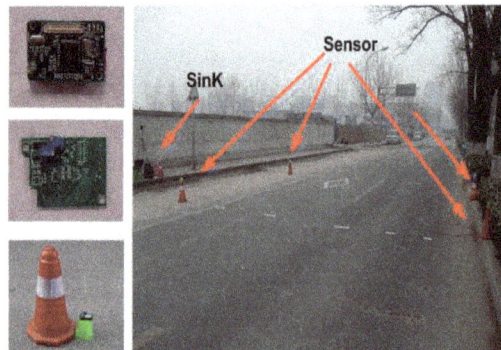

Figure 9. On-Road experiment scenario.

Table 2. On-road experiment results.

	Delivery Rate	Packet Delay	Retransmission Number
Group 1	82.54%	1.08s	0.71
Group 2	85.87%	0.84s	0.73
Group 3	94.11%	0.81s	0.61

5. Conclusions

In this paper, we propose a TPLE mechanism to solve the problem of low link quality caused by motor engines in the on-road traffic monitoring environment. TPLE estimates link quality by dynamically updating target motion status and scheduling radio communication tasks when there is no motor engine disturbance. Simulation and on-road experiments demonstrate the good communication quality of TPLE in on-road environments.

Acknowledgments: This work was supported in part by the National Natural Science Foundation of China under Grant No. 61379134.

Author Contributions: Rui Wang and Fei Chang conceived and designed the experiments; Suli Ren performed the experiments; Rui Wang and Fei Chang contributed analysis tools; Suli Ren analyzed the data; Rui Wang, Fei Chang, and Suili Ren all wrote the paper.

Conflicts of Interest: The authors declare no conflict of interest.

References

1. Lee, W.H.; Tseng, S.S.; Shieh, W.Y. Collaborative real time traffic information generation and sharing framework for the intelligent transportation system. *Inf. Sci.* **2010**, *180*, 62–70. [CrossRef]

2. Coleri, S.; Cheung, S.; Varaiya, P. Sensor networks for monitoring traffic. In Proceedings of the Allerton Conference on Communication, Control and Computing, Monticello, IL, USA, 29 September–1 October 2004; pp. 32–40.

3. He, T.; Krishnamurthy, S.; Stankovic, J.; Abdelzahber, T.; Luo, L.; Stoleru, R.; Yan, T.; Gu, L.; Hui, J.; Krogh, B. Energy efficient surveillance system using wireless sensor networks. In Proceedings of the 2nd International Conference on Mobile Systems Applications and Services, Boston, MA, USA, 6–9 June 2004; pp. 270–283.

4. Chen, W.J.; Chen, L.F.; Chen, Z.L.; Tu, S.L. A realtime dynamic traffic control system based on wireless sensor network. In Proceedings of the 34th International Conference on Parallel Processing Workshops, Oslo, Norway, 14–17 June 2005; pp. 258–264.

5. Gu, L.; Jia, D.; Vicaire, P.; Yan, T.; Luo, L.; Tirumala, A.; Cao, Q.; He, T.; Stankovic, J.; Abdelzaher, T.; et al. Light weight detection and classification for wireless sensor networks in realistic environments. In Proceedings of the 3rd International Conference on Embedded Networked Sensor Systems, San Diego, CA, USA, 2–4 November 2005; pp. 205–217.

6. Texas Instruments. *CC1000: Single Chip Very Low Power RF Transceiver*; Texas Instruments: Dallas, TX, USA, 2010.

7. Rayanchu, S.; Mishra, A.; Agrawal, D.; Saha, S.; Banerjee, S. Diagnosing wireless packet losses in 802.11: Separating collision from weak signal. In Proceedings of the 27th IEEE Conference on Computer Communications, Phonenix, AZ, USA, 13–18 April 2008; pp. 735–743.

8. Vutukuru, M.; Balakrishnan, H.; Jamieson, K. Cross layer wireless bit rate adaptation. In *ACM SIGCOMM Computer Communication Review*; ACM Publications: New York, NY, USA, 2009; Volume 39, pp. 3–14.

9. Wu, K.; Tan, H.; Ngan, H.L.; Liu, Y.; Ni, L.M. Chip error pattern analysis in IEEE 802.15.4. *IEEE Trans. Mob. Comput.* **2012**, *11*, 543–552.

10. Eu, Z.A.; Lee, P.; Tan, H.P. Classification of packet transmissions outcomes in wireless sensor networks. In Proceedings of the 2011 IEEE International Conference on Communications (ICC), Kyoto, Japan, 5–9 June 2011; pp. 1–5.

11. Wong, S.H.; Yang, H.; Lu, S.; Bharghavan, V. Robust rate adaptation for 802.11 wireless networks. In Proceedings of the 12th Annual International Conference on Mobile Computing and Networking, Los Angeles, CA, USA, 24–29 September 2006; pp. 146–157.

12. Kim, S.; Verma, L.; Choi, S.; Qiao, D. Collision aware rate adaptation in multi rate WLAN: Design and implementation. *Comput. Netw.* **2010**, *54*, 3011–3030. [CrossRef]

13. Khan, M.A.Y.; Veitch, D. Isolating physical per for s-mart rate selection in 802.11. In Proceedings of the IEEE INFOCOM 2009, Rio de Janeiro, Brazil, 19–25 April 2009; pp. 1080–1088.

14. Sugano, M.; Kawazoe, T.; Ohta, Y.; Murata, M. Indoor localization system using RSSI measurement of wireless sensor network based on ZigBee standard. In Proceedings of the 2006 Wireless and Optical Communication Conference, Banff, AB, Canada, 3–4 July 2006; pp. 1–6.

15. Viani, F.; Robol, F.; Giarola, E.; Benedetti, G.; De Vigili, S.; Massa, A. Advances in wildlife road-crossing early-alert system: New architecture and experimental validation. In Proceedings of the IEEE 8th European Conference on Antennas and Propagation (EuCAP 2014), The Hague, The Netherlands, 6–11 April 2014; pp. 3457–3461.

16. Suhonen, J.; Kohvakka, M.; Hannikainen, M.; Hannikainen, T.D. Design, implementation, and experiments on outdoor deployment of wireless sensor network for environmental monitoring. In *Embedded Computer Systems: Architectures, Modeling, and Simulation*; Springer: Berlin/Heidelberg, Germany, 2006; pp. 109–121.

17. Jin, M.; Zhou, X.; Luo, E.; Qing, X. Industrial-QoS-Oriented Remote Wireless Communication Protocol for the Internet of Construction Vehicles. *IEEE Trans. Ind. Electr.* **2015**, *62*, 7103–7113. [CrossRef]

18. Ghirmai, T. Distributed Particle Filter for Target Tracking: With Reduced Sensor Communications. *Sensors* **2016**, *16*, 1454. [CrossRef] [PubMed]

19. Zhang, Q.; Song, T.L. Improved Bearings-Only Multi-Target Tracking with GM-PHD Filtering. *Sensors* **2016**, *16*, 1469. [CrossRef] [PubMed]

20. Zhu, W.; Wang, W.; Yuan, G. An Improved Interacting Multiple Model Filtering Algorithm Based on the Cubature Kalman Filter for Maneuvering Target Tracking. *Sensors* **2016**, *16*, 805. [CrossRef] [PubMed]

21. Kalman, R. A new approach to linear filtering and prediction problems. *J. Basic Eng.* **1960**, *82*, 167–175. [CrossRef]

22. Niu, Q.; Yang, X.; Gao, S.; Chen, P.; Chan, S. Achieving Passive Localization with Traffic Light Schedules in Urban Road Sensor Networks. *Sensors* **2016**, *16*, 1662. [CrossRef] [PubMed]

23. Wang, R.; Zhang, L.; Sun, R.; Gong, J.; Cui, L. Easitia: A pervasive traffic information acquisition system based on wireless sensor networks. *IEEE Trans. Intell. Transp. Syst.* **2011**, *12*, 615–621. [CrossRef]

24. Levis, P.; Gay, D. *TinyOS Programming*; Cambridge University Press: Cambridge, UK, 2009.

25. Van Dam, T.; Langendoen, K. An adaptive energy-efficient MAC protocol for wireless sensor networks. In Proceedings of the International Conference on Embedded Networked Sensor Systems, Los Angeles, CA, USA, 5–7 November 2003; pp. 23–27.

26. Mi, X.; Zhao, H.; Zhu, J. Research on EWMA based link quality evaluation algorithm for WSN. In Proceedings of the IEEE Cross Strait Quad-Regional Radio Science and Wireless Technology Conference, Harbin, China, 26–30 July 2011; pp. 757–759.

9

Design of a Mobile Low-Cost Sensor Network Using Urban Buses for Real-Time Ubiquitous Noise Monitoring

Rosa Ma Alsina-Pagès [1,*], **Unai Hernandez-Jayo** [2,3], **Francesc Alías** [1] **and Ignacio Angulo** [2,3]

[1]   GTM—Grup de recerca en Tecnologies Mèdia, La Salle—Universitat Ramon Llull, Quatre Camins, 30, Barcelona 08022, Spain; falias@salleurl.edu

[2]   DeustoTech—Fundación Deusto, Avda. Universidades, 24, Bilbao 48007, Spain; unai.hernandez@deusto.es (U.H.-J.); ignacio.angulo@deusto.es (I.A.)

[3]   Facultad Ingeniería, Universidad de Deusto, Avda. Universidades, 24, Bilbao 48007, Spain

*   Correspondence: ralsina@salleurl.edu

Academic Editors: Andrea Zanella and Toktam Mahmoodi

Abstract: One of the main priorities of smart cities is improving the quality of life of their inhabitants. Traffic noise is one of the pollutant sources that causes a negative impact on the quality of life of citizens, which is gaining attention among authorities. The European Commission has promoted the Environmental Noise Directive 2002/49/EC (END) to inform citizens and to prevent the harmful effects of noise exposure. The measure of acoustic levels using noise maps is a strategic issue in the END action plan. Noise maps are typically calculated by computing the average noise during one year and updated every five years. Hence, the implementation of dynamic noise mapping systems could lead to short-term plan actions, besides helping to better understand the evolution of noise levels along time. Recently, some projects have started the monitoring of noise levels in urban areas by means of acoustic sensor networks settled in strategic locations across the city, while others have taken advantage of collaborative citizen sensing mobile applications. In this paper, we describe the design of an acoustic low-cost sensor network installed on public buses to measure the traffic noise in the city in real time. Moreover, the challenges that a ubiquitous bus acoustic measurement system entails are enumerated and discussed. Specifically, the analysis takes into account the feature extraction of the audio signal, the identification and separation of the road traffic noise from urban traffic noise, the hardware platform to measure and process the acoustic signal, the connectivity between the several nodes of the acoustic sensor network to store the data and, finally, the noise maps' generation process. The implementation and evaluation of the proposal in a real-life scenario is left for future work.

Keywords: END; noise mapping; hardware platform; connectivity; acoustic sensing; signal processing; dynamic measurement; ubiquitous; smart city

1. Introduction

Nowadays, more people live in urban than in rural areas, representing in 2010 50.5% of the world's population for the first time in history [1]. It is expected that this tendency will continue since all of the world's population growth will take place in urban areas in the next four decades according to the United Nations [1,2]. This reality poses new challenges to authorities so as to guarantee the efficient use of the resources of those urban areas and the quality of life of their inhabitants through better services management, which requires significant changes in governance, decision-making and developing specific action plans [3]. To this aim, a technological revolution is driving the change of most of the urban cities under the umbrella of the so-called smart city or smart region paradigms [4,5].

However, the transformation of any city into a smart city is a long and complex process [3], which should take most of the experiences and best practices of initiatives already developed previously.

In this sense, city managers should take note of the results obtained by a recent large-scale study in Europe, which has revealed significant adverse impact of environmental noise on the health and longevity of the inhabitants [6]. Since noise mapping has been evaluated as an answer to this problem [7], the acoustic pollution issue has been included within the smart city paradigm, due to its negative impact on the quality of life of citizens.

The smart city revolution, together with the definition of the Environmental Noise Directive 2002/49/EC (END) [8] and the consequent strategic noise mapping assessment (CNOSSOS-EU) [9], both from the European Commission, has encouraged national, regional and local authorities to address the negative impact of citizens' quality of life in terms of noise pollution. Several studies can be found in the literature about the implementation of the END directive and its impact on health. In a recent work [10], the road traffic noise exposure for 2012 is detailed based on the END. Other works focus on the END proposal for noise mapping, studying whether it is accurate enough for people exposure estimation [11] or for epidemiological studies [12]. For instance, the evaluation of the analyzed information can lead national road authorities to optimize the installation of noise-reducing pavements or noise barriers where required, as well as evaluating the noise exposure and reduction when these plans are implemented [13]. Despite several studies have already implemented the END, the first years of its application have been characterized by the lack of a common method to perform the noise mapping, which have made results' comparison almost unfeasible. This problem has been recently solved by the publication of the annex to the Commission Directive 2015/996 [14], which settles common methods that the Member States will have to use from 31 December 2018.

The European Commission (EC) has promoted the research in those topics, by funding working groups like FONOMOC [15] in the framework of the European cities association EUROCITIES [16], constituted to share experiences and knowledge across European cities. From these work groups, several conclusions and future work road-maps have been generated, e.g., as reported in [17], where the mobility in urban areas is studied from different perspectives and several suggestions of improvement are detailed through diverse technical measures and smart or innovative solutions.

In terms of European funding schemes focused on environmental issues, several LIFEprojects have been funded during the last few years to address the noise pollution consequences: from its diagnosis, through the improvement of citizens' involvement, to the development of specific solutions. The recently ended Harmonica project (LIFE10 ENV/FR/000211) [18] has developed new tools to give the public better information about environmental noise and to help local authorities make the right decisions in fighting noise pollution [19]. On the other hand, the DYNAMAPproject (LIFE13 ENV/IT/001254) [20] aims to develop a dynamic noise mapping system able to detect and represent in real time the acoustic impact of road infrastructure based on a static network of acoustic sensors, reducing the cost of the periodical actualization of the noise maps [21,22]. The latter builds on the conclusions obtained from the SENSEable PISA project [23,24], which proposed collecting the information describing not only the urban environment, but also some aspects of the social behavior of the citizens, to study possible relationships with public health, mobility or pollution. Regardless of the the dramatic improvement that these kinds of systems entail in terms of having a dynamic picture of the noise pollution in the city, their main drawback is that they are static, i.e., they are designed and adjusted once and deployed in specific locations from that moment on. Hence, if there is some significant change in the urban area environment (e.g., super-blocks in Barcelona [25,26]), the deployed system should be redesigned and adapted appropriately to the new operational context.

Nowadays, the European Union (EU) authorities are proposing cities to make a step beyond current approaches and use noise mapping to improve the strategy of urban sound planning [27]; this indication is slowly being acquired by local authorities and being integrated with their environmental noise policies. Noise mapping is used together with soundscape analysis and sound level measurements to follow the management guidelines of areas with good environmental noise quality [28]. To this aim, the SONORUS project [29] is focused on the training of professionals for the integration of urban sound planning in the urban development process of cities, with the final goal of reducing the noise levels.

In order to face the main drawback of static networks, we propose the development of a low-cost ubiquitous sensor network for real-time noise monitoring using urban bus routes. This proposal builds on the idea that urban traffic noise levels can be measured on certain city routes by means of the implementation of a mobile sensor network installed on public buses. From the gathered data, a noise map is dynamically calculated and published right after to inform citizens and authorities. After describing the design of the proposal, we discuss the main challenges it faces, as the mandatory noise cancellation of the sensing vehicle to avoid biasing the noise map computation, the development of low-cost ubiquitous hardware platform and its connectivity to provide the measured data to the cloud, the bus trajectory to collect the most suitable data and, finally, the data gathering in the cloud to tailor an integrated noise map. Finally, it is worth noting that the implementation and evaluation of the proposal in a real-life scenario are not addressed in this work, since these are left for future studies.

This paper is structured as follows. In Section 2, the state of the art of dynamic acoustic urban sensing applications, considering both static acoustic networks and participative sensing approaches, is presented. In Section 3, we compare several low-cost hardware platforms and their connectivity. Section 4 discusses the challenges of the proposal in terms of signal processing, hardware platform and network design. Finally, Section 5 resumes the key characteristics that should be taken into account for the design of a bus-based mobile acoustic urban sensor network to map the urban noise dynamically.

2. State of the Art of Dynamic Acoustic Urban Sensing

Traditional noise measurements in cities have been mainly carried out by professionals that record and analyze the data in a certain location typically using certified sound level meters [7]. This approach is hardly scalable when it comes to tackling the current demand for more frequent noise level measurements in both time and space (thus, more measures and in more places). In the last decade, several approaches for monitoring environmental noise have been proposed [30]. Their main goal has been developing less expensive and smaller hardware solutions, but assuring the reliability of the acoustic measure, as well as allowing the scalability of the system through the improvement of network data communications.

In this section, we describe representative approaches developed to automatically measure the noise levels of the cities in order to tailor noise maps [30]. The classification takes into account how the measure was performed, assuming that all of them have been designed to obtain a dynamic noise measure, where the term 'dynamic' refers in all this paper to a system that updates its measurement frequently in time. Static acoustic urban networks have been developed in some cities to provide automatic monitoring based on fixed sensor networks placed in specific locations. Finally, some mobile platforms designed to perform acoustic sensing are also described. On the one hand, some of them have conducted the measurements on vehicles, taking advantage of their ubiquity. On the other hand, some projects have asked citizen to become contributors to the city sensing system thanks to the democratization of technology; in particular, smartphones.

2.1. Static Acoustic Urban Sensing

In this section, several static noise sensing platforms and their applications are summarized, focusing exclusively on those whose main focus is the monitoring of environmental noise on specific locations in urban areas.

In [31], a project designed for the monitoring of the traffic noise in Xiamen City (China) is presented. The designed system considers noise meters, ZigBee technology and GPRS communication. In [32], the authors prefer to use a customized noise level meter, which is designed to remove the burden of computational and energy-expensive operations from the sensor node and process them in the cloud. In [33], the RUMEUR (Urban Network of Measurement of the sound Environment of Regional Use) network developed by Buitparif is detailed. The sensor network includes both high accuracy equipment for critical places, like airports, where the focus is to obtain detailed acoustic information due to the intense noise environment, together with less precise measuring equipment in other locations where the goal is only the updating of the noise map. In [34], the deployment of an acoustic sensor network based on the FIWARE platform is described. The information collected is used to create quasi-real-time dynamic noise and event maps, as well as for identifying the sound sources. The applications of the project are focused on surveillance, and they range from localizing fire to finding people in distress.

Another related field of study is Wireless Acoustic Sensor Networks (WASN) [35], enabled by the advances in the technology and the availability of small, low-cost and smart sensors that result in deployable nodes of WASN. They are usually applied to sensing remote areas where power connections are difficult or even impossible to reach. Despite most of the applications in this field being found in underwater [36] or environmental situations [37], several smart city applications can be also found in the literature. For instance, in [38], an ad hoc wireless sensor network system that detects and accurately locates shooters in the city is detailed. Thus, in this case, not only the specific type of sound is identified, but also the precise place of the shooter. In Section 2.3, another wireless sensor network is presented [39], but it is used as a mobile sensor network for participative sensing. For more references related to this field and, in particular, about signal processing-related issues, the reader is referred to [40] and the references therein.

The last group of static acoustic urban sensing systems is based on custom-built sensor networks usually designed to be very economical and autonomous, with low power consumption, so they can be installed pervasively. Two of the projects using this kind of network are the IDEA (Intelligent Distributed Environmental Assessment) project [41] and the MESSAGE (Mobile Environmental Sensing System Across Grid Environments) project [42]. They are based on a single-board computer with low computational capacity, working with a low-cost sound card and microphones. This hardware choice allows the deployment of large sensor networks due to the economic cost of each node, besides allowing the collection of relevant environmental data from several critical locations in the city. In the IDEA project [43], a cloud-based platform is developed integrating an environmental sensor network with an informative web platform and pretends to measure noise and air quality pollution levels in urban areas in Belgium. This is also the base idea of the DYNAMAP (DYNamic Acoustic MAPping) project [21,44], aimed at the deployment of two low-cost acoustic sensor networks, one in Rome and the other in Milan, as a step ahead from the preliminary results obtained by the SENSEable project deployed in the city of Pisa (all of them in Italy) [24].

2.2. Mobile Acoustic Urban Sensing

Some acoustic urban sensing approaches have built on mobile urban sensing systems during the last years. The mobile concept refers here to the execution of the measurement; thus, in this section we are focusing on those platforms that allow changing the place of measurement. The goal of these systems in comparison with the previous static urban sensing approaches is that the potential area to be sensed is widened by means of the sensor network ubiquity. One of the challenges that these systems face is the spatial resolution of urban noise maps when the measures are collected through mobile devices. In [45], the authors detail the relevance of different interpolation techniques in mobile sensing in comparison with the usual spatial interpolation techniques (such as inverse distance weighting [46] or kriging [47]).

One of the first experiences for this goal is detailed in [48], where a Mobile Sensing Unit (MSU) associated with a GPS position is used, allowing the ubiquitous sensing of larger areas. The Seoul ubiquitous sensing project conducted a wide range of tests across several city locations with a reduced set of sensors. The MSUs were even installed in cars and buses moving around the city following repetitive circuits. These nodes measured temperature and humidity and also noise level; but no signal processing is detailed in the description of the system oriented toward the cancellation of any noise, but traffic noise; especially the one produced by the MSU car or bus.

In [49], an array of sensors mounted on a vehicle driving along the streets of the city is proposed in order to acquire measurements from different locations quickly. The goal of that piece of research is estimating the locations and the power of the stationary noise sources of the places of interest. For this purpose, the data gathered by the array is post-processed before plotting the several sources in the noise map.

The work in [50] focuses on the study of low density roads. The introduced approach includes mobile [51] and fixed noise monitoring platforms, obtaining data about traffic, noise and air pollution. The proposal is based on performing the mobile measurements by bicycle, which provides a new view of the local variability of noise and air pollution based on computing the differences of measurements along road segments [52]. This proposal is easily applicable to other cities to monitor noise and air pollution. One of the strong assets of this proposal is that it is prepared to measure more parameters, but audio; so, a wide amount of environmental information can be collected. In [51], also black carbon concentration measurements are performed, so potential health risks for citizens can also be estimated. Not only estimating environmental parameters, but mobile acoustic sensors can also be used for other applications, but noise monitoring, e.g., for soundscape identification [53], where they derived geo-information from audio data, to determine the position of a mobile device in motion by means of the audio data recorded by its built-in microphone.

2.3. Participative Urban Sensing

Finally, the last urban sensing method used nowadays is based on participative measurements; that is, considering the so-called citizens as sensors paradigm [54]. In this case, tools and protocols have to be predefined to allow information integration of all of the individual measurements (e.g., when uploading geolocalized photos in Google Maps, reporting traffic accidents in the Waze application, etc.).

When it comes to measuring city noise levels, two main participative sensing approaches can be found in the literature, depending on the tools used by citizens: (i) those considering citizens' tools, usually through mobile applications installed in their smartphones, as described in [55]; and (ii) those where citizens are provided with specific tools to be used for conducting the measurements, as in the project CITI-SENSE [56]. These tools and observation protocols are defined to allow citizens to collect simultaneous objective and subjective data of the environmental sound by themselves; each project develops products and services to perform the tests. For example, in the CITI-SENSE project [57], it includes a mobile application, an external microphone with a windscreen to protect the microphone of the smartphone and other external equipment, connected via Bluetooth to the smartphone, to measure climatic conditions.

In [39], the authors reviewed the approaches already presented by the wireless sensing research community to assess noise pollution using both acoustic sensor nodes and mobile phones. Since then, several works have been published dealing with the integration of the results of the individual measurements. This approach increases the number of potential measurements to be evaluated in the integration of the data, which is usually non-equally distributed in the city and not stable in time either, so it makes the final noise mapping integration a complex problem. The NoiseSPY project [58] designed a sound sensing system considering a mobile phone as a low-cost data logger to monitor environmental noise; citizens were allowed to visualize real-time noise levels in different places of the city. The NoiseTube project [59] provides a low-cost solution for the citizens to measure their personal exposure to noise in their everyday life by means of a mobile application that evaluates noise using the smartphone, allowing them to participate in the generation of a collective noise map using the geo-localized measurements. In [60], an application named NoiseMap is presented, which gathers loudness data and transfers them to an open urban sensing central platform. Afterwards, the data become public by means of a web-based service. In [61], the participative sensing contributes to updating a previous noise map of the city in order to dynamically refine the granularity of the noise patterns on different places. The frequency of updating depends on the level of participation of citizens in each road segment. Finally, [55] presents the OnoM@p system, which collects crowd noise data recorded by inexperienced volunteers by means of a smartphone application; the data are afterwards processed and filtered from outliers in the cloud, before being finally mapped and published on the web.

The Sounds Of The City project [62] is focused on the measurement of the noise level that surrounds a citizen. The citizen can measure the loudness of his/her environment using a simple smartphone application, which sends the data to a central server where all of the data are aggregated before computing and plotting the visual noise map. This idea was also exploited in a platform that modeled the noise situation of New York City considering three dimensions [63]: the region, the type of noise and a time stamp. From the collected data, a noise pollution indicator is inferred from the intelligent composition of noises measured by citizens.

Nowadays, this approach is gaining in importance since almost any smartphone can be used as a tool to sense citizens' environment. On the one hand, authorities are interested in having real-time data measured in the city, especially in noise critical places, and on the other hand, citizens are interested in reporting noise excesses in their surroundings. One of the most recent trends in analyzing the acoustic data obtained from citizens is distinguishing between pleasant and unpleasant sounds, analyzing the relationship between soundscapes and emotions, as well as the relationship between soundscapes and people's perceptions [64].

However, besides the complexity of the gathering and integration of such an amount of data, several voices [65] in the literature point out that the reliability of the derived noise maps depends on the quality of the smartphone's microphone, plus the employed application, and the nature of the measure itself, which is conducted by any citizen at any place and at any time. Therefore, the accuracy of the measures cannot be guaranteed as if it were performed by experts using calibrated equipment. Other voices [66], supported by several studies comparing participative noise mapping with standard noise mapping techniques, maintain that the same accuracy can be achieved if the participative sensing is conducted properly.

2.4. Hybrid Urban Sensing

Nowadays, many projects are already designed to collect samples of acoustic signals in the city using the combination of more than one method. The most usual procedure is to take the basis to build the map from a static network in the city, with a small number of nodes of measurement. Then, they complement the measures with either mobile acoustic sensors or participative measurements made by volunteers.

In [67], they work in a Rotterdam urban neighborhood to detect noticeable sound events, because the perception of one's acoustic environment in which they live is mainly driven by those. They used twelve intelligent sound measure devices, active for several months. At the same time, they collected citizens opinion about the sound in their home with a continuous survey. The final goal of the project was to report the relationship of between the detected events and the sound quality of the citizens. In [68], the same authors worked on a project in the 13th district of Paris using the data from twenty-four fixed sound measurement devices, while mobile sound measurements are performed in regular periods in the same area. The map is basically generated with the data of the fixed sensors, and it is changed with the mobile measurements accordingly.

3. Mobile Measure Platforms and Their Connectivity

There are two outstanding challenges of systems in charge of monitoring sensors that are located in remote areas or a net of sensors where nodes are moving in a dynamic topology: the first one it is to design the system in charge of collecting the information recorded by the sensors, reducing its price as much as possible. This issue would not be a problem when only a few sensors are needed, there are no power supply restrictions and the information processed by the sensors can be easily accessible by wired and inexpensive solutions, as happens normally in industrial applications. However, when the application requires a certain number of sensors and also a level of in situ real-time signal processing, this can be an important issue in the application design process.

The second challenge is to receive all of this information in a central system to post-process, store and take actions according to a predefined protocol of operation or to provide applications and functionalities to final users based on these data. As was mentioned before, industrial sensors are usually wired to Programmable Logic Controllers (PLC) or other complex systems, which deploy different communication interfaces. However, when the nodes are moving, the connectivity with the central system could be a great challenge if there are restrictions about radio frequency coverage and the cost of the communications (mainly when mobile communications are used to create these links with the nodes).

In the following subsections, these issues are analyzed taking into consideration the requirement of an embedded electronic system that is required to control the noise sensors used to create a low-cost network for monitoring in real time the acoustic environment in an urban scenario.

3.1. Hardware Platforms

Attending to the requirements mentioned previously, the electronic device that better satisfies them is a microcontroller-based embedded system. Nowadays, embedded systems can be found everywhere, from simply electronic applications to complex automation systems. Basically, an embedded system is a special purpose computer that has been designed for a specific function combining a software application and hardware controlled by a real-time operative system. These systems normally include soft-core processing units as microcontrollers, microprocessors, FPGAs, Digital Signal Processors (DSPs) and Application-Specific Integrated Circuits (ASICs) [69].

When talking about embedded systems, there is a huge variety of options in the market, but the selection can start by the core of the system, that is the microprocessor or the Central Processing Unit (CPU) of the embedded system. Nowadays, 32-bit microcontrollers are offered at a really competitive price and are included in almost all of the commercial embedded systems, as is shown in Table 1.

Table 1. Embedded systems' comparison.

Embedded System	Processor Core	Price
The chipKIT^{TM} MX3	Microchip® PIC32MX320F128H Microcontroller (80-MHz 32-bit MIPS 128 KB Flash, 16 KB SRAM)	44.99$
STM32VLDiscovery	ARM® Cortex-M3 (24-MHz 32-bit 128 KB Flash memory, 8 KB RAM)	9.90$
FRDM-KL25Z	ARM® Cortex®-M0+ (48-MHz 32-bit MIPS 128 KB Flash 16 KB SRAM)	13.25$
BeagleBone Black	Sitara^{TM} ARM® Cortex-A8 (2x PRU 32-bit microcontrollers, 512 MB DDR3 RAM)	51.15$
Raspberry Pi 3 Model B	1.2-GHz Quad-Core ARM Cortex-A53	37.00$
CYPRESS PSoC® 4 CY8C4245AXI	32-bit ARM® Cortex^{TM}-M0 48-MHz CPU	24.31$

Table 1 only gathers a set of the most representative embedded systems used nowadays. All of the prices were obtained from common dealers, such as Farnell, Digikey and RS-online. Most of the embedded systems that are available on the market have a processor based on an ARM architecture, but the ARM® Cortex® series of cores considers a very wide range of options, offering designers a great deal of choice and the opportunity to use the best-fit core for their application [70]. There are three main categories of the Cortex Series:

- Cortex-A: basically used by embedded systems that need a high level of operational system computing capabilities as low-cost handsets to smartphones or tablets.
- Cortex-R: this series is the smallest ARM processor and is commonly used in automotive, networking and data storage applications.

- Cortex-M: being the most popular of the ARM family, this series is being used for all types of low-cost and low consumption applications, from real-time signal processing to industrial control.

Based on this analysis, the FRDM-KL25Z designed by NXP is considered as a good choice for the topic under discussion; that is, the processing of the acoustic environment in an urban scenario. This selection is done basically due to its low price, capability of processing analog signals in real time thanks to its 12-bit ADCs and its low power consumption (down to 47 μA/MHz), which makes it possible that the system could be powered by a portable battery pack. Furthermore, applications for this ultra-low-cost development platform can be designed using two very simple development environments:

- mbed: this is a platform that provides free libraries, hardware designs and online tools for rapid prototyping of 32-bit ARM-based microcontroller products. This framework includes a standards-based C/C++ SDK, a microcontroller HDK and supported development boards, an online compiler and online developer collaboration tools [71,72].
- MATLAB: MathWorks offers the Embedded Coder Support Package for the Freescale FRDM-KL25Z Board to run the Simulink® model on an FRDM-KL25Z board. The support package includes a library of Simulink blocks for configuring and accessing Freescale FRDM-KL25Z peripherals and communication interfaces. Then, it is quite simple to build applications using the block-based interface of Simulink®, which generates also the code for the Freescale FRDM-KL25Z board and runs the generated code on the board [73].

Joined to the basic electronic elements of the embedded system, the application to which this paper is devoted requires an audio module to monitor the acoustic environment. Basically, in an embedded system, an audio module is based on a set of blocks, each one with a specific function [74]:

- Signal conditioning module, in charge of accommodating the analog signal captured by the microphone to the next block; that is, the Analog to Digital Converter (ADC). This conditioning stage adjusts the analog signal levels of voltage and current to the ADC input. Besides, this module also matches the output impedance of the microphone to the input impedance of the ADC.
- Analog to digital audio converter. Obviously, this block converts the analog signal captured by the microphone to a digital signal that can be later processed. To select the ADC, the main characteristics that should be analyzed are: the number of bits used in the conversion (as many bits are used, a bigger resolution will be obtained in the conversion, but a longer time of conversion), the speed of the conversion, distortion performance, sensitivity and errors in the conversion. The ADC can be implemented inside the System on Chip (SoC) on which the embedded system is based or also it can be placed externally, and it is controlled via an I2C or SPI bus.
- CODEC block: Although this block is optional, it can be used to reduce the number of bits needed to save the digitalized signal in the memory of the embedded system to its post-process. A huge variety of codecs can be used, but all of them are focused on compressing the audio stream with the maximum fidelity and quality.
- Memory: after the CODEC, data are dumped into a memory block from which the data can be retrieved for further processing by the application program.

In this application, the sensor needed to monitor the acoustic environment is a microphone, an acoustic-to-electric transducer that converts sound into an electrical signal. To select the best solution, one more advantage is found if the FRDM-KL25Z board is chosen, due to it featuring Arduino-compatible headers. Then, a great set of microphones can be used in the project. However, as it wanted to collect and log noise pollution, the sensor has to satisfy a set of characteristics, as:

- Directivity: as it is desired to monitor the noise generated in all directions, the microphone must be omnidirectional; that is, it must be able to capture noise from all directions, and then, it must be omnidirectional.

- Sensitivity: understood as the ratio of the analog output voltage or digital output value to the input noise pressure. This value, combined with the signal to noise ratio, is quite important to obtain a high quality monitoring system in order to record the sound with the maximum fidelity.
- Signal to noise ratio: it specifies the ratio of a reference signal to the noise level of the microphone output. Measured in decibels, it is the difference between the noise level and a standard 1-kHz, 94-dB SPL (Sound Pressure Level) reference signal. This specification is typically specified as an A-weighted value (dBA), which means that it includes a correction factor that corresponds to the human ear's sensitivity to sound at different frequencies. Combined with the sensitivity, these factors will be important to be able to discriminate background noises during the monitoring of the acoustic environment.
- Operating frequency: this is the range of frequencies that can be collected by the microphone. As the application under design is to control the noise in an urban scenario and analyze its impact on humans, the frequency range should be from 20 to 20 kHz; that is, the dynamic range in with human ears work.

Taking into account these premises, the recommended solution is the elected microphone with auto again control based on the MAX9814 amplifier [75]. This microphone amplifier module is based on the CMA-4544PF-W omnidirectional capsule microphone, which provides a high sensitivity (-44 ± 2 dB), an operating frequency from 20 to 20 kHz and a signal to noise ratio of 60 dBA. Moreover, thanks to the MAX9814 amp, it performs automatic gain control, avoiding overwhelming and distortion of the amplifier when sound levels can change randomly, as in the scenario proposed in this paper.

3.2. Connectivity of the Platforms

Once the data are recorded and saved in the embedded system, they can be processed on-board or sent to a central system for being post-processed. Anyway, the embedded system must provide a wireless communication interface. Two alternatives are available to be deployed on the FRDM-KL25Z platform: a wireless and/or a 3G connection. With the adopted hardware, both solutions are possible at the same time, because the FRDM-KL25Z provides several universal asynchronous serial ports that can be used to communicate with both interfaces. The recommended hardware solutions are:

- WiFi ESP8266: this is a very low price and consumption WiFi module that implements a complete TCP/IP protocol stack. It provides a set of instructions and functionalities that make it very easy to control and start to work without any complicated configuration.
- Adafruit FONA 808: this is and all-in-one mobile communication interface plus a GPS module. Although it works with a 2 G SIM card, it is quite enough to allow remote control and data download, reducing the price of the whole system. It communicates with the controller through a serial port, making the deployment of applications easy and fast.

Thanks to these interfaces, the platform can be remotely accessed and controlled. The easiest and fastest way to implement this communication is deploying web services, which are transparent to the communication interface. The use of web servers is quite extended in the field of remote monitoring and controlling [76,77]. As in [78], the development of this application will be based on a Service-Oriented Architecture (SOA), in which most of the control is distributed in the server, which will access the remote sensing platform through SOAP messages to a web service developed for the control software solution.

The adoption of both interfaces increases the options of remote controlling. Basically, the WiFi interface will be used to upload stored data (acoustic samples) in the embedded system to the central system when a WiFi network is available. If the storage capacity of the embedded system is almost full, then the mobile interface would be used to send these data to the central system. The management of the communication interfaces and the selection of each one according to their availability will be done by the central processor of the embedded unit. Moreover, in all of the situations for which the central system wants to access the embedded system, the mobile communication interface assures this link.

Therefore and according to the suggestions of this section, the complete hardware solution for monitoring the acoustic environment in the frame of a mobility scenario in a city is shown in Figure 1.

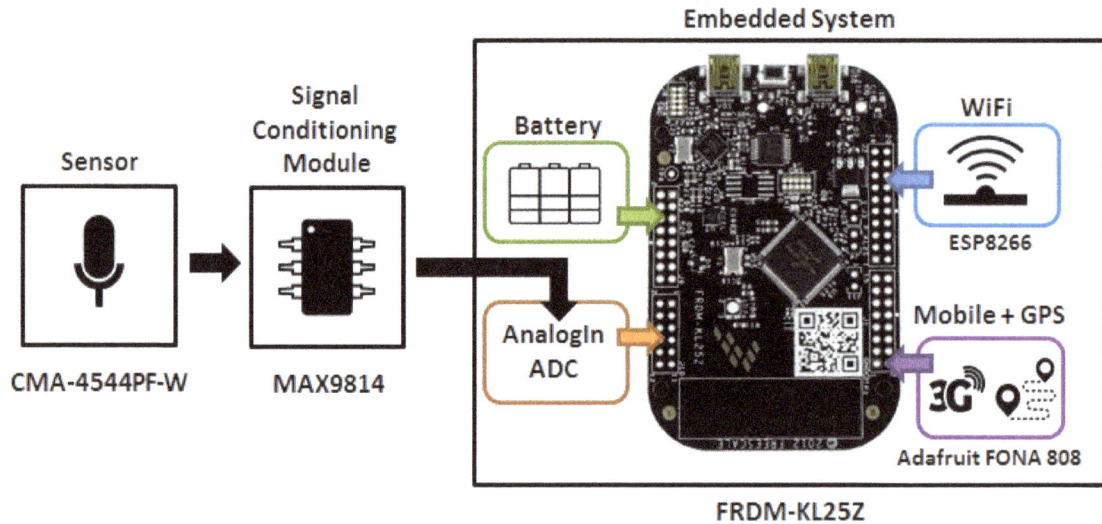

Figure 1. Suggested hardware platform. This is based on the FRDM-KL25Z embedded system. The microphone is the CMA-4544PF-W capsule followed by the MAX9814 amplifier. The platform also provides GPS, WiFi and mobile communications modules.

4. Mobile Bus Acoustic Measurement

The use of a public transport vehicle, such as a bus, to perform the measures presents evident advantages—power supply, mobility and connectivity—but also some specific drawbacks. In the literature, detailed in Section 2, there are several projects based on mobility to measure the noise in the city; on the one hand, deploying equipment temporally in vehicles (e.g., cars) to perform punctual measurements in specific locations of the city and, on the other hand, installing the sensors in non-motorized vehicles to get the vehicle to get the benefits of fast moving and maximum flexibility. In this paper, we propose a new scenario, in which a line bus carries the low-cost instruments, installed permanently for the realization of dynamic noise measurements in time. There are already previous experiences for the connectivity of urban buses [79,80], but mainly focused on fleet management, vehicle punctuality and the analysis of environmental parameters. Our proposal focuses on the possibility of evaluating dynamic noise maps in the city with a stable installation in their bus networks. This proposal faces new challenges, and for this purpose, in this section, we detail several challenges that a ubiquitous acoustic measurement system has to solve to allow a reliable dynamic acoustic mapping.

4.1. Signal Processing Challenges

There are several challenges in terms of signal processing when facing the mobile, bus acoustic measure of traffic noise levels. First, acoustic events different from the ones produced by traffic noise (hereafter, denoted as anomalous noise events) have to be detected and discarded to avoid biasing the L_{eq} measurement of the traffic noise. Second, the noise generated by the mobile vehicle used to perform the measure has to be appropriately considered as traffic noise, but adjusting its contribution according to its closer distance to the point of measurement. The integration of input data in the server of the mobile sensor network is also an issue to be solved by means of noise mapping generation techniques;

to tailor the most suitable noise map for every city, the chosen bus routes have to be previously studied so that they are representative enough and cover critical places in terms of noise. Finally, the classification of different type of vehicles is a real challenge in order to improve the information about traffic noise mapping, e.g., to distinguish between light and heavy traffic.

4.1.1. Reliability of the L_{eq} Measure

The goal of measuring the road traffic noise automatically faces inevitably the problem of dealing with non-traffic acoustic events (e.g., train, road work, bells, animals, human talking, etc.). An anomalous noise event detection algorithm has to be designed if we face the challenge of improving the robustness of the traffic noise mapping system. The challenge in this case is focused on identifying the most salient events to avoid their contribution to the L_{eq} computation value.

Several works have been already conducted about this topic. In [81], a two-stage recognition schema is presented to detect abnormal events that take place in an usually noisy environment. In [82], the work is focused on detecting only police sirens recorded together with traffic noise. Both algorithms are based on Hidden Markov Models (HMM) working with Mel-Frequency Cepstral Coefficients (MFCCs). In [83], MFCCs are combined with Support Vector Machines (SVM) to identify hazardous situations, e.g., tire skidding and car crashes, in order to detect accidents. These methods follow a classic classification approach since the number of anomalous events is limited to a certain universe. In [84], acoustic summaries of several quiet and noisy areas are constructed using an automatic method, based on Self-Organizing Maps (SOM), including a validation stage of the local residents. The map is tuned to the sounds that are likely to be heard in a certain location by means of unsupervised training.

However, one of the key challenges when dealing with anomalous noise event detection is the definition of anomalous events themselves, i.e., any sound source different from traffic noise; hence, it cannot be limited to a finite number of categories a priori. In [85,86], an anomalous event detection algorithm is presented as a binary classifier, detecting whether the input acoustic data come from road traffic noise or if it is an anomalous event by considering Gammatone Cepstral Coefficients (GTCC) for audio signal parametrization [87]. Both K-Nearest Neighbor (KNN) and Fisher's Linear Discriminant (FLD) performances are compared in [85] and in [86]. This algorithm can be performed in real time, just before the L_{eq} computation, in order to avoid the integration of the anomalous events L_{eq} in the noise map computation.

The main challenge of the anomalous event detection is to increase its accuracy when detecting the events, being capable of discarding those anomalous events to be detected that have not been yet observed during the training stage. As far as we know, this generalization has not been fully reached up to day, so it remains an open scientific challenge to be solved.

4.1.2. Mobile Vehicle Noise Contribution

Another challenge associated with audio signal processing for ubiquitous noise measurements is the correction that has to be applied to the noise generated by the mobile vehicle transporting the acoustic sensor; in our case, a bus. Obviously, the bus itself contributes to traffic noise. However, this noise source is very close to the point of measure. Therefore, its has to be detected before filtering its contribution accordingly. Nevertheless, this problem presents a counterpart: the process can take advantage of having the audio reference of the noise source (mainly produced by the bus engine). In the following paragraphs, we describe some related approaches that can be considered to address the problem at hand.

In [88], a signal processing system dealing with identification and the estimation of the contribution of each source to an overall noise level is presented. The proposal is based on Fisher's linear discriminant classifier and estimates the contribution based on a distance measure. Later, in [89], a system based on probabilistic latent component analysis is presented. This approach is based on a sound event dictionary where each element consists of a succession of spectral templates, controlled by class-wise HMMs.

In [90], the difficulties for appropriately measuring the performance of polyphonic sound event detection are stated before gathering several metrics specifically designed for this purpose. Sound event recognition complexity varies, and for the problem at hand of bus noise and traffic noise, the complexity is significant as, on the one hand, the sounds are continuously overlapped and, on the other hand, the type of signal to be detected is very similar. To this aim, a supervised model training for overlapping sound events based on an unsupervised source separation is presented in [91]. In [92], the authors detect sound events from real data using coupled matrix factorization of spectral representations and class annotations. Finally, in [93], the authors exploit deep learning methods to detect acoustic events by means of exploiting the spectro-temporal locality. For more references and detail, the reader is referred to [90].

In a nutshell, the most challenging issue for the problem at hand is integrating the bus and surrounding traffic noise levels to the noise map due to the similarity between the signal to be canceled and the signal to be processed to compute the L_{eq} value. This increases the complexity of the separation system significantly, which may be addressed thanks to having the reference signal (i.e., considering as input the noise generated by the bus for the sound source separation).

4.1.3. Classification of Road Traffic Vehicles

Finally, and despite it not being strictly necessary for dynamic noise mapping, there is the possibility of identifying the type and evolution of vehicles along time for a given area. This completes the information of the acoustic mapping, not only in terms of L_{eq}, but also in terms of type of traffic at each point of the city, e.g., by identifying the volume of heavy or light traffic [8].

In [94], a hierarchical classifier of urban sounds is proposed differentiating mechanical from non-mechanical sounds. The approach is based on MFCCs and HMM to perform the classification. In [95], the type of vehicle is also hierarchically classified besides extracting individual pass-by acoustic signatures of non-overlapping road traffic vehicles. In [96], the noise maps are refreshed using source separation from different types of vehicles (e.g., train, bus, car, etc.). In [97], an automatic sound recognition algorithm for the urban environment is detailed, reaching the best results for an SVM classifier. Finally, in [98], a complete review describing several methods to classify urban sounds is described. However, they are described under the umbrella of audio surveillance; the reviewed techniques are also applicable to the described challenge.

The challenge at this point is to classify a closed type of vehicle to characterize the type of traffic both at the high level (by means of noise maps) and low level (by means of sound source identification) across a particular urban area. These data may complete the information of the dynamic noise mapping services, allowing authorities to manage city traffic more precisely.

4.2. Challenges in Terms of Noise Mapping

In this section, two problems related to the design and tailoring of the noise maps are discussed. The first one deals with the problem of determining the most suitable locations that the mobile acoustic sensor should measure, i.e., the bus routes used to sense the city. The second one is related to the data gathering together with subsequent processing and integration to guarantee real-time noise mapping.

4.2.1. Mobile Trajectories Design and Data Collection

In order to collect the suitable data to build the noise map from the acoustic information gathered in some city buses, previous field studies have to be conducted so as to choose which bus lines better fit the expected goal depending on their trajectory, the traffic volume, the authorities criteria, etc.

In [99], the authors collect a total amount of 4200 h of measurements, in 24-h block periods. This information is used to analyze the time patterns of the noise levels in certain places, but most relevantly, under several weather conditions.

In [100], the authors study if urban noise can be stratified by measuring traffic noise after dividing the streets of a particular city depending on their use on the type of street. The approach is tested and validated in five medium-sized Spanish towns. Later, the same authors studied to what extent this preliminary approach could be generalized to larger towns, showing a significant predictive capacity [101].

In [102], a statistically-based method to choose the optimal number and location of static monitoring sites to improve the noise mapping is presented, being focused to the development of noise maps with data obtained from a low-cost network. They present some preliminary results of the application of the method to the streets of Milan, improving the spatial sampling considering legislative road divisions, which are notonly based on geometry, but on noise emission.

The project will consider the study of the streets of the city to conclude the most suitable bus routes to conduct the traffic audio measurements. The system will collect traffic noise measurements, but as the hardware platform allows us to collect data from more sensors, we also intend to gather meteorological data, such as temperature and humidity. These allow one to widen the study of traffic noise variations taking into account different weather situations, in addition to the time noise level variations that a study of the whole day already provides. All of this gathered information allows us to perform comparative studies between the legislative classification of streets and acoustic measurement tests, pretending to obtain a classification of streets depending only on noise.

4.2.2. Noise Mapping Real-Time Update

In the last years, several algorithms to address real-time noise mapping have been proposed in the literature, following three main approaches: (i) real-time calculation; (ii) map rescaling and sum and (iii) citizens' contributions; a similar approach to the method described in Section 2.3.

In the real-time calculation approach, the measured values are introduced to the simulation software to evaluate a new value for the noise maps; the server has to reevaluate the entire noise map continuously, with a heavy computational load associated with that process. Gdansk University developed a real-time calculation approach system [103], and Ghent simplified the algorithms to make the evaluations faster [104].

The map scaling and sum approach is based on a similar infrastructure, but the noise mapping server evaluates the new noise map with the sum of the prescaled precalculated map according to the measured noise (i.e., it follows an incremental process). Hence, the computational load of the server is lower than the real-time approach. ACCON [105] uses an external GIS software technique to scale and sum the partial maps, and so does SADMAM (*Sistema de Actualización Dinámica del Mapa Acústico de Madrid*) [106]. In [106], the details of the software are given: the measurement data and GIS, the measurements integration to improve the source model, calibration of multiple sources in complex environments and also integrating GPS in the data. Finally, IDASC [107] scales and sums the partial maps using a function implemented inside the noise model software, avoiding the use of an external GIS, and publishes on the web new updated maps together with the noise data measured.

Finally, the citizens' measurements using their smartphones also contribute to updating the information about noise levels in urban areas. In this case, the smartphone sends the data to the cloud with the GPS position. The aforementioned project NoiseTube [108] is an example of this type of platform, and Ear-Phone project [109] deals with the context-aware sensing, using classifiers to determine the sensing context of the phone; it allows to decide whether to sense or not automatically.

The proposed system, which is nowadays in a previous stage of development, will analyze the literature proposals, and the solution that fits best the dynamic of the measurements will be implemented to build the noise maps. However, the update time of the noise maps and how this is implemented are also open research issues. The noise maps should be refreshed only when there is a significant variation in the L_{eq} value; the noise mapping update rhythm is calculated for each system depending on the time variation of the noise measurements 24 h a day in the places of interest. This study has to be conducted for the planned routes after the study detailed in Section 4.2.1, and the variation of the noise measurements values depending on the hour of the day has to be evaluated to conclude with the periodical update needed.

4.3. Challenges in Terms of Hardware Platform Selection

From the hardware platform point of view, the application under analysis presents the following challenges:

- Low price: the main goal of the application being the collection of data about the noise in the city to generate a dynamic noise map, the more sensors can be placed, the better. Then, the use of a inexpensive hardware platforms is recommended, which can be easily deployed, and this should include all of the elements needed to achieve the objectives of the applications.
- Non-intrusive: that is, as the platform will be deployed on a public means of transport, such as urban buses, it is recommended not to require any special restrictions regarding its installation. Then, it must be auto-powered, small in size and easily integrated on the buses. Moreover, during the performance of the system, it has to have no interference with the electronic and communication systems of the vehicle.
- Low energy consumption: as the hardware platform cannot be connected to the energy system of the vehicle, it must have a power system itself. Then, the complete system should be low consumption and implemented with any software functionality able to control the waste of energy during the operation, for example disabling some subsystems if they are not needed.
- Communication interfaces: one of the main tasks of the platform is to send the recorded data to a central system. Then, different communication interfaces can be deployed on board, according to the characteristics of the environment. That is, if a WiFi network is available, then a free link can be established with the server, but in the case of being out of the WiFi coverage, a mobile communication link should be available to allow the data upload and also the remote control of the platform in case a remote reconfiguration or maintenance is necessary. At the same time, the platform must perform a geolocation interface in order to associate noise-data to location references.
- Communications management sub-system: as the platform needs different interfaces, a management software is needed to control the communications to the server. This functionality will be in charge of determining which interface can be used in each situation in order to save energy and to save money. For example, when free WiFi is available, this functionality will upload all of the saved data to the server automatically.
- Storage capacity: as the platform must save data about the noise in different locations, it must be able to save as much information as possible in order to keep it in memory until being in a free WiFi coverage area to send it to the server or just before overloading of the memory, through a mobile link.
- Real-time data processing capability: the processor provided on the embedded system must be able to process the noise captured by the microphone in real time to obtain the noise frequency and level. Then, this information is saved on the memory joined to the location provided by the GPS. Later, these data are sent to the central server. During this signal processing stage, the processor also must be able to discriminate the noise of the environment from the noise generated by the vehicle in which the hardware will be deployed. Moreover, the resolution of the processing must be enough to allow the classification and characterization of the traffic vehicles.

In summary, the hardware platform chosen for creating the dynamic noise map must be easily installed in a public means of transport like the buses of a city. It should be as inexpensive as possible to allow multiple instances of the platform in different vehicles, obtaining information about the noise in all of the areas of the city. It must deploy different communication interfaces with the aim of allowing the upload of recorded data to the central server. At the same time, the platform must perform real-time data processing to obtain as much information as possible about the noise and to detect and filter the vehicle audio contribution, in which the hardware will be deployed.

5. Conclusions

This work describes the design of a mobile approach for ubiquitous urban noise sensing using public buses. It widens the input data coverage with respect to static approaches, since it allows sensing the city by means of several bus lines carrying acoustic low-cost sensing platforms. Hence, due to the fact that the vehicle moves across the city, the noise map is fed with ubiquitous measurements. Moreover, when compared to other mobile acoustic sensing systems, it takes advantage, on the one hand, of the possibility of changing the measurement route by just selecting another bus line to carry the sensor and, on the other hand, of using a vehicle where the hardware can be installed and powered easily. In regards to participative sensing, a closed route repeated several times a day should guarantee enough data to build a reliable noise map, and it solves the heterogeneity and the accuracy problems when integrating data from citizens' measurements (i.e., different types of smartphones, under- or over-sampled locations, etc.).

Nevertheless, before implementing the proposed approach, several challenges should be addressed; signal processing challenges, to work with the bus engine noise and its contribution to the noise map, which nowadays is an open scientific topic; noise mapping challenges, to study which are the best bus routes for the measurements, and the proper building of the noise maps in the cloud from ubiquitous acoustic data. Finally, hardware challenges face the design of the sensor in a low-cost and low-power consumption platform, but with enough computational capacity to implement the necessary signal processing algorithms to perform the measures and afterwards send them to the cloud. These are key factors in all systems that will be deployed under the paradigm of smart cities, since these systems should be autonomous and transparent, by requiring easy and occasional maintenance and minimizing the interference with the other elements of the city, as in the scenario proposed in this work, with the public transport system. The long-term vision of the implementation of the proposed system is that the entire city, or at least its major roads, will be mapped through the proposed mobile noise sensing system.

In the near future, we plan to address the aforementioned challenges before implementing and evaluating the proposal in a real-life scenario.

Acknowledgments: The authors would like to thank Project ACM2016_06 entitled 'Towards the development of low-cost ubiquitous sensor networks for real-time acoustic monitoring in urban mobility', from the II Convocatoria del Programa de Ayudas a Proyectos de Investigación Aristos Campus Mundus 2016. Francesc Alías and Rosa Ma Alsina-Pagès also would like to thank the Secretaria d'Universitats i Recerca del Departament d'Economia i Coneixement (Generalitat de Catalunya) under Grant Ref. 2014 - SGR - 0590.

Author Contributions: Rosa Ma Alsina-Pagès has prepared the state-of-the art of acoustic signal processing applications and projects and worked on the project proposal's challenges. Unai Hernandez-Jayo has participated in the analysis of the available hardware platforms from the connectivity point of view and in the definition of the challenges of the hardware system. Francesc Alías has worked on the audio signal processing and noise mapping parts of the paper and its challenges and applications. Ignacio Angulo has contributed to the analysis of the available embedded systems.

Conflicts of Interest: The authors declare no conflicts of interest.

Abbreviations

The following abbreviations are used in this manuscript:

ADC	Analog to Digital Converter
ANED	Anomalous Noise Event Detection
ARM	Advanced RISC Machine
ASIC	Application-Specific Integrated Circuits
CPU	Central Processing Unit
EC	European Commission
END	Environmental Noise Directive
EU	European Union

DSP	Digital Signal Processor
FLD	Fisher's Linear Discriminant
FPGA	Field Programmable Gate Array
GPS	Global Positioning System
GTCC	Gammatone Cepstrum Coefficients
HMM	Hidden Markov Models
ICT	Information and Communication Technology
KNN	K-Nearest Neighbor
LMS	Least Mean Squares Filter
MFCC	Mel-Frequency Cepstrum Coefficients
MSU	Mobile Sensing Unit
PLC	Programmable Logic Device
RAM	Random Access Memory
RLS	Recursive Least Squares Filter
SOA	Service-Oriented Architecture
SoC	System on Chip
SOM	Self-Organizing Maps
SPL	reference signal
SVM	Support Vector Machine
WASN	Wireless Acoustic Sensor Network

References and Notes

1. World Demographics Profile 2012. Index Mundi. Available online: http://www.indexmundi.com/world/demographics_profile.html (accessed on 22 May 2016).

2. World Population 2015. United Nations, Department of Economic and Social Affairs, Population Division. Available online: https://esa.un.org/unpd/wpp/Publications/Files/World_Population_2015_Wallchart.pdf (accessed on 13 September 2016).

3. Bouskela, M.; Casseb, M.; Bassi, S.; De Luca, C.; Facchina, M. *La Ruta Hacia las SmartCities: Migrando de una Gestión Tradicional a una Ciudad Inteligente*; Banco Interamericano de Desarrollo (BID): Washington, DC, USA, 2016.

4. Morandi, C.; Rolando, A.; Di Vita, S. *From Smart City to Smart Region, Digital Services for an Internet of Places*; Politecnico de Milano; SpringerBriefs in Applied Sciences and Technology: Zurich, Switzerland, 2016.

5. Smart City Expo World Congress, Report 2015. Fira de Barcelona. Available online: http://media.firabcn.es/content/S078016/docs/Report_SCWC2015.pdf (accessed on 14 September 2016).

6. The European Environment, State and Outlook 2010 (Sythesis). European Environment Agency, 2010. Available online: http://www.eea.europa.eu/soer/synthesis/synthesis (accessed on 27 August 2016).

7. Ripoll, A. *State of the Art of Noise Mapping in Europe*; Internal Report; Universitat Autònoma de Barcelona, European Environment Agency: Copenhagen, Denmark, 2005.

8. Environmental Noise Directive (END). 2002/49/EC of the European parliament and the Council of 25 June 2002 relating to the assessment and management of environmental noise. *Off. J. Eur. Communities L* **2002**, *189*, 2002.

9. Common Noise Assessment Methods in Europe (CNOSSOS-EU) for strategic noise mapping following Environmental Noise Directive 2002/49/EC, European Commission, Joint Research Centre—Institute for Health and Consumer Protection, 2012.

10. Alberts, W.; Faber, N.; Roebben, M. Road Traffic Noise Exposure in Europe in 2012 based on END data. In Proceedings of the INTERNOISE, Hamburg, Germany, 21–24 August 2016; pp. 1236–1247.

11. Licitra, G.; Palazzuoli, D.; Ascari, E. END Noise Mapping for a Sufficiently Accurate People Exposure Estimation in Epidemiological Studies. In Proceedings of the INTERNOISE, Hamburg, Germany, 21–24 August 2016; pp. 5687–5698.

12. Burden of Disease from Environmental Noise. Quantification of Healthy Life Years Lost in Europe. World Health Organization, Regional Office for Europe—European Comission, 2011. Available online: http://www.euro.who.int/__data/assets/pdf_file/0008/136466/e94888.pdf (accessed on 12 September 2016).

13. Murphy, E.; King, E.A. *Environmental Noise Pollution, Noise Mapping, Public Health and Policy*; Elsevier: San Diego, CA, USA, 2014.

14. Commission Directive (EU) 2015/996 Establishing Common Noise Assessment Methods According to Directive 2002/49/EC of the European Parliament and of the Council. Available online: http://eur-lex.europa.eu/legal-content/EN/TXT/?uri=CELEX%3A32015L0996 (accessed on 5 December 2016).

15. FONOMOC, Subgroup of Working Group Noise in EUROCITIES. Available online: https://workinggroupnoise.com/fonomoc/ (accessed on 27 August 2016).

16. EUROCITIES, the Network of Major European Cities. Available online: www.eurocities.eu (accessed on 27 August 2016).

17. Wolfert, H. Towards New Less Noisy Mobility Patterns in Cities. In Proceedings of the Internoise, Melbourne, Australia, 16–19 November 2014.

18. Harmonica, An Innovative Noise-pollution Index and a Platform to Inform and to Assist Decision-Making. Available online: http://www.harmonica-project.eu/en (accessed on 27 August 2016).

19. Mietlicki, C.; Mietlicki, F.; Ribeiro, C.; Gaudibert, P.; Vincent, B. The HARMONICA Project, New Tools to Assess Environmental Noise and Better Inform the Public. In Proceedings of the Forum Acusticum, Krákow, Poland, 7–12 September 2014.

20. DYNAMAP, Dynamic Acoustic Mapping—Development of Low Cost Sensor Networks for Real Time Noise Mapping. LIFE+. Available online: http://www.life-dynamap.eu/ (accessed on 27 August 2016).

21. Sevillano, X.; Socoró, J.C.; Alías, F.; Bellucci, P.; Peruzzi, L.; Radadelli, S.; Coppi, P.; Nencini, L.; Cerniglia, A.; Bisceglie, A.; et al. DYNAMAP—Development of Low Cost Sensors Networks for Real Time Noise Mapping. *Noise Mapp.* **2016**, *3*, 172–189.

22. Zambon, G.; Angelini, F.; Salvi, D.; Zanaboni, W.; Smiraglia, M. Traffic Noise Monitoring in the City of Milan: Construction of a Representative Statistical Collection of Acoustic Trends. In Proceedings of the 22nd International Congress on Sound and Vibration, Florence, Italy, 12–16 July 2015,

23. Progetto SENSEable PISA. Sensing The City. Description of the Project. Available online: http://senseable.it/ (accessed on 10 September 2016). (In Italian)

24. Nencini, L.; De Rosa, P.; Ascari, E.; Vinci, B.; Alexeeva, N. SENSEable Pisa—A Wireless Sensor Network for Real-Time Noise Mapping. In Proceedings of the EURONOISE, Prague, Czech Republic, 10–13 June 2012,

25. Superblocks Project, a Sustainable Strategy for Regenerating the City. Ajuntament de Barcelona. Available online: http://smartcity.bcn.cat/en/superblocks.html (accessed on 14 September 2016).

26. Urban Mobility Plan of Barcelona, PMU 2013–2018. Ajuntament de Barcelona. 2014. Available online: http://www.bcnecologia.net/sites/default/files/proyectos/pmu_angles.pdf (accessed on 12 September 2016).

27. Alves, S.; Scheuren, J.; Altreuther, B. Review of recent EU funded research projects from the perspective of urban sound planning: Do the results cope with the needs of Europe's noise policy? *Noise Mapp.* **2016**, *3*, 86–106.

28. Aletta, F.; Kang, J. Soundscape approach integrating noise mapping techniques: A case study in Brighton, UK. *Noise Mapp.* **2015**, *2*, 1–12.

29. Alves, S.; Estévez-Mauriz, L.; Aletta, F.; Echevarria-Sanchez, G.M.; Puyana Romero, V. Towards the integration of urban sound planning in urban development processes: The study of four test sites within the SONORUS project. *Noise Mapp.* **2015**, *2*, 57–85.

30. Basten, T.; Wessels, P. An Overview of Sensor Networks for Environmental Noise Monitoring. In Proceedings of the 21st International Congress on Sound and Vibration, Beijing, China, 13–17 July 2014.

31. Wang, C.; Chen, G.; Dong, R.; Wang, H. Traffic Noise Monitoring and Simulation Research in Xiamien City based on the Environmental Internet of Things. *Int. J. Sustain. Dev. World Ecol.* **2013**, *20*, 248–253.

32. Filiponni, L.; Santini, S.; Vitaletti, A. Data Collection in Wireless Sensor Networks for Noise Pollution Monitoring. In Proceedings of the 4th IEEE International Conference on Distributed Computing in Sensor Systems, Santorini Island, Greece, 11–14 June 2008,

33. Mietlicki, C.; Mietlicki, F.; Sineau, M. An Innovative Approach for Long-term Environmental Noise Measurement: RUMEUR Network. In Proceedings of the Euronoise 2015, Maastrich, The Netherlands, 31 May–3 June 2015; pp. 2309–2314.

34. Paulo, J.P.; Fazenda, P.; Oliveira, T.; Carvalho, C.; Felix, M. Framework to Monitor Sound Events in the City Supported by the Fiware Platform. In Proceedings of the 46º Congreso Español de Acústica, Valencia, Spain, 21–23 October 2015.

35. Rawat, P.; Singh, K.D.; Chaouchi, H.; Bonnin, J.M. Wireless Sensor Networks: A Survey on Recent Developments and Potential Synergies. *J. Supercomput.* **2013**, *68*, 1–48

36. Heidemann, J.; Li, Y.; Syed, A.; Wills, J.; Ye, W. Underwater Sensor Networking: Research Challenges and Potential Applications. In Proceedings of the IEEE Wireless Communications and Networking Conference, Las Vegas, NV, USA, 3–6 April 2006.

37. Werner-Allen, G.; Lorincz, K.; Welsh, M.; Marcillo, O.; Johnson, J.; Ruiz, M.; Lees, J. Deploying a Wireless Sensor Network on an Active Volcano. *IEEE Internet Comput.* **2006**, *10*, 18–25.

38. Simon, G.; Maróti, M.; Lédeczi, A.; Balogh, G.; Kusy, B.; Nádas, A.; Pap, G.; Sallai, J.; Frampton, K. Sensor Network-based Countersniper System. In Proceedings of the 2nd International Conference on Embedded Networked Sensor Systems, Baltimore, MD, USA, 3–5 November 2004; ACM: New York, NY, USA, 2004; pp. 1–12.

39. Santini, S.; Ostermaier, B.; Adelmant, R. On the Use of Sensor Nodes and Mobile Phones for the Assessment of Noise Pollution Levels in Urban Environments. In Proceedings of the Sixth International Conference on Networked Sensing Systems (INSS), Pittsburgh, PA, USA, 17–19 June 2009.

40. Bertrand, A.; Doclo, S.; Gannot, S.; Ono, N.; van Waterschoot, T. Special Issue on Wireless Acoustic Sensor Networks and ad hoc Microphone Arrays. *Signal Process.* **2015**, *107*, 1–3.

41. Botteldooren, D.; De Coensel, B.; Oldoni, D.; Van Rentherghem, T.; Dauwe, S. Sound Monitoring Networks New Style. In Proceedings of the Acoustics 2011, Gold Coast, Australia, 2–4 November 2011.

42. Bell, M.C.; Galatioto, F. Novel Wireless Pervasive Sensor Network to Improve the Understanding of Noise in Street Canyons. *Appl. Acoust.* **2013**, *74*, 169–180.

43. Domínguez, F.; Dauwe, S.; The Cuong, N.; Cariolaro, D.; Touhafi, A.; Dhoedt, B.; Steenhaut, K. Towards an Environmental Measurement Cloud: Delivering Pollution Awareness to the Public. *Int. J. Distrib. Sens. Netw.* **2014**, doi:10.1155/2014/541360.

44. Nencini, L. DYNAMAP Monitoring Network Hardware Development. In Proceedings of the 22nd International Congress on Sound and Vibration, Florence, Italy, 12–16 July 2015.

45. Can, A.; Dekoninck, L.; Botteldooren, D. Measurement network for urban noise assessment: Comparison of mobile measurements and spatial interpolation approaches. *Appl. Acoust.* **2014**, *83*, 32–39.

46. Shepard, D. A two-dimensional interpolation function for irregularity-spaced data. In Proceedings of the 23rd National Conference of the ACM, New York, NY, USA, 27–29 August 1968; pp. 517–524.

47. Krige, D.G. A Statistical Approach to Some Mine Valuations and Allied Problems at the Witwatersrand. Master's Thesis, University of Witwatersrand, Johannesburg, South Africa, 1951.

48. Hong, P.D.; Lee, Y.W. A Grid Portal for Monitoring of the Urban Environment Using the MSU. In Proceedings of the International Conference on Advanced Communication Technology, Phoenix Park, Korea, 15–18 February 2009,

49. Zhao, S.; Nguyen, T.N.T.; Jones, D.L. Large Region Acoustic Source Mapping using Movable Arrays. In Proceedings of the International Conference on Acoustic, Speech and Signal Processing, Brisbane, Australia, 19–24 April 2015; pp. 2589–2593.

50. Dekonick, L.; Botteldoren, D.; int Panis, L. Sound Sensor Network Based Assessment of Traffic, Noise and Air Pollution. In Proceedings of the EURONOISE, Maastrich, The Netherlands, 31 May–3 June 2015; pp. 2321–2326.

51. Can, A.; Van Renterghem, T.; Botteldooren, D. Exploring the Use of Mobile Sensors for Noise and Black Carbon Measurements in an Urban Environment. In Proceedings of the Acoustics, Nantes, France, 23–27 April 2012.

52. Dekoninck, L.; Botteldooren, D.; Int Panis, L.; Hankey, S.; Jain, G.; Marshall, J. Applicability of a Noise-based Model to Estimate in-Traffic Exposure to Black Carbon and Particle Number Concentrations in Different Cultures. *Environ. Int.* **2015**, *74*, 89–98.

53. Fiser, M.; Pokorny, F.B.; Graf, F. Acoustic Geo-sensing Using Sequential Monte Carlo Filtering. In Proceedings of the 6th Congress of the Alps Adria Acoustics Association, Graz, Austria, 16–17 October 2014.

54. Mostashari, A.; Arnold, F.; Maurer, M. Citizens as Sensors: The Cognitive City Paradigm. In Proceedings of the 8th International Conference & Expo on Emerging Technologies for a Smarter World, Hauppauge, NY, USA, 2–3 November 2011.

55. Guillaume, G.; Can, A.; Petit, G.; Fortin, N.; Palominos, S.; Gauvreau, B.; Bocher, E.; Picaut, J. Noise Mapping Based on Participative Measurements. *Noise Mapp.* **2016**, *3*, 140–156.

56. Aspuru, I.; García, I.; Herranz, K.; Santander, A. CITI-SENSE: Methods and Tools for Empowering Citizens to Observe Acoustic Comfort in Outdoor Public Spaces. *Noise Mapp.* **2016**, *3*, 37–48.

57. Aspuru, I.; García, I.; Herranz-Pascual, K.; Santander, A. Empowering People on the Assessment of the Acoustic Confort of Urban Places: CITI-SENSE Project. In Proceedings of the INTERNOISE, Hamburg, Germany, 21–24 August 2016.

58. Kanjo, E. NoiseSPY: A Real-Time Mobile Phone Platform for Urban Noise Monitoring and Mapping. *Mob. Netw. Appl.* **2010**, *15*, 562–574.

59. Maisonneuve, N.; Stevens, M.; Ochab, B. Participatory noise pollution monitoring using mobile phones. *Inf. Polity* **2010**, *15*, 51–71.

60. Schweizer, I.; Bartl, R.; Schulz, A.; Probst, F.; Muhlhauser, M. NoiseMap—Real-time Participatory Noise Maps. In Proceedings of the 2nd International Workshop on Sensing Applications on Mobile Phones, Seattle, WA, USA, 1–4 November 2011.

61. Hu, M.; Che, W.; Zhang, Q.; Luo, Q.; Lin, H. A Multi-Stage Method for Connecting Participatory Sensing and Noise Simulations. *Sensors* **2015**, *15*, 2265–2282.

62. Ruge, L.; Altakrouri, B.; Schrader, A. SoundOfTheCity—Continuous Noise Monitoring for a Healthy City. In Proceedings of the 5th International Workshop on Smart Environments and Ambient Intelligence, San Diego, CA, USA, 18–22 March 2013; pp. 670–675.

63. Zheng, Y.; Liu, T.; Wang, Y.; Zhu, Y.; Liu, Y.; Chang, E. Diagnosing New York City's Noises with Ubiquitous Data. In Proceedings of the 2014 ACM International Joint Conference on Pervasive and Ubiquitous Computing, Seattle, WA, USA, 13–17 September 2014; pp. 715–725.

64. Aiello, L.M.; Schifanella, R.; Quercia, D.; Aletta, F. Chatty Maps: Constructing sound maps of urban areas from social media data. *R. Soc. Open Sci.* **2016**, *3*, 150690, doi:10.1098/rsos.150690.

65. Loreto, V.; Haklay, M.; Hotho, A.; Sevedio, V.; Stumme, G.; Theunis, J.; Tria, F. Participatory Sensing, Opinions and Collective Awareness. In *Collection Understanding Complex Systems*; Springer: Cham, Switzerland, 2017.

66. D'Hondt, E.; Stevens, M.; Jacobs, A. Participatory noise mapping works! An evaluation of participatory sensing as an alternative to standard techniques for environmental monitoring. *Pervasive Mob. Comput.* **2013**, *9*, 681–694.

67. De Coensel, B.; Botteldooren, D. Smart Sound Monitoring for Sound Event Detection and Characterisation. In Proceedings of the INTERNOISE, Melbourne, Australia, 16–19 November 2014.

68. De Coensel, B.; Sun, K.; Wei, W.; Van Renterghem, T.; Sineau, M.; Ribeiro, C.; Can, A.; Aumond, P.; Lavandier, C.; Botteldooren, D. Dynamic Noise Mapping based on Fixed and Mobile Sound Measurements. In Proceedings of the Euronoise, Maastrich, The Netherlands, 31 May–3 June 2015; pp. 2339–2344.

69. Malinowski, A.; Yu, H. Comparison of Embedded System Design for Industrial Applications. *IEEE Trans. Ind. Inf.* **2011**, *7*, 244–254.

70. Which ARM Cortex Core Is Right for Your Application: A, R or M? Available online: http://www.silabs.com (accessed on 26 May 2016).

71. Macías, M.M.; Agudo, J.E.; Orellana, C.J.G.; Velasco, H.M.G.; Manso, A.G. The "mbed" platform for teaching electronics applied to product design. In Proceedings of the Tecnologias Aplicadas a la Ensenanza de la Electronica (Technologies Applied to Electronics Teaching) (TAEE), Bilbao, Spain, 11–13 June 2014; pp. 1–6.

72. Toulson, R.; Wilmshurst, T. *Fast and Effective Embedded Systems Design Applying the ARM Mbed*; Newnes: Oxford, UK, 2012.

73. Getting Started with Embedded Coder Support Package for Freescale FRDM-KL25Z Board. Available online: http://es.mathworks.com/ (accessed on 26 May 2016).

74. Wu, X.; Obeng, M.; Wang, J.; Kulas, D. A survey of techniques to add audio module to embedded systems In Proceedings of the IEEE Southeastcon, Orlando, FL, USA, 15–18 March 2012; pp. 1–5.

75. Adafruit. Available online: https://www.adafruit.com/products/1713 (accessed on 31 May 2016).

76. Magdaleno, E.; Rodríguez, M.; Pérez, F.; Hernández, D.; García, E.A. FPGA Embedded Web Server for Remote Monitoring and Control of Smart Sensors Networks. *Sensors* **2014**, *14*, 416–430.

77. Ozer, E.; Feng, M.Q.; Feng, D. Citizen Sensors for SHM: Towards a Crowdsourcing Platform. *Sensors* **2015**, *15*, 14591–14614.

78. Moreno, A.; Angulo, I.; Perallos, A.; Landaluce, H.; Zuazola, I.J.G.; Azpilicueta, L.; Astrain, J.J.; Falcone, F.; Villadangos, J. IVAN: Intelligent Van for the Distribution of Pharmaceutical Drugs. *Sensors* **2012**, *12*, 6587–6609.

79. Mitchell, S.; Wagener, W. The Connected Bus: Connected and Sustainable Mobility Pilot. A Partnership between the City and County of San Francisco and Cisco. *Pilot Results*, Connected Urban Development, 2009. Available online: http://www.cisco.com/c/dam/en_us/about/ac79/docs/cud/The_Connected_Bus_Pilot.pdf (accessed on 5 December 2016).

80. Central Policy Unit, The Government of the Hong Kong Special Administrative Region. Research Report on Smart City. September 2015. Available online: http://www.cpu.gov.hk/doc/en/research_reports/CPU%20research%20report%20-%20Smart%20City(en).pdf (accessed on 5 December 2016).

81. Ntalampiras, S.; Potamitis, I.; Fakotakis, N. On Acoustic Surveillance of Hazardous Situations. In Proceedings of the International Conference on Acoustic, Speech and Signal Processing, Taipei, Taiwan, 19–24 April 2009; pp. 165–168.

82. Schröder, J.; Goetze, S.; Grützmacher, V.; Anemüller, J. Automatic Acoustic Siren Detection in Traffic Noise by Part-Based Models. In Proceedings of the International Conference on Acoustic, Speech and Signal Processing, Vancouver, BC, Canada, 26–31 May 2013; pp. 493–497.

83. Foggia, P.; Petrov, N.; Saggese, A.; Strisciuglio, N.; Vento, M. Audio Surveillance of Roads: A System for Detecting Anomalous Sounds. *Trans. Intell. Transp. Syst.* **2016**, *17*, 279–288.

84. Oldoni, D.; De Coensel, B.; Bockstael, A.; Boes, M.; De Baets, B.; Botteldooren, D. The Acoustic Summary as a Tool for Representing Urban Sound Environments. *Landsc. Urban Plan.* **2015**, *144*, 34–48.

85. Socoró, J.C.; Ribera, G.; Sevillano, X.; Alías, F. Development of an Anomalous Noise Event Detection Algorithm for Dynamic Road Traffic Noise Mapping. In Proceedings of the 22nd International Congress on Sound and Vibration (ICSV22), Florence, Italy, 12–16 July 2015.

86. Socoró, J.C.; Albiol, X.; Sevillano, X.; Alías, F. Analysis and Automatic Detection of Anomalous Noise Events in Real Recordings of Road Traffic Noise for the LIFE DYNAMAP Project. In Proceedings of the INTERNOISE, Hamburg, Germany, 21–24 August 2016; pp. 6370–6379.

87. Valero, X.; Alías, F. Gammatone Wavelet Features for Sound Classification in Surveillance Applications. In Proceedings of the 20th European Signal Processing Conference, Bucharest, Romania, 27–31 August 2012; pp. 1658–1662.

88. Creixell, E.; Haddad, K.; Song, W.; Chauhan, S.; Valero, X. A Method for Recognition of Coexisting Environmental Sound Sources based on the Fisher's Linear Discriminant Classifier. In Proceedings of the INTERNOISE, Innsbruck, Austria, 15–18 September 2013.

89. Benetos, E.; Lafay, G.; Lagrange, M.; Plumbley, M. Detection of Overlapping Acoustic Events using a Temporally-Constrained Probabilistic Model. In Proceedings of the International Conference on Acoustic, Speech and Signal Processing, Shanghai, China, 20–25 March 2016; pp. 6450–6454.

90. Mesaros, A.; Heittola, T.; Virtanen, T. Metrics for Polyphonic Sound Event Detection. *Appl. Sci.* **2016**, *6*, 162.

91. Heittola, T.; Mesaros, A.; Virtanen, T.; Gabbouj, M. Supervised Model Training for Overlapping Sound Events based on Unsupervised Source Separation. In Proceedings of the 38th International Conference on Acoustics, Speech, and Signal Processing (ICASSP 2013), Vancouver, BC, Canada, 26–31 May 2013; pp. 8677–8681.

92. Mesaros, A.; Dikmen, O.; Heittola, T.; Virtanen, T. Sound Event Detection in Real Life Recordings Using Coupled Matrix Factorization of Spectral Representations and Class Activity Annotations. In Proceedings of the IEEE International Conference on Acoustics, Speech and Signal Processing (ICASSP), Brisbane, Australia, 19–24 April 2015; pp. 151–155.

93. Espi, M.; Fujimoto, M.; Kinoshita, K.; Nakatani, T. Exploiting Spectro-temporal Locality in Deep Learning Based Acoustic Event Eetection. *EURASIP J. Audio Speech Music Process.* **2015**, doi:10.1186/s13636-015-0069-2.

94. Ntalampiras, S.; Potamitis, I.; Fakotakis, N. Automatic Recognition of Urban Environmental Sounds Events. In Proceedings of the EURASIP—Workshop on Cognitive Information Processing, Santorini, Greece, 9–10 June 2008; pp. 110–113.

95. Valero, X.; Alías, F. Automatic Classification of Road Vehicles Considering their Pass-by Acoustic Signature. In Proceedings of the International Conference on Acoustics (ICA), Montréal, QC, Canada, 2–7 June 2013.

96. Geréb, G. Real-time updating of noise maps by source-selective noise monitoring. *Noise Control Eng. J.* **2013**, *61*, 228–239.

97. Theodorou, T.; Mporas, I.; Fakotakis, N. *Automatic Sound Recognition of Urban Environment Events*; Speech and Computer Vol. 9319 of the series Lecture Notes in Computer Science; Springer: Zurich, Switzerland, pp. 129–136.

98. Crocco, M.; Cristani, M.; Trucco, A.; Murino, V. Audio Surveillance: A Systematic Review. *ACM Comput. Surv. (CSUR)* **2016**, *48*, 52.

99. Garcia, A.; Faus, L.J. Statistical Analysis of Noise Levels in Urban Areas. *Appl. Acoust.* **1991**, *34*, 227–247.

100. Morillas, J.M.; Escobar, V.G.; Sierra, J.A.; Vilchez-Gomez, R.; Vaquero, J.M.; Carmona, J.T. A Categorization Method Applied to the Study of Urban Road Traffic Noise. *J. Acoust. Soc. Am.* **2005**, *117*, 2844–2852.

101. Del Río, F.J.C.; Escobar, V.G.; Carmona, J.T.; Vílchez-Gómez, R.; Sierra, J.A.M.; Gozalo, G.R.; Morillas, J.M.B. A Street Categorization Method to Study Urban Noise: The Valladolid (Spain) study. *Environ. Eng. Sci.* **2011**, *28*, 811–817.

102. Zambon, G.; Benocci, R.; Brambilla, G. Cluster Categorization of Urban Roads to Optimize their Noise Monitoring. *Environ. Monit. Assess.* **2016**, *188*, 26.

103. Andrzej, C.; Maciej, S. Software for Calculation of Noise Maps Implemented on Supercomputer. *Prac. Poligr.* **2009**, *TQ4131*, 363–378.

104. Wei, W.; Botteldooren, D.; Van Renterghem, T. Monitoring Sound Exposure by Real Time Measurement and Dynamic Noise Map. In Proceedings of the Forum Acusticum, Krakow, Poland, 7–12 September 2014.

105. Cerniglia, A. State of the Art on Real Time Noise Mapping System and Related Software Development. In Proceedings of the 22nd International Congress on Sound and Vibration, Florence, Italy, 12–16 July 2015.

106. Manvell, D.; Ballarin Marcos, L.; Stapelfeldt, H.; Sanz, R. SADMAM—Combining Measurements and Calculations to Map Noise in Madrid. In Proceedings of the INTERNOISE, Prague, Czech Republic, 22–25 August 2004.

107. Brambilla, G.; Cerniglia, A.; Verardi, P. New Potential of Long Term Real Time Noise Monitoring Systems. In Proceedings of the EuroNoise, Tampere, Finland, 30 May–1 June 2006.

108. Maisonneuve, N.; Stevens, M.L.L.; Niessen, M.; Steels, L. NoiseTube: Measuring and Mapping Noise Pollution with Mobile Phones. In Proceedings of the Information Technologies in Environmental Engineering, Thessaloniki, Greece, 28–29 May 2009.

109. Rana, R.; Chou, C.T.; Bulusu, N.; Kanhere, S.; Hu, W. Ear-Phone: A Context-Aware Noise Mapping using Smart Phones. *Pervasive Mob. Comput.* **2015**, *17*, 1–22.

10

A Canopy Density Model for Planar Orchard Target Detection Based on Ultrasonic Sensors

Hanzhe Li [1], Changyuan Zhai [1,2,*], Paul Weckler [2], Ning Wang [2], Shuo Yang [1] and Bo Zhang [1]

[1]   College of Mechanical and Electronic Engineering, Northwest A&F University, Yangling 712100, China; lihanzhe187@163.com (H.L.); yangshuosjz@163.com (S.Y.); zhangbo_609@163.com (B.Z.)

[2]   Department of Biosystems and Agricultural Engineering, Oklahoma State University, Stillwater, OK 75078, USA; paul.weckler@okstate.edu (P.W.); ning.wang@okstate.edu (N.W.)

*   Correspondence: zhaichangyuan@nwsuaf.edu.cn

Academic Editors: Dipen N. Sinha and Cristian Pantea

Abstract: Orchard target-oriented variable rate spraying is an effective method to reduce pesticide drift and excessive residues. To accomplish this task, the orchard targets' characteristic information is needed to control liquid flow rate and airflow rate. One of the most important characteristics is the canopy density. In order to establish the canopy density model for a planar orchard target which is indispensable for canopy density calculation, a target density detection testing system was developed based on an ultrasonic sensor. A time-domain energy analysis method was employed to analyze the ultrasonic signal. Orthogonal regression central composite experiments were designed and conducted using man-made canopies of known density with three or four layers of leaves. Two model equations were obtained, of which the model for the canopies with four layers was found to be the most reliable. A verification test was conducted with different layers at the same density values and detecting distances. The test results showed that the relative errors of model density values and actual values of five, four, three and two layers of leaves were acceptable, while the maximum relative errors were 17.68%, 25.64%, 21.33% and 29.92%, respectively. It also suggested the model equation with four layers had a good applicability with different layers which increased with adjacent layers.

Keywords: precision spray; target detection; canopy density model; ultrasonic sensor; orthogonal regression central composite experiment

1. Introduction

Prevention of insects and diseases of crops is a crucial factor in orchard management. In conventional agriculture chemical spray application is still the main way to insure high yields at a low cost. However, excessive pesticide application results in residues on fruits and soil, which pollute the environment and threaten the safety of agricultural products. Precision target-oriented variable spraying is an effective method to reduce pesticide residue. To achieve this goal, real-time acquisition of the orchard targets' characteristic information is the key.

The characteristic information of the orchard target includes the tree's diameter, volume, Leaf Area Index (LAI), canopy density, etc. Many researchers have applied ultrasonic techniques, digital photographic techniques, optical sensors, high-resolution radar images, high-resolution X-ray computed tomography, stereo vision and LIDAR (light detection and ranging) sensors for target characteristic information acquisition [1–4]. Crop management plans including spraying, irrigation and fertilization have benefited from the application of targets' characteristic information [1,2]. Especially in air-assisted variable-rate spraying, a controller adjusts the spraying parameters which include pesticide flow rate and airflow rate based on targets' characteristic information to improve the performance of spraying [3,4]. Light interception and aerial photogrammetry have been used to measure the shape and size of trees, which were sufficient for plant protection [5,6]. The computerized spraying control system with ultrasonic measurement arrays and GPS (global positioning system) or DGPS (differential global positioning system), can automatically adjust pesticide flow rate according to real-time sensing, monitoring, calculation, storage and mapping of tree canopy volume and height [7–10]. The distance from a sprayer to orchard targets at different heights can be measured using several ultrasonic ranging sensors, and the trees' volumes can be estimated based on a neural network algorithm [11–13]. But the sound cone determination, angle errors, crosstalk errors and field measurements were affected by surroundings [14]. LIDAR has been widely used in measuring 3D (three-dimensional) structural characteristics of trees including the target's height, width, volume, leaf area index and leaf area density [15]. The unstructured point cloud is obtained from LIDAR scanning, and then the computer processes the point cloud data and rebuilds the 3D digital model of the target. This method allows fast and nondestructive measurement of a target's parameters and also has a high correlation with actual measurement [16–19]. LAI is an important indicator in determining the growth status of plants. Some researchers also use the digital photographic techniques to estimate LAI. Compared to the other methods this estimating method observably expand the spatial area and frequency of analysis [20–22]. The LIDAR also has been used in drift detection and crop discrimination to guide spraying [23,24].

Integrating many different types of target characteristic information, Walklate compared spray volume deposition based on different models including a vertical wall area model, cylindrical wall area model, tree row volume model, tree area index model, tree area density model and light interception flux model. The result suggested that the tree area density is one of the most important parameters of a single tree target [25]. Ultrasonic techniques, digital photographic techniques and LIDAR have been used in detecting orchard target canopy density, but they still lack quantitative/parametric equations. Palleja estimated canopy density using ultrasonic envelope signals [26]. The results showed that ultrasound's wave intensity can be used as an indicator of canopy density, however it could only reflect the change of density and lacks a quantitative relationship between density and ultrasound's wave intensity. It could not provide the control basis for real-time mathematical equations in variable spraying. This paper aims to explore the quantitative relationship between ultrasounds' wave intensity energy and canopy density, and establish the orchard target canopy density model.

2. Materials and Methods

2.1. Target Canopy Density Detection Method

In order to measure the intensity of ultrasonic echo, an ultrasonic sensor XL-MaxSonar MB7092 (MaxBotix Inc., Brainerd, MN, USA) was used. The sensor operated on 3.0 V–5.5 V with five functional pins. It could output analog voltage of range measurement at pin3, and output the analog voltage envelope of the acoustic waveform at pin2. The other three pins were controlling pins. The echo analog voltage of ultrasonic sensor was recorded from pin2 to analyze echo intensity. Echo intensity was influenced not only by the target distance and the target spatial dimension, but also by the canopy density.

Time-domain energy analysis is a common method of signal analysis. The time-domain energy calculation method is as follows:

$$E = \int_{-\infty}^{+\infty} x^2(t)dt \tag{1}$$

$$E = \sum_{k=0}^{n} x^2(k) \tag{2}$$

where E is the energy of signal, $x(t)$ is the analog signal and $x(k)$ is the sequence of digital signal.

The output signal voltage representing the ultrasonic wave is shown in Figure 1. This graph displays the transmitted wave and echo wave. The ultrasonic energy was analyzed based on these waves.

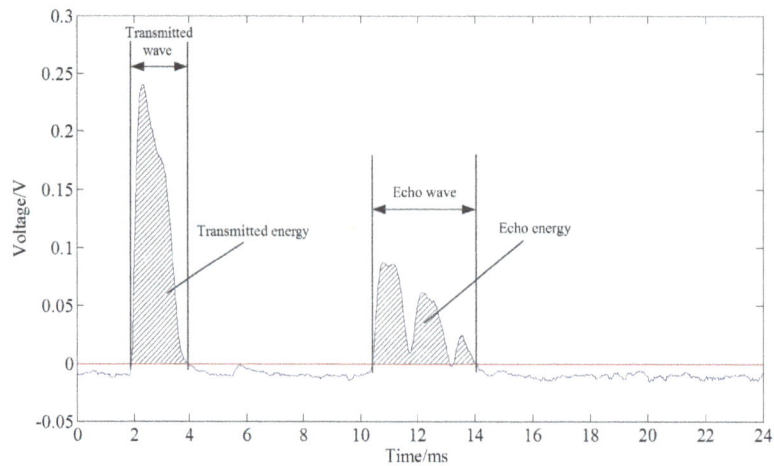

Figure 1. Ultrasonic transmitted wave and echo wave.

The ultrasonic transmitted energy and echo energy were calculated using MATLAB software (MathWorks, Natick, MA, USA). The ultrasonic signal was recorded by an oscilloscope and a computer. The signals of the transmitted and echo waves should not be negative in theory. The negative data was treated as zero. Then the signal was smoothed using the smooth function in MATLAB. The transmitted energy and echo energy were calculated after the signal processing.

2.2. Target Density Detection System

The target density detection system included a test bench, an ultrasonic sensor, a fixed mount, a DC power supply, an oscilloscope and a computer. The detection system was developed as shown in Figure 2. It provided a controllable test environment in which density and detecting distance could be accurately adjusted.

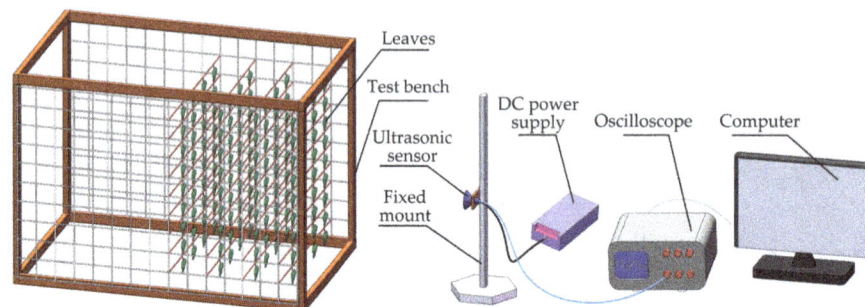

Figure 2. Diagram of target density detection system.

The test bench consists of a wooden frame, fishing lines and wire fencing. The size of the wooden frame (length × width × height) was 150 cm × 103 cm × 103 cm. The fishing line with a diameter of 0.234 mm had almost no effect on ultrasonic echo waves. The wire fencing was fixed to two sides of the wooden frame to attach fishing lines. The grid size of the wire fencing was 1 cm × 1 cm. Fishing line crossed the wire fencing grid in the same plane to constitute a layer, the spacing of fishing line was 5 cm. In each layer there were nine rows of two fishing lines. The leaves could be clamped on each row using clips. The spacing of layers was 20 cm, therefore the volume of each layer was 0.188 m^3. The leaves of the Chinese glossy privet (*Ligustrum lucidum*) were chosen for the experiment. The weight of each leaf was between 1.0 and 2.0 g, while the size of the leaf was about 10 cm × 6 cm. Under such conditions, the maximum weight of leaves that could be arranged in each layer was 212 g, while the maximum density of each layer was 1127.66 g/m^3. The minimum density was set as 112.77 g/m^3, which was 10% of the maximum density. In the test bench several layers of leaves could be combined to simulate canopies with different thicknesses. The density of each layer was the same. The leaves were evenly fixed in each row with interspersed arrangements in adjacent rows. In the adjacent layers, the arrangements were interspersed as well.

The DC power supply was S-25-5 5V DC power supply (Weiming Power, Qidong, China), whose actual voltage output was 5.69 V. The oscilloscope used was a RIGOL DS1062E-EDU (Beijing RIGOL Technology Co. Ltd., Beijing, China), which recorded the waveform from an ultrasonic sensor. The Ultrascope for DS1000E Series software was used to read the waveform of oscilloscope on the screen. This software could save the waveform as a BMP picture and an Excel file to a computer through an RS-232 to USB converter.

2.3. Experiment for the Relationship between the Ultrasonic Energy and the Power Supply Voltage

During practice use of the ultrasonic sensor, it was found that the ultrasonic energy would change with its power supply voltage. An experiment was designed to establish the relationship between the ultrasonic energy and the power supply voltage. The sensor was powered by an MPS-3003L-3 laboratory power supply (Matrix Technology Inc., Shenzhen, China), whose voltage range was 0–30 V with the regulation precision of 0.1 V. Since the ultrasonic sensor accepted a power of 3.0–5.5 V, the voltage of the power supply was set between 3.0 and 6.0 V with a current of 200 mA. A smooth solid wall was used as a test target, which was 1.0 m away from the surface of the sensor. An oscilloscope and a computer with a DS1000E Series software Ultrascope recorded the waveform when the power supply was set from 3.0 V to 6.0 V with increments of 0.1 V. In each treatment, the waveform data was recorded three times. The averages of transmitted energy and echo energy were calculated using MATLAB to analyze the relationship between the ultrasonic energy and the supply voltage.

It was meaningful to analyze the echo energy under a unified transmitted energy, but it was difficult to keep the supply voltage constant. Thus normalization of transmitted energy through a mathematical method was determined. The fitting equation between the correction coefficient and supply voltage was obtained using the CFTool in MATLAB.

2.4. Experiment for Beam Width of Ultrasonic Sensor

The beam width of the ultrasonic sensor is an important parameter which determines the detecting range. The diagram of the measuring method to obtain the beam width at different detecting distances is shown in Figure 3, where S is the detecting distance between the ultrasonic sensor and the test plate edge; W_R is the distance between the center line and the right test plate; and W_L is the distance between the center line and the left test plate. The value of S was calculated in the orthogonal regression central composite experiment (will be mentioned in Section 2.5).

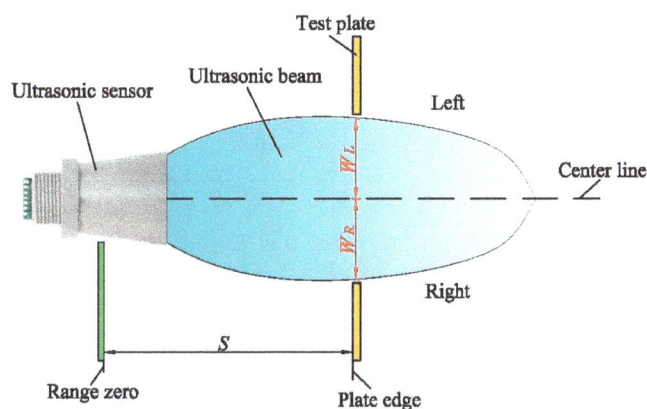

Figure 3. Diagram of measuring method for the beam width of an ultrasonic sensor.

In the measurement experiments, the ultrasonic sensor was placed in an empty space where the sensor couldn't receive any echoes. The oscilloscope read the waveform output of the ultrasonic sensor in real-time. A test plate was moved slowly from right (or left), to the center line until the ultrasonic sensor received echoes. Then, the distance W_R or W_L was manually measured between the test plate and the center line. Each measurement was conducted 3 times. The final value of W_R or W_L was the average of the 3 repetitions. The beam width was the sum of W_R and W_L which must be measured at the same detecting distance S.

2.5. Orthogonal Regression Central Composite Experimental Design

A central composite design is the most commonly used response surface designed experiment. Central composite designs are a factorial or fractional factorial design with center points, augmented with a group of axial points (also called star points) that allow estimation of curvature. The orthogonal regression central composite experimental design is an effective method to obtain mathematical relationships between factors and variables [27,28]. Only the representative test points are chosen from the comprehensive full-scale tests based on orthogonality, which makes this method more efficient by reducing test times. Canopy density models were designed to be obtained based on orthogonal regression central composite experiments. The factors were the density and the distance, while the result was the echo energy of the ultrasonic sensor. The parameter γ which was used to determine factors levels, was calculated by the following equations:

$$m_c - \frac{m_c^2}{n} - \frac{4m_c}{n}\gamma^2 - \frac{4}{n}\gamma^4 = 0, \gamma = \sqrt{-2^{p-1} + (2^p + 2p + m_0)^{\frac{1}{2}} \times 2^{\frac{p}{2}-1}} \tag{3}$$

where p is the number of factors; m_c is the number of orthogonal tests; m_0 is the number of the zero level repeat tests; n is the number of the total tests; γ is star test point parameter; and m_0 is the number of zero level repeat tests. In these orthogonal regression experiments, parameter m_0 was set as: $m_0 = 3$. The values of the other parameters were: $p = 2$, $m_c = 4$, $n = 11$, $\gamma = 1.15$:

$$z_{0j} = \frac{\left(z_{lj} + z_{uj}\right)}{2}, \Delta_j = \frac{\left(z_{uj} - z_{lj}\right)}{2r}, x_j = \frac{\left(z_j - z_{0j}\right)}{\Delta_j} \tag{4}$$

where: Z_{lj}, Z_{uj} and Z_{0j} are the lower level, upper level and zero level of the factor j respectively; Δ_j is the range radius; Z_j is the value of factor j; and x_j is the factor level code. The factor levels coding is shown in Table 1.

Table 1. Factor levels coding.

Factor	Z_{lj}	Z_{uj}	Z_{0j}	Δ_j	$-\gamma$	-1	0	1	γ
Density (Z_1) [g/m^3]	112.77	1127.66	620.21	440.91	112.77	179.31	620.21	1061.12	1127.66
Distance (Z_2) [m]	0.5	1.5	1.0	0.43	0.5	0.57	1.0	1.43	1.5

In orthogonal experiments, the detection points distributed on the test bench were set due to the results of the beam width experiment (will be mentioned in Section 3.2). The distribution diagram of detection points is shown in Figure 4.

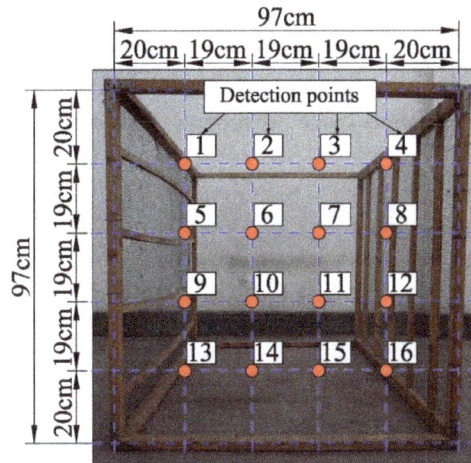

Figure 4. Distribution diagram of detection points.

According to the results of the beam width experiments, the maximum of the beam width was 20 cm. The detection points were set 20 cm inside from the test bench boundary, while the spacing of adjacent detection points was 19 cm. At each of the 16 detection points, an average of echo energy was obtained by three replicated measurements. In the 16 echo energies, the three maximum and the three minimum ones were removed, and the 10 left were averaged to generate the final echo energy data to establish orthogonal regression equations. The regression equations were calculated by the following computational process:

$$y = b_0 + \sum_{j=1}^{p} b_j x_j + \sum_{j<k} b_{jk} x_j x_k + \sum_{j=1}^{p} b_{jj} x_j' \tag{5}$$

where b is the coefficient of regression equation; and y is the echo energy calculated by regression equation. The calculation of coefficients was omitted, but the detailed calculation process can be obtained from [27,28].

For the sake of confirming the reliability of the equation established, the regression equation and its parameters were hypothesis tested using the following expressions:

$$F_j = \frac{\frac{S_j}{f_j}}{\frac{S_e}{f_e}}; F = \frac{\frac{S_T}{f_T}}{\frac{S_R}{f_R}}; \tag{6}$$

where S_j are the sums of partial regression squares; f_j is the degree of freedom of S_j; S_e was the sum of error squares within repeat test group; f_e was the degree of freedom of S_e; S_T was the sum of regression squares; f_T was the degree of freedom of S_T; S_R was the sum of residual squares; f_R was the degree of freedom of S_R; F_j was the F distribution statistic of parameter j; and F was the F distribution statistic of the regression equation. The significant coefficients will be selected to build the regression equations based on the F-test.

In case of the repeated measurement data, the model can be evaluated by test for lack of fit. The test for lack of fit of canopy density model was calculated by the following equation:

$$F_{lf} = \frac{\frac{S_{lf}}{f_{lf}}}{\frac{S_e}{f_e}} \tag{7}$$

where S_{lf} is the sum of lack of fit squares; f_{lf} is the degrees of freedom of S_{lf}; and F_{lf} is the F distribution statistic used in the test for lack of fit.

In order to establish a canopy density model, orthogonal experiments were conducted with three and four layers of leaves, and two canopy density models were obtained based on three layers and four layers. A verification experiment was conducted to select a better model. The experiments for establishing canopy density models were performed indoors. Canopy model experiments with four layers are shown in Figure 5. Each experimental datapoint was recorded three times, and the average data was used as the result. The final result was a decuple result to reduce the round-off error. During the experiments, the temperature was 25–29 °C, and the humidity was 32%–53%.

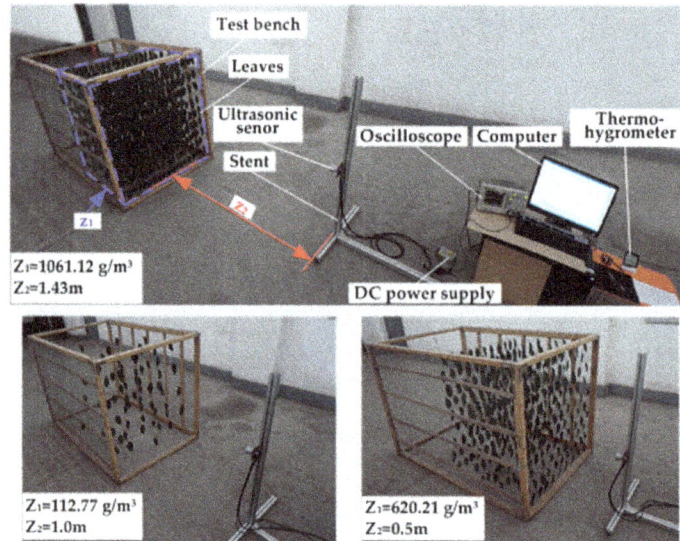

Figure 5. Canopy model experiments performance with four layers of leaves.

2.6. Verification Test Design

In order to verify the universality of canopy density models, this paper selected different layers with different density which weren't used in establishing the density model. The actual value was calculated using MATLAB, while the model value was calculated based on the selected canopy density model. The relative errors between model value and actual value were used to analyze the universality of the canopy density model. The verification test was conducted with the same density values and detecting distances with different layers.

3. Results and Discussion

3.1. Relationship between the Ultrasonic Energy and the Power Supply Voltage

The relationship between the ultrasonic energy and the supply voltage is shown in Figure 6a. It shows that both the transmitted energy and the echo energy went up as the supply voltage increased. Therefore, the stability of supply voltage has an important influence on the time-domain energy analysis. In order to calibrate the transmitted energy, it should be normalized by the correction coefficient. The reference voltage was set as 5.0 V, while the reference transmitted energy was 1.1130 J. The correction coefficient vs. supply voltage curve is shown in Figure 6b. The mathematical equation was obtained as follows, and the value of R^2 was 0.9984:

$$c = 0.0894U^4 - 1.7704U^3 + 13.1537U^2 - 43.7620U + 56.3865, \ 3.0 \leq U \leq 6.0 \tag{8}$$

where U is the supply voltage in V, and c is the correction coefficient.

The normalized transmitted energy was multiplied by transmitted energy and correction coefficient, normalized echo energy was echo energy multiplied by echo energy and correction coefficient, and the formula was obtained as follows:

$$E_N = c \times E \tag{9}$$

where E_N is the normalized energy, c is the correction coefficient, and E is the calculation energy.

The normalized echo energy corrected by this coefficient can reduce the deviation caused by supply voltage variation, but it cannot totally eliminate the deviation. Figure 6a shows that the slopes of the transmitted energy variation and the echo energy variation were different, thus the normalized echo energy still has deviation. In order to obtain a uniform reference, it is necessary to ensure the stability of the sensor supply voltage.

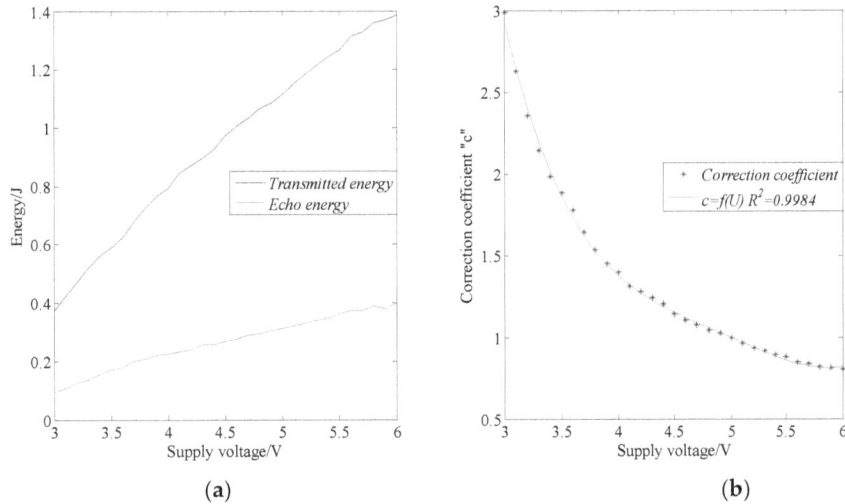

(a)   (b)

Figure 6. (a) Relationship between ultrasonic energy and supply voltage; and (b) fitting equation of correction coefficient and supply voltage.

3.2. Beam Width of the Ultrasonic Sensor

Table 2 shows the result of the beam width experiment. The results showed that the beam width was different at different detection distances. In order to avoid detecting the boundary of the test bench, the detection points were set at 20 cm inside the test bench.

Table 2. Results of beam width experiments.

Tests	S [m]	W_L [cm]	W_R [cm]	Average of W_L [cm]	Average of W_R [cm]
1		6	7		
2	0.5	6.5	6.5	6.3	6.8
3		6.5	7		
4		8	7		
5	0.57	7.5	7	7.7	7.0
6		7.5	7		
7		11	12		
8	1.0	10.5	11	10.5	11.2
9		10	10.5		
10		12	13		
11	1.43	11	12	12.0	12.3
12		13	12		
13		16	14		
14	1.5	15	14.5	15.0	14.2
15		14	14		

3.3. Canopy Density Model

Results of the canopy model experiments with 4 layers are shown in Table 3. The equation coefficients and statistical parameters were calculated (Table 4).

Table 3. Results of canopy density model experiments with 4 layers of leaves.

Z_1 [g/m³] (x_1)	Z_1 [m] (x_2)	x_1x_2	x_1'	x_2'	Transmitted Energy [J]	Echo Energy [J]	Normalized Echo Energy [J]	Decuple Normalized Echo Energy [J]
1061.12(1)	1.43(1)	1	0.396	0.396	1.2738	0.1815	0.1586	1.586
1061.12(1)	0.57(−1)	−1	0.396	0.396	1.2601	0.4878	0.4309	4.309
179.31(−1)	1.43(1)	−1	0.396	0.396	1.2601	0.1528	0.1350	1.350
179.31)(−1)	0.57(−1)	1	0.396	0.396	1.3354	0.4230	0.3526	3.526
1127.66(r)	1.0(0)	0	0.716	−0.604	1.3347	0.3338	0.2784	2.784
112.77(−r)	1.0(0)	0	0.716	−0.604	1.3524	0.1818	0.1496	1.496
620.21(0)	1.5(r)	0	−0.604	0.716	1.3291	0.2265	0.1897	1.897
620.21(0)	0.5(−r)	0	−0.604	0.716	1.3036	0.5774	0.4930	4.930
620.21(0)	1.0(0)	0	−0.604	−0.604	1.3211	0.3670	0.3092	3.092
620.21(0)	1.0(0)	0	−0.604	−0.604	1.3249	0.3189	0.2679	2.679
620.21(0)	1.0(0)	0	−0.604	−0.604	1.3363	0.3126	0.2603	2.603

Table 4. Equation coefficients and statistical parameters of canopy density model.

Regression Equation Parameters		Test for Lack of Fit of Density Model		Equation Parameter Hypothesis Test	
b_0	2.750	S_R	0.301	F_1	13.601
b_1	0.376	S_T	13.243	F_2	153.131
b_2	−1.262	S_{Lf}	0.163	F_{12}	0.037
b_{12}	−0.137	S_e	0.138	F_{11}	14.276
b_{11}	−0.533	F_{Lf}	0.786	F_{22}	9.481
b_{22}	0.434	f_R	5	F	43.971
		F_T	5		
		f_{Lf}	3		
		f_e	2		

As $F_{Lf} < 1$, and $F > F_{0.90}(5,5) = 3.45$, the flowing model was acceptable. The value of F_{12} was less than $F_{0.90}(1,2) = 8.53$, so the term x_1x_2 could be ignored. The canopy density model equation with four layers was obtained as follows:

$$10y = 2.750 + 0.376x_1 - 1.262x_2 - 0.533x_1' + 0.434x_2' \tag{10}$$

$$x_1 = \frac{z_1 - 620.21}{440.91}, x_2 = \frac{z_2 - 1}{0.43} \tag{11}$$

$$y = -2.742 \times 10^{-7}z_1^2 + 0.2348z_2^2 + 4.225 \times 10^{-4}z_1 - 0.7831z_2 + 0.6609 \tag{12}$$

where z_1 is the canopy density in g/m³, z_2 is the distance in m, and y is the echo energy. Similar experiments were conducted to establish canopy density models with three layers (Table 5). The equation coefficients and statistical parameters were calculated (Table 6).

Table 5. Results of canopy density model experiments with three layers of leaves.

Z_1 [g/m³] (x_1)	Z_1 [m] (x_2)	x_1x_2	x_1'	x_2'	Transmitted Energy [J]	Echo Energy [J]	Normalized Echo Energy [J]	Decuple Normalized Echo Energy [J]
1061.12(1)	1.43(1)	1	0.396	0.396	1.3381	0.2098	0.1745	1.745
1061.12(1)	0.57(−1)	−1	0.396	0.396	1.3362	0.5665	0.4718	4.718
179.31(−1)	1.43(1)	−1	0.396	0.396	1.3506	0.1185	0.0977	0.977
179.31)(−1)	0.57(−1)	1	0.396	0.396	1.3301	0.3277	0.2742	2.742
1127.66(r)	1.0(0)	0	0.716	−0.604	1.3468	0.3936	0.3253	3.253
112.77(−r)	1.0(0)	0	0.716	−0.604	1.3531	0.1693	0.1393	1.393
620.21(0)	1.5(r)	0	−0.604	0.716	1.3524	0.2002	0.1648	1.648
620.21(0)	0.5(−r)	0	−0.604	0.716	1.3424	0.5427	0.4499	4.499
620.21(0)	1.0(0)	0	−0.604	−0.604	1.3512	0.3146	0.2591	2.591
620.21(0)	1.0(0)	0	−0.604	−0.604	1.3569	0.2879	0.2361	2.361
1061.12(1)	1.43(1)	0	−0.604	−0.604	1.3426	0.3253	0.2697	2.697

Table 6. Equation coefficients and statistical parameters of Canopy density model.

Regression Equation Parameters		Test for Lack of Fit of Density Model		Equation Parameter Hypothesis Test	
b_0	2.602	S_R	0.144	F_1	121.882
b_1	0.735	S_T	14.193	F_2	328.565
b_2	−1.207	S_{Lf}	0.085	F_{12}	0.182
b_{12}	−0.302	S_e	0.059	F_{11}	9.270
b_{11}	−0.280	F_{Lf}	0.963	F_{22}	9.915
b_{22}	0.290	f_R	5	F	95.589
		f_T	5		
		f_{Lf}	3		
		f_e	2		

As $F_{Lf} < 1$, and $F > F_{0.90}(5,5) = 3.45$, the model was acceptable. The value of F_{12} was less than $F_{0.90}(1,2) = 8.53$, so the term $x_1 x_2$ could be ignored. The canopy density model equation with three layers was obtained as follows:

$$10y = 2.602 + 0.735x_1 - 1.207x_2 - 0.28x_1' + 0.29x_2' \tag{13}$$

$$x_1 = \frac{z_1 - 620.21}{440.91}, x_2 = \frac{z_2 - 1}{0.43} \tag{14}$$

$$y = -1.440 \times 10^{-7}z_1^2 + 0.1596z_2^2 + 3.454 \times 10^{-4}z_1 - 0.5945z_2 + 0.5386 \tag{15}$$

where z_1 is the canopy density in g/m^3, z_2 is the distance in m, and y is the echo energy.

3.4. Model Equation Selection

With the purpose of selecting a better model equation to simplify the application in practice, experimental data with four layers and three layers were used to contrast the two different model equations. The results are shown in Tables 7 and 8.

The model echo energy was calculated based on canopy density models for four layers and three layers (Equations (12) and (15)). Table 7 shows that the relative errors of model echo energy and actual normalized echo energy with three layers and four layers were different. The maximum relative error was 53.47%, and the average relative error was 16.07%. The maximum relative error of the model with four layers was 19.57% with the average relative error of 8.80%. Table 8 shows that the maximum relative error of the model with three layers was 26.14%, and the average relative error was 8.26%. The maximum relative error of the model with four layers was 26.83% and the average relative error was 10.76%. More importantly, the variance of relative error for the model with four layers was smaller than the model with three layers in those comparisons. The result of the model equation comparison showed that the canopy density model with four layers was more universal than the canopy density model with three layers, thus this paper selected the canopy density model with four layers as the optimal equation.

Table 7. Result of model equation contrast with 4 layers of leaves.

Density [g/m^3]	Distance [m]	Normalized Echo Energy [J]	Model Equation with Three Layers		Model Equation with Four Layers	
			Calculated Value [J]	Relative Error [%]	Calculated Value [J]	Relative Error [%]
1061.12	1.43	0.1586	0.2099	32.35	0.1896	19.57
1061.12	0.57	0.4309	0.4513	4.74	0.4420	2.59
179.31	1.43	0.1350	0.0628	53.47	0.1168	13.50
179.31	0.57	0.3526	0.3042	13.71	0.3692	4.71
1127.66	1.0	0.2784	0.3036	9.06	0.2606	6.38
112.77	1.0	0.1496	0.1343	10.23	0.1768	18.14
620.21	1.50	0.1897	0.1549	18.33	0.2012	6.08
620.21	0.50	0.4930	0.4356	11.64	0.4947	0.34
620.21	1.0	0.3092	0.2560	17.19	0.2892	6.45
620.21	1.0	0.2679	0.2560	4.43	0.2892	7.96
620.21	1.0	0.2603	0.2560	1.66	0.2892	11.10

Table 8. Result of model equation contrast with 3 layers of leaves.

Density [g/m^3]	Distance [m]	Normalized Echo Energy [J]	Model Equation with Three Layers		Model Equation with Four Layers	
			Calculated Value [J]	Relative Error [%]	Calculated Value [J]	Relative Error [%]
1061.12	1.43	0.1745	0.2192	25.60	0.1608	7.87
1061.12	0.57	0.4718	0.4560	3.37	0.4304	8.78
179.31	1.43	0.0977	0.0721	26.14	0.0882	9.75
179.31	0.57	0.2742	0.3089	12.64	0.3478	26.83
1127.66	1.0	0.3253	0.3101	4.67	0.2504	23.02
112.77	1.0	0.1393	0.1408	1.09	0.1568	12.54
620.21	1.50	0.1648	0.1648	0.01	0.1711	3.85
620.21	0.50	0.4499	0.4401	2.19	0.4846	7.71
620.21	1.0	0.2591	0.2625	1.32	0.2692	3.88
620.21	1.0	0.2361	0.2625	11.18	0.2692	14.00
620.21	1.0	0.2697	0.2625	2.65	0.2692	0.19

3.5. Model Equation Verification

The canopy density model universal analysis with five layers of leaves is shown in Table 9. The model value was calculated based on the canopy density model with four layers (Equation (12)). Table 9 shows that the relative errors of the model value and actual normalized echo energy were small. The maximum relative error was 17.68%, the minimum relative error was 1.46% and the average relative error was 8.33%.

Table 9. Canopy density model universal analysis with 5 layers of leaves.

Density [g/m^3]	Distance [m]	Transmitted Energy [J]	Echo Energy [J]	Normalized Echo Energy [J]	Model Value [J]	Relative Error [%]
319.15	0.8	1.3330	0.3631	0.3032	0.3076	1.46
319.15	1.2	1.3330	0.2002	0.1672	0.1902	13.78
478.72	0.8	1.3215	0.3886	0.3273	0.3402	3.92
478.72	1.2	1.3271	0.2540	0.2130	0.2228	4.59
744.68	0.8	1.3087	0.4354	0.3703	0.3634	1.88
744.68	1.2	1.3267	0.2500	0.2098	0.2460	17.26
904.26	0.8	1.3153	0.3995	0.3381	0.3587	6.10
904.26	1.2	1.3285	0.2448	0.2050	0.2413	17.68

Canopy density model universal analysis with four layers is shown in Table 10, which shows that the relative errors of model value and actual normalized echo energy still were small. The maximum relative error was 25.64%, the minimum relative error was 1.23% and the average relative error was 12.61%.

Table 10. Canopy density model universal analysis with four layers of leaves.

Density [g/m^3]	Distance [m]	Transmitted Energy [J]	Echo Energy [J]	Normalized Echo Energy [J]	Model Value [J]	Relative Error [%]
319.15	0.8	1.2742	0.3378	0.2951	0.3076	4.26
319.15	1.2	1.3317	0.1985	0.1659	0.1902	14.63
478.72	0.8	1.3245	0.4098	0.3444	0.3402	1.23
478.72	1.2	1.3256	0.2112	0.1773	0.2228	25.64
744.68	0.8	1.3112	0.3698	0.3139	0.3634	15.78
744.68	1.2	1.3274	0.2372	0.1989	0.2460	23.68
904.26	0.8	1.3192	0.3754	0.3168	0.3587	13.24
904.26	1.2	1.3211	0.2936	0.2474	0.2413	2.46

Canopy density model universal analysis with 3 layers is shown in Table 11. The results showed that the relative errors of model value and actual normalized echo energy still were small.

Table 11. Canopy density model universal analysis with 3 layers of leaves.

Density [g/m³]	Distance [m]	Transmitted Energy [J]	Echo Energy [J]	Normalized Echo Energy [J]	Model Value [J]	Relative Error [%]
319.15	0.8	1.3285	0.3267	0.2737	0.3076	12.41
319.15	1.2	1.3235	0.2340	0.1967	0.1902	3.31
478.72	0.8	1.3184	0.3805	0.3213	0.3402	5.88
478.72	1.2	1.3272	0.2189	0.1836	0.2228	21.33
744.68	0.8	1.3155	0.3546	0.3000	0.3634	21.13
744.68	1.2	1.3260	0.2416	0.2028	0.2460	21.31
904.26	0.8	1.3077	0.3939	0.3352	0.3587	6.99
904.26	1.2	1.3221	0.2365	0.1991	0.2413	21.17

The maximum relative error was 21.33%, the minimum relative error was 3.31% and the average relative error was 14.19%. Canopy density model universal analysis with two layers is shown in Table 12. The results showed that the relative errors of model value and actual normalized echo energy were acceptable. The maximum relative error was 29.92%, the minimum relative error was 2.32% and the average relative error was 17.98%.

Table 12. Canopy density model universal analysis with two layers of leaves.

Density [g/m³]	Distance [m]	Transmitted Energy [J]	Echo Energy [J]	Normalized Echo Energy [J]	Model Value [J]	Relative Error [%]
319.15	0.8	1.3148	0.3293	0.2788	0.3076	10.35
319.15	1.2	1.3806	0.2306	0.1859	0.1902	2.32
478.72	0.8	1.3110	0.3378	0.2867	0.3402	18.63
478.72	1.2	1.3852	0.2232	0.1793	0.2228	24.21
744.68	0.8	1.3248	0.3541	0.2975	0.3634	22.15
744.68	1.2	1.3363	0.2343	0.1951	0.2460	26.04
904.26	0.8	1.3202	0.3275	0.2761	0.3587	29.92
904.26	1.2	1.3277	0.2611	0.2188	0.2413	10.26

As a consequence of model equation verification, the model equation had a good applicability with different layers, but a higher relative error was experienced with two layers.

4. Conclusions

A method for estimating canopy density of a planar orchard target based on ultrasonic echo energy was studied. Testing indicated that there were strong relationships among the ultrasonic echo energy, detecting distance and canopy density. Two canopy density models with three layers and four layers of leaves were established and compared. The model with four layers was selected as optimal. The verification test results using the optimal model showed that the maximum relative error of model value and actual value with different layers was 17.68%, 25.64%, 21.33% and 29.92%, respectively. The data also suggested the canopy density model with four layers would provide reasonable estimates for different layers. Therefore, it could be used as a control basis in precision sprayers to adjust liquid flow rate and airflow rate.

The relationship between the ultrasonic energy and the power supply voltage showed that the slopes of transmitted energy variation and echo energy variation were different, so normalized echo energy calculated still deviated from the actual echo energy. If supply voltage could be stabilized, the errors can be further reduced without normalization. Future work will focus on field experiments in combination with the real situation of orchard targets.

Acknowledgments: This work is supported by the National Natural Science Foundation of China (31201128), the "Young Faculty Study Abroad Program" of Northwest A&F University Scholarship Fund, Shaanxi science and technology overall planning and innovation project (2014KTCL02-15) and science and technology project of Northwest A&F University (Z222021560).

Author Contributions: Hanzhe Li, Changyuan Zhai, Paul Weckler, and Ning Wang conceived and designed the algorithms and experiments; Hanzhe Li, Shuo Yang and Bo Zhang performed the experiments and analyzed the data; Hanzhe Li and Changyuan Zhai wrote the paper; and Paul Weckler and Ning Wang reviewed and revised the paper.

Conflicts of Interest: The authors declare no conflict of interest.

References

1. Rosell, J.R.; Sanz, R. A review of methods and applications of the geometric characterization of tree crops in agricultural activities. *Comput. Electron. Agric.* **2012**, *81*, 124–141. [CrossRef]
2. Gil, E.; Arno, J.; Llorens, J.; Sanz, R.; Llop, J.; Rosell-Polo, J.R.; Gallart, M.; Escola, A. Advanced technologies for the improvement of spray application techniques in Spanish viticulture: An overview. *Sensors* **2014**, *14*, 691–708. [CrossRef] [PubMed]
3. Miranda-Fuentes, A.; Rodriguez-Lizana, A.; Gil, E.; Aguera-Vega, J.; Gil-Ribes, J.A. Influence of liquid-volume and airflow rates on spray application quality and homogeneity in super-intensive olive tree canopies. *Sci. Total Environ.* **2015**, *537*, 250–259. [CrossRef] [PubMed]
4. Song, Y.; Sun, H.; Li, M.; Zhang, Q. Technology Application of Smart Spray in Agriculture: A Review. *Intell. Autom. Soft Comput.* **2015**, *21*, 319–333. [CrossRef]
5. Meron, M.; Cohen, S.; Melman, G. Tree shape and volume measurement by light interception and aerial photogrammetry. *Trans. ASAE* **2000**, *43*, 475–481. [CrossRef]
6. Sinoquet, H.; Sonohat, G.; Phattaralerphong, J.; Godin, C. Foliage randomness and light interception in 3-D digitized trees: An analysis from multiscale discretization of the canopy. *Plant Cell Environ.* **2005**, *28*, 1158–1170. [CrossRef]
7. Schumann, A.W.; Zaman, Q.U. Software development for real-time ultrasonic mapping of tree canopy size. *Comput. Electron. Agric.* **2005**, *47*, 25–40. [CrossRef]
8. Zaman, Q.U.; Schumann, A.W.; Miller, W.M. Variable rate nitrogen application in Florida citrus based on ultrasonically-sensed tree size. *Appl. Eng. Agric.* **2005**, *21*, 331–335. [CrossRef]
9. Escola, A.; Planas, S.; Rosell, J.R.; Pomar, J.; Camp, F.; Solanelles, F.; Gracia, F.; Llorens, J.; Gil, E. Performance of an ultrasonic ranging sensor in apple tree canopies. *Sensors* **2011**, *11*, 2459–2477. [CrossRef] [PubMed]
10. Llorens, J.; Gil, E.; Llop, J.; Escola, A. Ultrasonic and LIDAR sensors for electronic canopy characterization in vineyards: Advances to improve pesticide application methods. *Sensors* **2011**, *11*, 2177–2194. [CrossRef] [PubMed]
11. Maghsoudi, H.; Minaei, S.; Ghobadian, B.; Masoudi, H. Ultrasonic sensing of pistachio canopy for low-volume precision spraying. *Comput. Electron. Agric.* **2015**, *112*, 149–160. [CrossRef]
12. Jejcic, V.; Godesa, T.; Hocevar, M.; Sirok, B.; Malnersic, A.; Lesnik, M.; Strancar, A.; Stajnko, D. Design and Testing of an Ultrasound System for Targeted Spraying in Orchards. *Stroj. Vestnik J. Mech. Eng.* **2011**, *57*, 587–598. [CrossRef]
13. Stajnko, D.; Berk, P.; Lesnik, M.; Jejcic, V.; Lakota, M.; Strancar, A.; Hocevar, M.; Rakun, J. Programmable ultrasonic sensing system for targeted spraying in orchards. *Sensors* **2012**, *12*, 15500–15519. [CrossRef] [PubMed]
14. Gamarra-Diezma, J.L.; Miranda-Fuentes, A.; Llorens, J.; Cuenca, A.; Blanco-Roldan, G.L.; Rodriguez-Lizana, A. Testing accuracy of long-range ultrasonic sensors for olive tree canopy measurements. *Sensors* **2015**, *15*, 2902–2919. [CrossRef] [PubMed]
15. Rosell, J.R.; Llorens, J.; Sanz, R.; Arnó, J.; Ribes-Dasi, M.; Masip, J.; Escolà, A.; Camp, F.; Solanelles, F.; Gràcia, F.; et al. Obtaining the three-dimensional structure of tree orchards from remote 2D terrestrial LIDAR scanning. *Agric. For. Meteorol.* **2009**, *149*, 1505–1515. [CrossRef]
16. Osterman, A.; Godeša, T.; Hočevar, M.; Širok, B.; Stopar, M. Real-time positioning algorithm for variable-geometry air-assisted orchard sprayer. *Comput. Electron. Agric.* **2013**, *98*, 175–182. [CrossRef]
17. Sanz, R.; Rosell, J.R.; Llorens, J.; Gil, E.; Planas, S. Relationship between tree row LIDAR-volume and leaf area density for fruit orchards and vineyards obtained with a LIDAR 3D Dynamic Measurement System. *Agric. Forest Meteorol.* **2013**, *171–172*, 153–162. [CrossRef]
18. Méndez, V.; Rosell-Polo, J.R.; Sanz, R.; Escolà, A.; Catalán, H. Deciduous tree reconstruction algorithm based on cylinder fitting from mobile terrestrial laser scanned point clouds. *Biosyst. Eng.* **2014**, *124*, 78–88. [CrossRef]

19. Miranda-Fuentes, A.; Llorens, J.; Gamarra-Diezma, J.L.; Gil-Ribes, J.A.; Gil, E. Towards an optimized method of olive tree crown volume measurement. *Sensors* **2015**, *15*, 3671–3687. [CrossRef] [PubMed]

20. Zarate Valdez, J.L.; Whiting, M.L.; Lampinen, B.D.; Metcalf, S.; Ustin, S.L.; Brown, P.H. Prediction of leaf area index in almonds by vegetation indexes. *Comput. Electron. Agric.* **2012**, *85*, 24–32. [CrossRef]

21. Liu, C.; Kang, S.; Li, F.; Li, S.; Du, T. Canopy leaf area index for apple tree using hemispherical photography in arid region. *Sci. Hortic.* **2013**, *164*, 610–615. [CrossRef]

22. Zarate Valdez, J.L.; Metcalf, S.; Stewart, W.; Ustin, S.L.; Lampinen, B. Estimating light interception in tree crops with digital images of canopy shadow. *Precis. Agric.* **2015**, *16*, 425–440. [CrossRef]

23. Gil, E.; Llorens, J.; Llop, J.; Fabregas, X.; Gallart, M. Use of a terrestrial LIDAR sensor for drift detection in vineyard spraying. *Sensors* **2013**, *13*, 516–534. [CrossRef] [PubMed]

24. Andujar, D.; Rueda-Ayala, V.; Moreno, H.; Rosell-Polo, J.R.; Escola, A.; Valero, C.; Gerhards, R.; Fernandez-Quintanilla, C.; Dorado, J.; Griepentrog, H.W. Discriminating crop, weeds and soil surface with a terrestrial LIDAR sensor. *Sensors* **2013**, *13*, 14662–14675. [CrossRef] [PubMed]

25. Walklate, P.J.; Cross, J.V.; Richardson, G.M.; Murray, R.A.; Baker, D.E. Comparison of Different Spray Volume Deposition Models Using LIDAR Measurements of Apple Orchards. *Biosyst. Eng.* **2002**, *82*, 253–267. [CrossRef]

26. Palleja, T.; Landers, A.J. Real time canopy density estimation using ultrasonic envelope signals in the orchard and vineyard. *Comput. Electron. Agric.* **2015**, *115*, 108–117. [CrossRef]

27. Yang, Z.P.; Yan, X.L. *Experimental Optimization Technique*, 1st ed.; Northweat A&F University Press: Yangling, China, 2003; pp. 74–100, 142–166. (In Chinese)

28. Zhai, C.Y.; Wang, X.; Liu, D.Y.; Ma, W.; Mao, Y.J. Nozzle flow model of high pressure variable-rate spraying based on PWM technology. *Adv. Mater. Res.* **2011**, *422*, 208–217. [CrossRef]

11

Characteristics of the Fiber Laser Sensor System Based on Etched-Bragg Grating Sensing Probe for Determination of the Low Nitrate Concentration in Water

Thanh Binh Pham [1,*], Huy Bui [1], Huu Thang Le [2] and Van Hoi Pham [1]

[1] Institute of Materials Science, Vietnam Academy of Science and Technology, 18 Hoang Quoc Viet Rd, Cau giay District, Hanoi 100000, Vietnam; buihuy@ims.vast.ac.vn (H.B.); hoipv@ims.vast.ac.vn (V.H.P.)

[2] Small and Medium Enterprise Development and Support Center 1, Directorate for Standards, Metrology and Quality, 8 Hoang Quoc Viet Rd, Cau giay District, Hanoi 100000, Vietnam; lhthang2001@gmail.com

* Correspondence: binhpt@ims.vast.ac.vn

Academic Editor: Vamsy P. Chodavarapu

Abstract: The necessity of environmental protection has stimulated the development of many kinds of methods allowing the determination of different pollutants in the natural environment, including methods for determining nitrate in source water. In this paper, the characteristics of an etched fiber Bragg grating (e-FBG) sensing probe—which integrated in fiber laser structure—are studied by numerical simulation and experiment. The proposed sensor is demonstrated for determination of the low nitrate concentration in a water environment. Experimental results show that this sensor could determine nitrate in water samples at a low concentration range of 0–80 ppm with good repeatability, rapid response, and average sensitivity of 3.5×10^{-3} nm/ppm with the detection limit of 3 ppm. The e-FBG sensing probe integrated in fiber laser demonstrates many advantages, such as a high resolution for wavelength shift identification, high optical signal-to-noise ratio (OSNR of 40 dB), narrow bandwidth of 0.02 nm that enhanced accuracy and precision of wavelength peak measurement, and capability for optical remote sensing. The obtained results suggested that the proposed e-FBG sensor has a large potential for the determination of low nitrate concentrations in water in outdoor field work.

Keywords: nitrate; etched-Fiber Bragg Grating; fiber laser; optical sensor

1. Introduction

Nitrate (NO_3^-) is considered to be one of the important substances to measure in water, because of its potential environmental and human health implications. The main anthropogenic sources of nitrates in the environment are municipal and industrial waste, artificial fertilizers, septic systems, animal feedlots, and food processing waste such as food preservatives, especially to cure meats. Nitrates can cause eutrophication of surface waters. Nitrates are not directly toxic to human health, but their possible reduction to nitrites and a next reaction of nitrites with secondary or tertiary amines present in the body can result in the formation of carcinogenic N-nitrosamines. Moreover, the nitrite oxidizes iron in the hemoglobin of the red blood cells to form methemoglobin, which lacks the oxygen-carrying ability of hemoglobin. This creates a condition known as methmoglobinemia, wherein blood iron in hemoglobin (Fe^{+2}) is reduced to its oxidized form Fe^{+3}. Different methods, including ultraviolet-visible spectroscopy (UV-VIS), electrophoresis, electrochemical detection, chromatography, mass spectroscopy, and potentiometry coupled with sequence injection analysis are adopted in finding the concentration of nitrate [1–5]. These methods are expensive and/or inconvenient for field work.

Optical fiber sensors offer very attractive solutions over conventional technologies due to some unique characteristics such as multiplexing capabilities, high sensitivity, fast response, and immunity to electromagnetic interference. The small physical size of optical fiber allows the development of very small and flexible fiber sensors, and enables the remote in-situ sensing of species in difficult or hazardous environments [6,7]. Optical fiber sensors based on colorimetric technique [8,9] and evanescent wave absorption [10] for in-situ nitrate detection in water have been proposed. These methods can detect the nitrate concentration in the range from ppb to ppm, but the measurement response time is some tens of minutes.

Fiber Bragg gratings (FBGs) have been demonstrated as optical sensors for various applications [11–13], especially for chemical and biochemical sensing [14–17]. In various chemical and biochemical applications, refractive index sensing is important, since several substances can be detected by the measurement of the refractive indices. The FBG sensing operation principle relies on the dependence of the Bragg resonance wavelength on the grating pitch and effective refractive index. Normal FBGs are intrinsically insensitive to the ambient refractive index. However, if the fiber cladding diameter is reduced along the grating region, the effective refractive index is significantly affected by the external refractive index. Among different kinds of FBG, the Tilt-FBG and the Long Period Fiber Grating (LPFG) have shown a large potential for chemical and bio-sensing applications with high sensitivity and low cost. However, their multiple resonance peaks limit their multiplexing capabilities. Moreover, the measurement accuracy of LPFG is limited due to its broad line-width at full-width at half maximum (FWHM) [18].

The aim of our study is investigation of characteristics of etched-fiber Bragg grating (e-FBG) sensing probe integrated in fiber laser structure as lasing wavelength selected element for determination of low nitrate concentrations in water. The e-FBG sensing probe is designed and fabricated by wet chemical etch-erosion and put into a fiber cavity laser using Er^{+3}-doped silica fiber. In the interaction between the evanescent wave of the fundamental core mode and the surrounding medium, a small variation of the refractive index of the medium rounding the e-FBG will induce a significant change in the Bragg wavelength according to the Bragg condition, and the response time of measurement is less than a milli-second. The e-FBG sensing probe can be used to detect the nitrate in water samples at a low concentration range of 0–80 ppm. The line-width spectrum of lasing emission from a fiber laser is much narrower than that of reflected FBG spectra, thus enhancing the detection accuracy and capability for remote sensing.

2. Experiment

The FBGs used in our experiments were fabricated with the standard single-mode photosensitive fiber (Model: PS 1250/1500, Fiber-core, Southampton, UK) by the Talbot interferometric technique with exposure to the KrF Excimer Laser source of 248 nm wavelength (ASX-750 Excimer Laser, MPB Technology Inc., Montreal, QC, Canada). The Bragg resonant wavelength was 1550 nm with 12 mm long FBG, and the reflection line width at FWHM was 0.2 nm [12]. The e-FBG was fabricated by wet chemical etching the FBG region in hydrofluoric acid (HF) solution to increase the interaction of the propagating optical field in the fiber core with the surrounding medium. The etching technique was performed in two steps: the first step used a 30% HF solution to speed up the etching fiber cladding layer process. After an etching process for 75 min, the fiber diameter was below 15 μm. The etching solution was then replaced with a 15% HF solution for the second step. The purpose of the second step is to slow down the etching process and smoothing the etched fiber surface (for 20 min). A schematic diagram of the experimental setup for fiber-mount design and for measuring the reflected Bragg wavelength shift of e-FBG in the etching process is shown in Figure 1. A broadband light source from amplified spontaneous emission (ASE) of Erbium-doped fiber amplifier, an optical circulator, and an optical spectrum analyzer (OSA: Advantest Q8384 with a resolution of 0.01 nm) are used for monitoring the wavelength shift. The input ASE signal passes through a circulator before being reflected by the FBG and directed to an OSA. There is a need to have a protective mount for the fragile e-FBG with small diameter of micrometers, so the design of the FBG-mount is also shown in Figure 1. Before the corrosion process, the FBG is mounted and fixed at two ends with epoxy on Teflon V-groove mount, which is non-reactant to HF solution and decreased mechanical vibration for e-FBG.

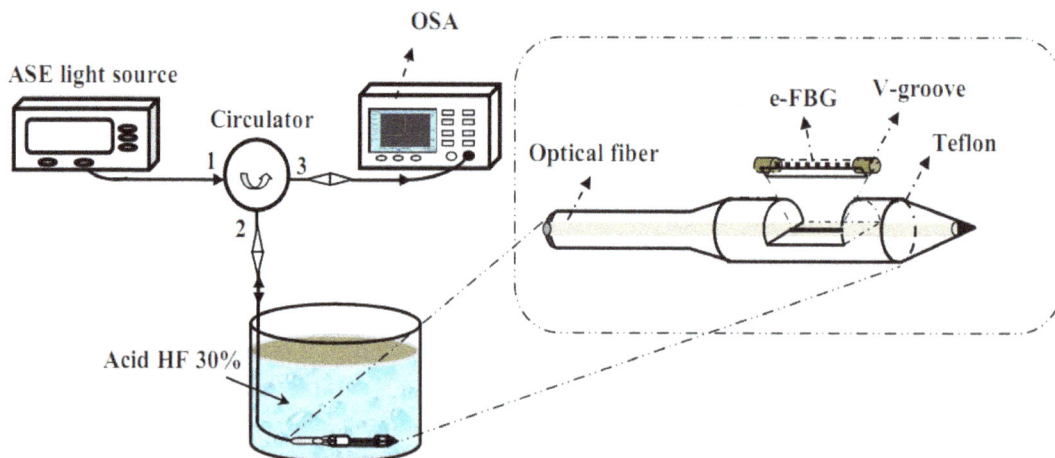

Figure 1. Experimental setup and mount design for making etched-fiber Bragg grating (e-FBG). ASE: amplified spontaneous emission; OSA: optical spectrum analyzer.

The fiber laser using e-FBG as a reflector (which operates as an optical sensor) was proposed for the determination of low nitrate concentration in water. The optical gain medium was an erbium-doped silica fiber (Model: EDF-HCO-4000, Core-active, Quebec, QC, Canada) with length of 3 m, and the optical pump was a 980 nm-laser diode with output optical power up to 170 mW in single-mode emission (SDLO-2564-170). The pumped light was through a 980/1550 nm wavelength division multiplexer (WDM) to Er-doped silica fiber for excitation of the erbium ions. The other end of the erbium doped fiber was connected to an e-FBG sensing element as a mirror of fiber laser system. A fiber-optic circulator was used to couple the light into the cavity and through an optical coupler 10/90 in order to extract 10% of the light from the cavity to the acquisition system, and 90% of the light comes back to the cavity. This fiber laser configuration will give narrow line-width of lasing emission and high optical signal-to-noise ratio. The spectral characteristics of the lasing emission were analyzed by the OSA. The e-FBG sensing probe was immersed in solvent, and the reflection spectrum from the e-FBG sensing element was changed by different solutions of nitrate concentration varying from 0 to 80 ppm. The potassium nitrate stock solution was prepared by dissolving 0.4075 g of anhydrous KNO_3 (Merck) in purified water (250 mL) to obtain a sample of 1000 ppm nitrate in water. One hundred milliliters of this solution was diluted to one litre of water to produce the stock 100 ppm solution. By this dilution method, we obtained nitrate solutions with low concentration from 0 to 80 ppm for use in our experiment. All the measurements were done at constant temperature of 25 °C.

3. Results and Discussion

During the etching process, the shift of Bragg wavelength was monitored at regular intervals by recording the reflected spectrum peak from the FBG in different moments (shown in Figure 2a). As time progressed, the Bragg wavelength shifted to the shorter wavelength range (blue shift) due to the reduction of the cladding diameter, since the fundamental mode is less confined in the fiber core region, leading to a higher evanescent field, and thus to a more efficient interaction with the surrounding medium. For the mechanical strength and the durability of e-FBG for practical use, we have limited the etched fiber diameter to 6–8 μm.

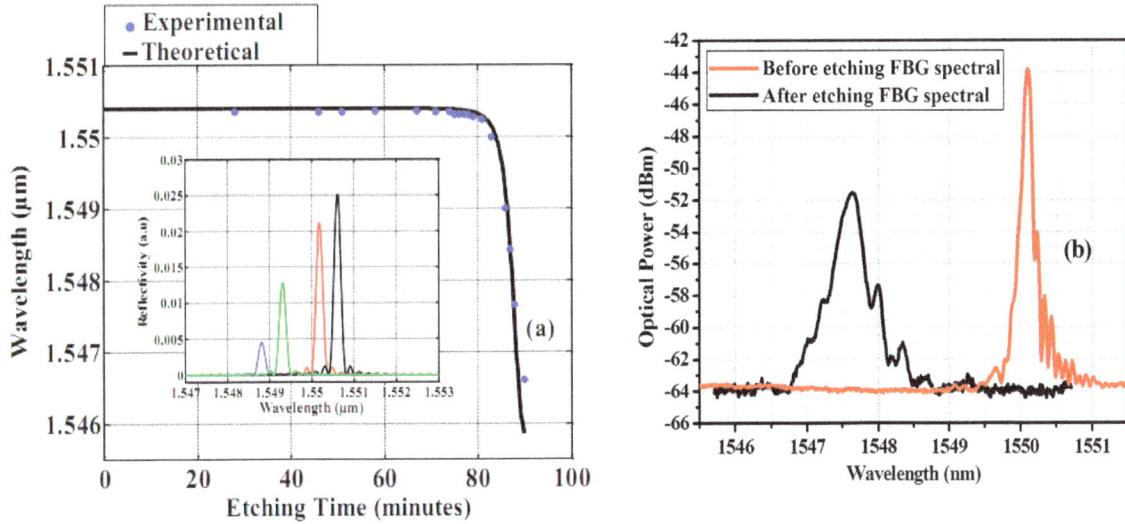

Figure 2. (a) Wavelength shift of FBG versus the etching time; and (b) reflected spectra of FBG before and after etching process.

In common FBG, the effective refractive index of the fundamental mode does not practically depend on the refractive index of the medium surrounding the fiber. However, if the cladding diameter is reduced, this effective refractive index shows a nonlinear dependence on the external refractive index and leads to a shift in the reflected wavelength. The effective refractive index in an e-FBG is evaluated by numerically resolving the dispersion equation of a double-clad fiber model. According to the theory of the fiber Bragg grating, the Bragg wavelength λ_B is as follows:

$$\lambda_B = n_{eff}\Lambda \tag{1}$$

where n_{eff} and Λ are the effective refractive index and periodic spacing of FBG, respectively. According to the optical fiber Coupled-Mode Theory, the relationships between the effective refractive index of the e-FBG, fiber diameter, and the normalized frequency V_{ext} of the etched single-mode fiber are follows [16]:

$$n_{eff}^2 = n_{co}^2 - \left(\frac{U}{V_{ext}}\right)^2 \left(n_{co}^2 - n_{cl}^2\right) U = a\sqrt{k_0^2 n_{co}^2 - \beta^2}, \ V_{ext} = \frac{\pi d}{\lambda}\sqrt{n_{co}^2 - n_{ext}^2} \tag{2}$$

where a and d are the fiber core radius and the e-FBG diameter, respectively; n_{co}, n_{cl}, and n_{ext} are the refractive indexes of the fiber core, cladding, and external medium, respectively; $k_0 = 2\pi/\lambda$ is vacuum wave number; and β is a propagation constant. The reflection wavelength shift of e-FBG is only related to the effective refractive index. The simultaneous differential equation from Equations (1) and (2) is as follows:

$$\frac{\Delta\lambda_B}{\lambda_B} = \frac{\Delta n_{eff}}{n_{eff}} = \frac{U^2\left(n_{co}^2 - n_{cl}^2\right)}{2V_{ext}^3\left[n_{co}^2 - \left(\frac{U}{V_{ext}}\right)^2\left(n_{co}^2 - n_{ext}^2\right)\right]} \tag{3}$$

In our calculation, the fiber parameters were chosen as: $n_{co} = 1.45$, $n_{cl} = 1.4464$, $d = 4.5$–125 µm, $\Lambda = 0.53472$ µm. The response simulation spectra of e-FBG (in the inset of Figure 2a), numerical predictions, and experimental values of the reflection wavelength shift of e-FBG during the etching process are shown in Figure 2a. It is observed that the calculated result fits very well to the experimental one. The reflected spectra of the original FBG and e-FBG obtained experimentally are shown in Figure 2b. When the diameter of fiber etched was 6.55 µm, the wavelength shift was 2.56 nm—this value is very close to the one provided by the numerical analysis by Equation (3) in the case of quasi-full etching. The experimentally-obtained spectral line-width increased from 0.2 nm to 0.7 nm between before and after etching grating. This ensures that the thickness of the FBG has reached core level, as shown in Scanning Electron Microscopy (SEM) images for produced e-FBGs with diameters of 33.9, 10, and 6.55 μm. Figure 3a shows an SEM image of e-FBG as it performed corrosion of the first step. It is observed that the e-FBG surface has large roughness, enabling it to induce adjacent modes and decrease intensity of the FBG reflection spectrum due to light scattering. Figure 3b,c shows SEM images of the final two produced e-FBGs after corrosion by the second step to be completed with a smoothed surface of the e-FBG. Figure 3d shows an SEM image of the e-FBG surface with roughness of about 7.94 nm corresponding to $\lambda/194$ (λ: 1550 nm wavelength of light transmitted in sensor system). This fine smoothness of the fiber surface will decrease evanescent wave scattering at the fiber surface, and we can obtain the high intensity of lasing emission.

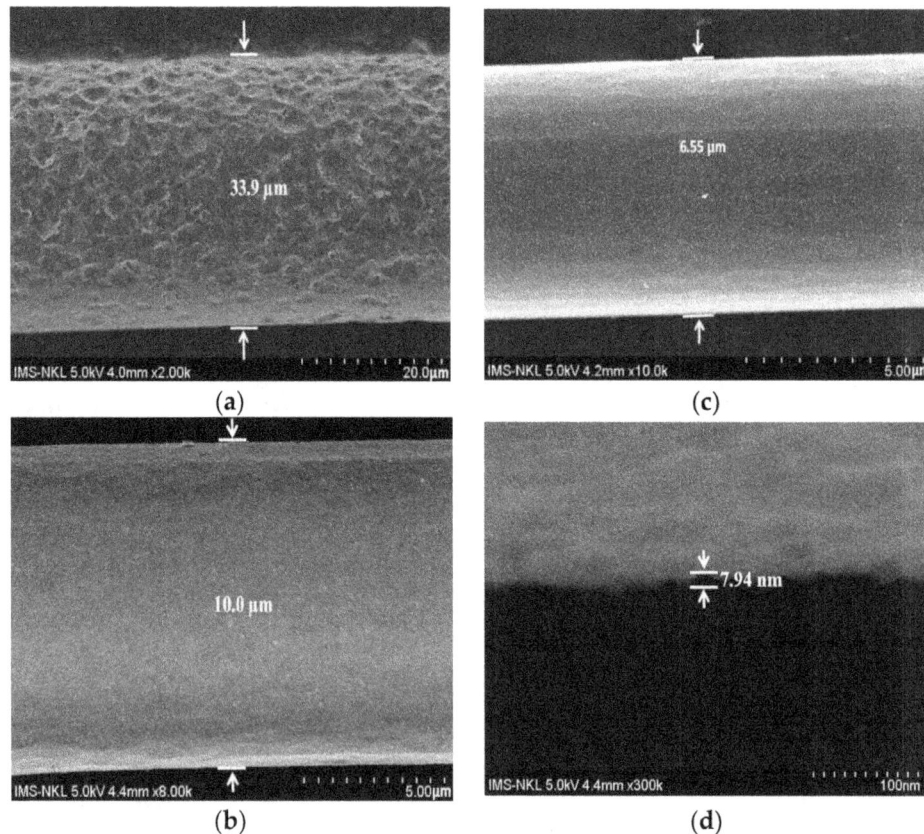

Figure 3. SEM images of e-FBGs with diameters of: (**a**) 33.9 μm; (**b**) 10 μm; (**c**) 6.55 μm; and (**d**) of e-FBG surface with roughness of 7.94 nm.

The spectra of reflection light from e-FBG sensing element and of lasing emission from fiber laser used the same e-FBG element as reflector experimentally obtained for different solutions of nitrate concentration in water measured on the OSA are shown in Figure 4a,b, respectively. The lasing emission from e-FBG integrated erbium-doped fiber laser has an optical signal-to-noise ratio (OSNR) higher than 40 dB and spectral line-width of 0.02 nm at −3 dB, whereas OSNR and spectral line-width of reflection light from e-FBG sensing element are about 3 dB and 0.55 nm, respectively. These narrow line-width and high OSNR characteristics of fiber laser emission will give high accuracy and high sensitivity for wavelength peak measurement method. In addition, the high optical intensity from the laser can be transmitted through the fiber for a long distance, which is required for remote sensing systems.

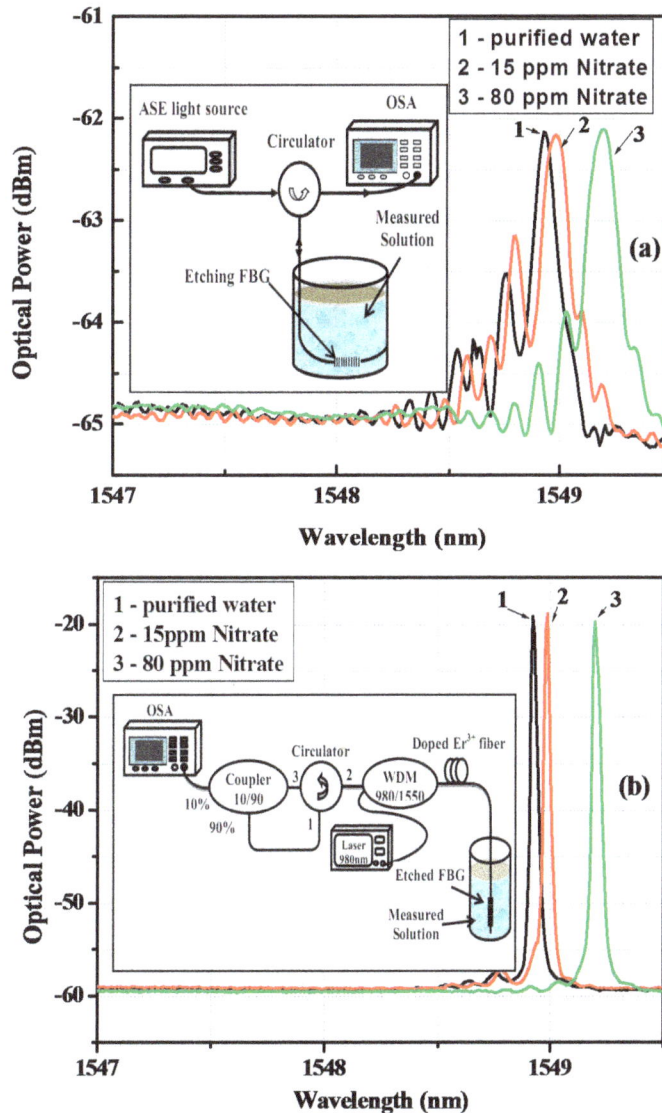

Figure 4. The spectral responses of (**a**) e-FBG from reflected configuration; and (**b**) of e-FBG integrated fiber laser configuration. The −3 dB-bandwidths of spectra decreased from 0.55 nm to 0.02 nm, and optical signal-to-noise ratio (OSNR) increased from 3 dB to 40 dB.

From the characteristic spectral response of the e-FBG sensing element, the signal-to-noise ratio (SNR) of the e-FBG sensor can be assumed to be inversely proportional to e-FBG spectral line-width, defined as [19]:

$$SNR(n_s) = \left[\frac{\Delta\lambda_{res}}{\Delta\lambda_{SW}} \right]_{n_s} \tag{4}$$

where $\Delta\lambda_{res}$ is the resonance wavelength shift induced by the e-FBG sensing element, and $\Delta\lambda_{SW}$ can be calculated as the full width at half maximum (FWHM) of the spectral response of the e-FBG sensing element. The *SNR* of the e-FBG sensor depends on how accurately and precisely the e-FBG sensor can detect the resonant wavelength shift of the sensing element. Therefore, the characteristic spectral response of the fiber laser-based sensor system has shown much narrower spectral line-width in comparison with e-FBG-based sensors, such that the fiber laser using e-FBG sensing element sensor system has strongly enhanced detection accuracy when the wavelength peak measurement is used.

In order to demonstrate in detail the determination of nitrate concentration in water using an e-FBG integrated fiber laser sensor, the proposed sensor has been performed for the detection of nitrate in the concentration range of 0–80 ppm with step changes of 5–10 ppm. The designed e-FBG sensing probe was cleaned off the residual test sample solution by the de-ionized water before replacing the different test sample to avoid contamination and to ensure the accuracy of the measurement results. All the measurements were done at constant temperature 25 °C, and a standard deviation of the wavelength shift was obtained from the average value of five experimental data runs. The experimental results are depicted in Figure 5.

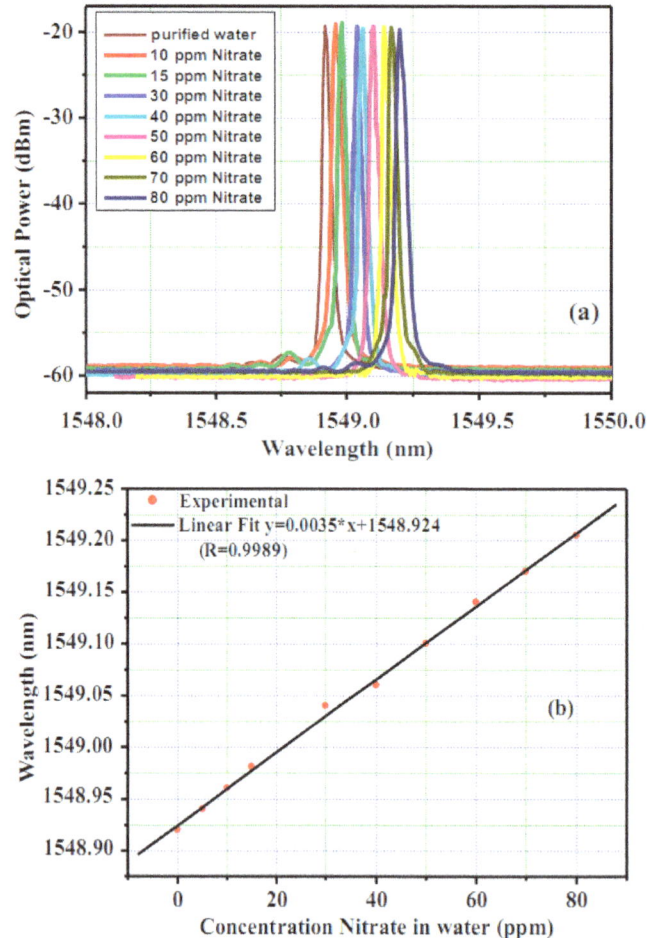

Figure 5. (**a**) Spectral response for different concentrations of nitrate solutions from fiber laser sensor; and (**b**) Bragg wavelength shift as a linear function of nitrate concentrations in water.

The lasing spectral response of the e-FBG sensing element corresponding to different nitrate concentrations in water is shown in Figure 5a. The spacing between the peaks of the lasing spectra can easily distinguish with OSA resolution of 0.01 nm. It is observed that the increase of nitrate concentration of the water environment caused the lasing wavelength to shift to longer wavelength range (red shift). From Figure 5b, the wavelength shift of the lasing emission was 0.3 nm when the nitrate concentration in water changed from 0 ppm to 80 ppm. The calculated slope of linearly fitted data can be used as the effective sensitivity of the sensor; it can be deduced that the sensitivity of this sensor is achieved to 3.5×10^{-3} nm/ppm. Therefore, assuming that the detectable spectral resolution of OSA is 0.01 nm, the optical sensor can measure nitrate concentrations in water with a detection limit of 3 ppm.

Table 1 shows the results of detection limit and response time of nitrate measurement using different sensing techniques, such as electrochemical sensors with different electrodes, colorimetric, evanescent wave absorption fiber sensors, and e-FBG integrated to fiber laser.

Table 1. Comparison of nitrate-in-water detection limits and response times using different sensors.

Type of Sensors	Limit of Detection (ppm)	Measured Response Time	References
Electrochemical sensor	1.35	Tens of minutes	Liang et al. [20]
Graphene oxide sensor	0.05	30 min	Ren et al. [21]
Disposable Electrochemical sensor	8.6	Not reported	Bui et al. [1]
Evanescent wave absorption Fiber sensor	0.06	Not reported	Kumar et al. [10]
Colorimetric sensor	4.0	30 min	Kunduru et al. [9]
Lopine sensitive layer Fiber sensor	1.0	40 milliseconds	Camas-Anzueto et al. [22]
e-FBG in fiber laser sensor	3.0	Milliseconds	This work

It is remarkable that the proposed e-FBG sensor is a typical physical sensor without functionalized materials on the surface, which has the fast response time of in-situ refractive index measurement of aqueous environment, and it is suitable for reversible use by easily cleaning the glass surface. In addition, the silica glass-based sensor has good repeatability and reproducibility in the aqueous medium from its non-corrosion and stable properties in this environment. The limit of detection of current sensor (3 ppm) is far below the maximum nitrate level allowed in drinking water by the United States Environmental Protection Agency (EPA) [23], giving it large potential for application in monitoring drinking water.

4. Conclusions

We successfully designed and prepared the fiber laser using e-FBG sensing element as reflector and used this device to detect nitrate concentration in water. This sensor provides a new approach for real-time in-situ measurement with high accuracy. Moreover, this proposed sensor system shows many advantages including a high resolution for wavelength shift identification, high OSNR and narrow band-width that enhances accuracy and precision of wavelength peak measurement and improves capability for remote sensing application. To confirm the feasibility of the determination of nitrate concentration in water, experimental results demonstrate the usefulness of the e-FBG sensor in measuring nitrate compounds in water with good sensitivity in the low concentration range of 0–80 ppm with detection limit of 3 ppm. This established that the proposed sensor can be used for monitoring the water quality in field work. This sensor has a large potential for applications in agriculture, industrial fluids, and the food industry.

Acknowledgments: This work is financially supported by the National Foundation for Science and Technology Development of Vietnam (NAFOSTED) under grant No. 103.03-2015.23 and the project of State Key Laboratory for Electronic Materials and Devices, Institute of Materials Science, Vietnam Academy of Science and Technology under grant No. CSTD.01.16.

Author Contributions: Thanh Binh Pham and Huu Thang Le conceived, designed and performed the experiments, interpreted the results and drafted the manuscript; Huy Bui and Van Hoi Pham supervised the experimental data analysis; Thanh Binh Pham supervised the interpretation of results and wrote the manuscript.

Conflicts of Interest: The authors declare no conflict of interest.

References

1. Bui, M.-P.N.; Brockgreitens, J.; Ahmed, S.; Abbas, A. Dual detection of nitrate and mercury in water using disposable electrochemical sensors. *Biosens. Bioelectron.* **2016**, *85*, 280–286. [CrossRef] [PubMed]

2. De Perre, C.; McCord, B. Trace analysis of urea nitrate by liquid chromatography-UV/fluorescence. *Forensic Sci. Int.* **2011**, *211*, 76–82. [CrossRef] [PubMed]

3. Tamiri, T. Characterization of the improvised explosive urea nitrate using electrospray ionization and atmospheric pressure chemical ionization. *Rapid Commun. Mass Spectrom.* **2005**, *19*, 2094–2098. [CrossRef] [PubMed]

4. Pavel, M.; Lukas, C.; Zbynek, V.; Ivan, K.; Josef, K. Photo-induced flow-injection determination of nitrate in water. *Int. J. Environ. Anal. Chem.* **2014**, *94*, 1038–1049.

5. Gajaraj, S.; Fan, C.; Lin, M.; Hu, Z. Quantitative detection of nitrate in water and wastewater by surface-enhanced Raman spectroscopy. *Environ. Monit. Assess.* **2013**, *185*, 5673–5681. [CrossRef] [PubMed]

6. Moo, Y.C.; Matjafri, H.S.; Tan, C.H.L. New development of optical fiber sensor for determination of nitrate and nitrite in water. *Optik* **2015**, *9*, 1–36.

7. Cennamo, N.; Massarotti, D.; Conte, L.; Zeni, L. Low cost sensors based on spr in a plastic optical fiber for biosensor implementation. *Sensors* **2011**, *11*, 11752–11760. [PubMed]

8. Singh, A.N.; Singh, K.R. Fiber optic interrogator based on colorimetry technique for in-situ nitrate detection in groundwater. *Opt. Appl.* **2008**, *4*, 727–735.

9. Kunduru, K.R.; Basu, A.; Abtew, E.; Tsach, T.; Domb, A.J. Polymeric sensors containg P-dimethylamininnamaldehyde: Colorimetric detection of urea nitrate. *Sens. Actuators B Chem.* **2017**, *238*, 387–391.

10. Kumar, P.S.; Vallabhan, C.P.G.; Nampoori, V.P.N.; Pillai, V.N.S.; Radhakrishnan, P.A. A fiber optic evanescent wave sensor used for the detection of trace nitrites in water. *J. Opt. A Pure Appl. Opt.* **2002**, *4*, 247–250.

11. Shivananju, B.N.; Yamdagni, S.; Fazuldeen, R.; Sarin Kumar, A.K.; Hegde, G.M.; Varma, M.M.; Asokan, S. CO_2 sensing at room temperature using carbon nanotubes coated core fiber Bragg grating. *Rev. Sci. Instrum.* **2013**, *84*, 065002. [CrossRef] [PubMed]

12. Pham, T.B.; Nguyen, T.V.; Tran, A.V.; Bui, H.; Pham, V.H. Simulation and fabrication of uniform fiber Bragg gratings. In Proceedings of the Conference on Solid State Physics & Materials Science, Danang, Vietnam, 8–10 November 2009; pp. 1205–1208.

13. Pham, T.B.; Seat, H.C.; Bernal, O.; Suleiman, M. A Novel FBG interrogation method for potential structural health monitoring applications. *IEEE Sens. J.* **2012**, *11*, 1341–1344.

14. Liang, W.; Huang, Y.; Xu, Y.; Lee, R.K.; Yariv, A. Highly sensitive fiber Bragg grating refractive index sensors. *Appl. Phys. Lett.* **2005**, *86*, 151122. [CrossRef]

15. Sun, A.; Wu, Z. A Hybrid LPG/CFBG for highly Sensitive Refractive Index Measurements. *Sensors* **2012**, *12*, 7318–7325. [CrossRef] [PubMed]

16. Chen, N.; Yun, B.; Wang, Y.; Cui, Y. Theoretical and experimental study on etched fiber Bragg grating cladding mode resonances for ambient refractive index sensing. *J. Opt. Soc. Am. B* **2007**, *24*, 439–445. [CrossRef]

17. Corotti, R.P.; Thaler, J.; Kalnowski, H.J.; Muller, M.; Fabris, J.L.; Kamikawachi, R.C. Etched FBG written in multimode fibers: Sensing characteristics and application in the liquid fuels sector. *J. Microw. Opt. Electron. Appl.* **2015**, *14*, 51–59. [CrossRef]

18. Sang, X.; Yu, C.; Mayteevarunyoo, T.; Wang, K.; Zhang, Q.; Chu, L.P. Temperature-insensitive chemical sensor based on a fiber Bragg grating. *Sens. Actuators B Chem.* **2007**, *120*, 754–757. [CrossRef]

19. Kanso, M.; Cuenot, S.; Louarn, G. Sensitivity of optical fiber sensors based on surface plasmon resonance: Modeling and experiments. *Plasmonics* **2008**, *3*, 49–57. [CrossRef]

20. Liang, J.; Zheng, Y.; Liu, Z. Nanowire-based Cu electrode as electrochemical sensor for detection of nitrate in water. *Sens. Actuators B Chem.* **2016**, *232*, 336–344. [CrossRef]

21. Ren, W.; Mura, S.; Irudayaraj, J.M.K. Modified graphene oxide sensors for ultra-sensitive detection of nitrate ions in water. *Talanta* **2015**, *143*, 234–239. [CrossRef] [PubMed]

22. Camas-Anzueto, J.L.; Aguilar-Castillejos, A.E.; Castanon-Gonzalez, J.H.; Lujpan-Hidalgo, M.C.; Hernandez de Leon, H.R.; Mota Grajales, R. Finer sensor based on Lopine sensitive layer for nitrate detection in drinking water. *Opt. Lasers Eng.* **2014**, *60*, 38–43. [CrossRef]

23. United States Environmental Protection Agency. Available online: http://www.epa.gov/your-drinking-water/table-regulated-drinking-water-contaminants (accessed on 6 October 2016).

12

Experimental Investigations of a Precision Sensor for an Automatic Weapons Stabilizer System

Igor Korobiichuk

Industrial Research Institute for Automation and Measurements PIAP, Jerozolimskie 202, 02-486 Warsaw, Poland; kiv_Igor@list.ru

Academic Editor: Seung-Bok Choi

Abstract: This paper presents the results of experimental investigations of a precision sensor for an automatic weapons stabilizer system. It also describes the experimental equipment used and the structure of the developed sensor. A weapons stabilizer is designed for automatic guidance of an armament unit in the horizontal and vertical planes when firing at ground and air targets that are quickly maneuvering, and at lower speeds when firing anti-tank missiles, as well as the bypass of construction elements by the armament unit, and the automatic tracking of moving targets when interacting with a fire control system. The results of experimental investigations have shown that the error of the precision sensor developed on the basis of a piezoelectric element is 6×10^{-10} m/s^2 under quasi-static conditions, and $\sim 10^{-5}$ m/s^2 for mobile use. This paper defines metrological and calibration properties of the developed sensor.

Keywords: accelerometer; weapons stabilizer system; mobile objects

1. Introduction

Weapons stabilizer is an automatic control system that provides combat vehicles with weapon targeting and stabilization in the target fire direction during oscillation of a moving armoured vehicle (AV) [1]. To increase the effectiveness of fire during movement in all modern tanks and other combat vehicles, the main armament is stabilized by a special device referred to as a weapons stabilizer.

AV oscillations are random and never dampen in motion. The amplitude of angular oscillations and oscillation frequencies are quite high [2]. The accuracy of shooting is mostly influenced by AV oscillations in the longitudinal plane, changing the angle of gun elevation, and angular oscillations in the horizontal plane, changing the traverse [3]. This leads to a significant displacement of the aiming mark towards the target and does not allow the gunner to keep it on target even with the most advanced power drives. Lateral angular oscillations causing gun trunnion tilt have less impact, but increase with an extension of the firing range.

First of all, these causes increase vectoring errors by 10–30 times due to firing on the move, as compared with firing from a halt. Dispersion of projectiles increases by about 10–12 times [4]. Precision of fire on the move also decreases due to continuous changes in the target range. AV oscillations in motion significantly worsen firing conditions [5].

To increase the effectiveness of fire on the move, all modern combat vehicles are equipped with a special automatic device referred to as a weapons stabilizer system (WSS) [6]. Providing the accuracy improvement of measuring the mobile object's acceleration, such as those used in a weapons stabilization system (WSS), is a great challenge [6]. WSS effectiveness is mostly dependent on the accuracy and performance of the sensitive stabilizer elements and accelerometers. The modern stabilization systems, using spring, string, quartz, magnetic, and gyroscopic accelerometers cannot provide the required speed of response and accuracy [7,8].

Therefore, the urgent scientific and technical challenge is to improve the accuracy and speed of response when measuring the acceleration values by experimental investigations of a piezoelectric sensor (PS) for the automatic weapons stabilization system [9–11]. The recent literature concerning this subject [12–20] contain no information on the analysis of experimental investigations of precision sensors of an automated weapons stabilizer system.

The task given by the Ministry of Education and Science of Ukraine grant No. 0115U000210 was to improve the accuracy and error characteristics of the precision accelerometer sensor for WSS. The WSS under consideration was the SVU-500 (Figure 1), produced by the G. Petrovsky Kyiv Automatics Plant (Kyiv, Ukraine).

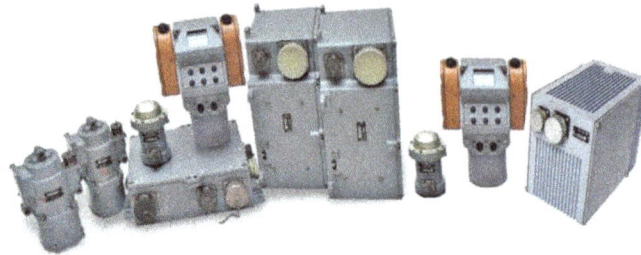

Figure 1. SVU-500 Weapon stabilising system (from left to right: SVU-500, SVU-500-01, SVU-500-3TS, SVU-500-4TS, SVU-500-4TS-01, SVU-500-10P, SVU-500-3TS-01).

The WSS of this type are used in the fire control systems of the "Shkval", on BTR 3E, on infantry fighting vehicles BMP IM (SVU-500-3TS), and in "Shturm" and "Parus" systems for APCs BTR1 3E and BMP 4 (SVU-500-4TS) [21].

The modernized version of SVU-500 is intended to be mounted on the AVs, such as the BMP-2 (BMP-2, BTR, BMP, BMD), to stabilize targeting and improve fire control during motion on land and on water. The existing WSS was using obsolete vibrating string accelerometers; therefore, a new approach was chosen, based on piezoelectric elements. The new sensor had to be compatible with existing WSS systems, which gave initial assumptions for the development process.

2. Measurement Stand

Experimental device was created for experimental investigations of the sensing element (SE). Its schematic diagram is shown in Figure 2, and its photo is presented in Figure 3. The test stand includes the following devices: a mechanical vibration generator GMK-1 (vibration table) with two on-board induction transducers converting electrical signals into mechanical displacement; an SE placed directly on the vibration table; an input/output module; an SE output signal amplifier unit; a personal computer (PC); an AC generator and voltmeters for logging voltage levels of the generator and of the induction transducers.

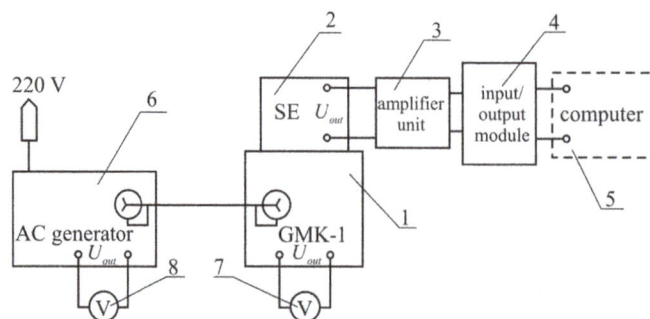

Figure 2. Schematic diagram of the test stand for the experimental investigation of piezoelectric SE: 1—GMK-1; 2—SE; 3—SE output signal amplifier unit; 4—input/output module; 5—PC; 6—AC generator; 7—voltmeter for logging generator voltage; and 8—voltmeter for logging induction transducer voltage.

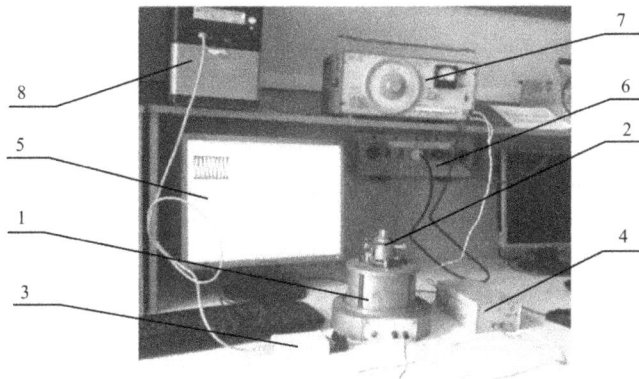

Figure 3. Appearance of the device for SE experimental investigations: 1—GMK-1; 2—SE; 3—SE output signal amplifier unit; 4—input/output module; 5—PC; 6—AC generator; 7—voltmeter for logging generator voltage; and 8—voltmeter for logging induction transducer voltage.

The basis of the experimental test stand is the vibration table GMK-1 (Figure 4). It is a mechanical vibration generator, structurally designed as two magnetic cores (8, 10) (see Figure 5). These magnetic cores are tightly fastened and form a single structure of solenoid type. The rod (7) can move in the middle of a solenoid created by two magnetic cores (8, 10).

Figure 4. Mechanical vibration generator GMK-1.

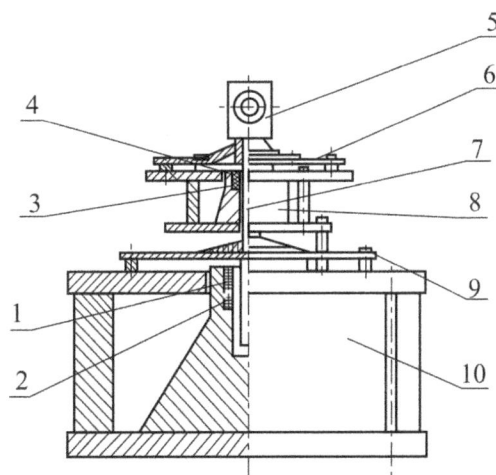

Figure 5. Mechanical vibration generator: 1, 3—generator windings; 2, 4—control winding; 5—working table with piezoelectric sensing element; 6, 9—flexible membranes; 7—rod; and 8, 10—magnetic cores.

Driving force for moving the rod (7) is generated by induction transducers. Windings (1, 3) perform the excitation function and (2, 4) perform the control function.

Induction transducers shown in the diagram are designed to convert input electrical excitation signals into the output mechanical rod movement.

The core (7) with windings (1, 2) and (3, 4) is restrained by flexible supports in the form of special membranes (6, 9) which combine sufficient stiffness with great linear power curve values.

Fixation of the rod by membranes on both sides allows minimization of movement in directions that do not coincide with the longitudinal axis. This will provide the rod with only one degree of freedom in the required direction of the vertical axis. Therefore, if a current is passed through the generator winding, the power generated by the winding will result in a vertical movement of the rod.

Thus, the vibration table GMK-1, creating oscillatory rod acceleration, affects the working table (5), where the SE is found.

If current, varying in time sinusoidally, is passed through the generator winding (1) of the vibration table, the power generated by the winding will result in rod movement h, which is also sinusoidal:

$$h = H \sin \omega t, \tag{1}$$

where H, $\omega = 2\pi f$ are the amplitude and frequency of the oscillation movement of the rod, respectively.

Oscillating rod movement rate h is connected with values of rod oscillation velocity h_c and acceleration h_z affecting the SE by the following relations:

$$h_z = -H\omega \sin \omega = H_z \sin \omega t, \tag{2}$$

$$h_c = H\omega \cos \omega = H_c \cos \omega t. \tag{3}$$

Only amplitude values of the rod oscillation velocity and acceleration are measured during experimental investigations. As a result we obtain:

$$H_c = Hn\omega, \tag{4}$$

$$H_z = -H_c\omega = Hn\omega^2. \tag{5}$$

Measurement transducer (2) has much smaller dimensions than the generator's and is designed to measure acceleration amplitudes of the vibration table rod oscillation.

Amplitude of output voltage U_{mt} of measurement transducer (2) winding is connected with the acceleration amplitude H_z of rod (7) movement by the following dependence:

$$U_{mt} = S_{mt} H_z, \tag{6}$$

where S_{mt} is the induction transducer sensitivity (S_{mt} = 8.8 mV/mm).

Investigated SE is shown in Figure 6. It is located on the working table of the mechanical vibration generator GMK-1. The sensing element SE operates on the basis of tension-compression strain.

There is acceleration due to gravity g_z influencing the inertial mass (IM); as a result, IM is displaced by the value x:

$$x \equiv f(g_z). \tag{7}$$

The movement of IM causes compression or tension of the piezoelectric element (PE) and the appearance of electric charge Q (direct piezoelectric effect phenomenon) on its surface, which is directly proportional to g_z. There is a measurement of the PE voltage U rather than charge Q:

$$U \equiv \frac{Q(g_z)}{C_{PE}}, \tag{8}$$

where C_{PE} is the PE capacity.

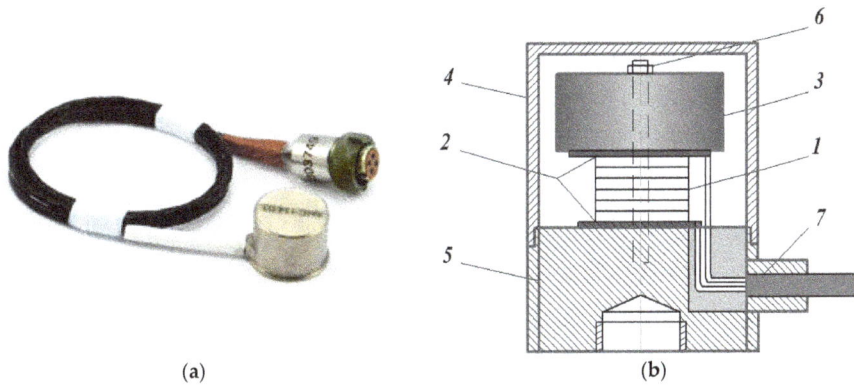

Figure 6. Investigated piezoelectric SE: (**a**) general view; (**b**) construction diagram: 1—PE; 2—insulators; 3—IM; 4—shielding; 5—base; 6—screw; 7—cable.

Software for the experimental investigation of SE characteristics, i.e., displaying the SE output signal on the PC, is developed on the LabVIEW platform and has the form of a virtual oscilloscope.

3. Experimental Investigation of Piezoelectric SE

A piezoelectric accelerometer AHC 114-08 [22], with its natural frequency $\omega_0 = 0.1$ rad/s, achieved by increasing the total resistance ($\tau = 1/C_\Sigma R_\Sigma$), has been chosen for experimental research. The investigated frequency range was chosen based on the sensor's natural frequency and typical vibration range experienced by APCs [12].

Dependence of the SE U_{SE} output voltage amplitude on the vibration table oscillation frequency ω for the generator's voltage amplitude $U_{gen} = 5, 7$ and 8 V was investigated. The 7 V signal amplitude of the generator was equivalent to the mean amplitude of vibrations experienced by APCs in normal conditions [12]. Lower and higher amplitude was checked for comparison purposes during the sensor's frequency dependence characterization. The experimental data are shown in the Table 1. Graphs of $U_{SE} = \psi(\omega)$ for $U_{gen} = 5, 7$ and 8 V are given in Figure 7.

Table 1. The experimental data.

	ω **(Rad/s)**	0.01	0.033	0.05	0.1	0.15	0.20	0.25	0.28	0.30	0.35
	At $U_{gen} = 5$ V	52.4	56.7	74	121.9	53.8	87	89.6	6.1	40.0	9.8
U_{SE} (mV)	At $U_{gen} = 7$ V	71	75.5	107.1	174.1	70.7	125.8	113	10.8	72.6	13.2
	At $U_{gen} = 8$ V	79	81.3	118	192	77.8	154.6	143	14	65.2	17

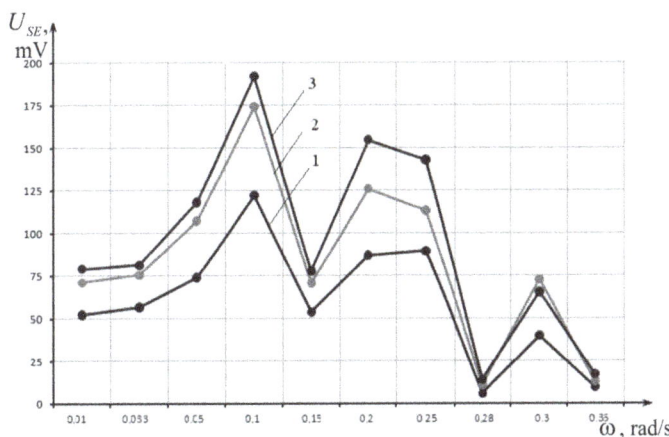

Figure 7. Dependence of SE output voltage on vibration table oscillation frequency at different excitation voltages: 1—$U_{SE} = \psi(\omega)$ at $U_{gen} = 5$ V; 2—$U_{SE} = \psi(\omega)$ at $U_{gen} = 7$ V; and 3—$U_{SE} = \psi(\omega)$ at $U_{gen} = 8$ V.

Table 1 and graphs in Figure 7 show that the maximum output voltage amplitude of the SE investigated takes place when the vibration table oscillation frequency values is $\omega = 0.1$ rad/s for $U_{gen} = 5$, 7 and 8 V, which equals the frequency of the natural oscillations of the investigated SE ($\omega = \omega_0 = 0.1$ rad/s). This is a case of the so-called "main resonance". This finding coincides with the findings of analytical studies and PC simulations.

With increasing vibration table oscillation frequency voltage U_{SE} decreases. The experimentally obtained characteristic $U_{SE} = \psi(\omega)$ is confirmed by the formula $U_{SE} = \frac{k_1 k_2 F_x}{\omega S_x}$, where $k_1 = d_{ij}$ is the piezoelectric modulus, and k_2 is the lithium niobate proportionality coefficient.

Secondly, research of the dependence of the induction transducer (IT) U_{it} output voltage amplitude on the vibration table oscillation frequency ω for the generator's voltage amplitude $U_{gen} = 5$, 7 and 8 V was conducted. The experimental data is shown in Table 2 and dependency graphs of $U_{it} = \psi(\omega)$ for $U_{gen} = 5$, 7 and 8 V (Figure 8).

According to the graphs from Figure 8 we conclude that U_{it} does not depend on the vibration table oscillation frequency and is directly proportional to U_{gen}. It has also been found that there is deviation from linearity characteristics in the area $\omega \leq 0.033$ rad/s, caused by technological errors of transducer production.

Table 2. Dependence of the IT output voltage on the vibration table oscillation frequency at different excitation voltages.

	ω (Rad/s)	0.01	0.033	0.05	0.1	0.15	0.20	0.25	0.28	0.30	0.35
	$U_{gen} = 5$ V	2.120	1.820	1.780	1.782	1.786	1.785	1.779	1.784	1.782	1.787
U_{it} (V)	$U_{gen} = 7$ V	2.480	1.920	1.850	1.830	1.829	1.822	1.821	1.818	1.819	1.813
	$U_{gen} = 8$ V	2.50	2.250	2.240	2.230	2.230	2.240	2.240	2.240	2.240	2.240

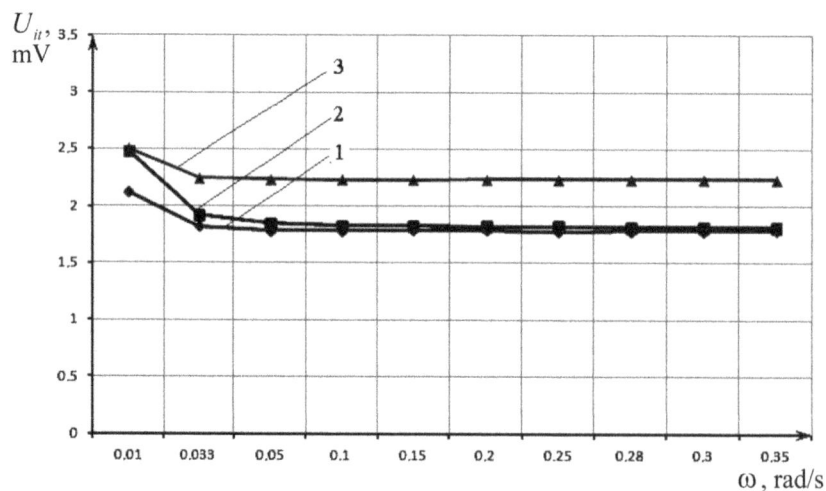

Figure 8. Dependence of the IP output voltage on the vibration table oscillation frequency at different excitation voltages: 1—$U_{it} = \psi(\omega)$ at $U_{gen} = 5$ V; 2—$U_{it} = \psi(\omega)$ at $U_{gen} = 7$ V; and 3—$U_{it} = \psi(\omega)$ at $U_{gen} = 8$ V.

Calibration is a metrological operation that provides measurement instrument (meter or gauge) with a scale or calibration table (curve). For this purpose, we use the device (Figure 9), which consists of an optical dividing head (1), SE (2), mounted on a bracket (3), amplifier unit (4), input/output module (5), and computer (6).

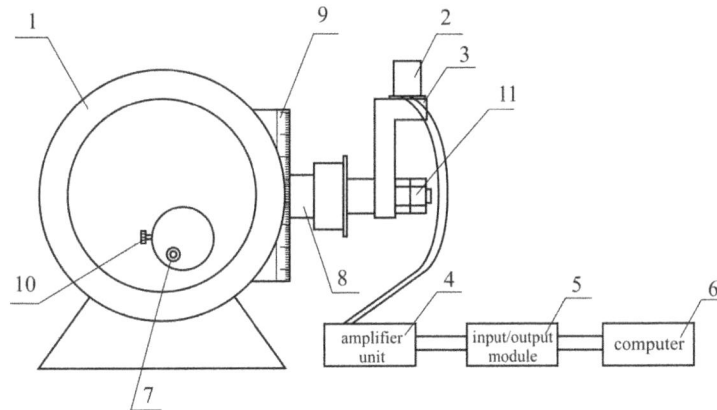

Figure 9. Experimental device for SE calibration 1—optical dividing head; 2—SE; 3—bracket; 4—amplifier unit; 5—input/output module; 6—PC; 7—10: knobs; 8—axle; 9—readout scale; and 11—mounting nuts.

Calibration of the SE takes place when its measurement axis OZ is tilted by the optical dividing head at an angle α_z (Figure 10). SE calibration is effected by the twist handle (7) of the optical dividing head (1). This brings the axle (8), the bracket (3), and SE (2) mounted on the bracket into rotation. Rotation angle α_z is adjusted according to the readout scale (9). The output signal of the SE (2) is displayed on the computer (6).

The results g_{zEXP} of the SE calibration obtained experimentally are displayed in Table 3 and compared with analytical calculations ($g_{zANL} = g \cdot \cos \alpha_z$). Figure 11 shows the constructed graphs of dependency of the SE signal g_z on rotation angle α_z.

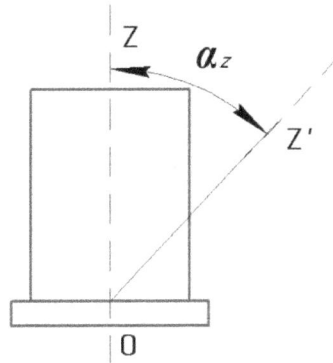

Figure 10. Tilt angle of the SE WSS.

Table 3. Calibration table of the SE WSS.

No.	α_z (°)	g_{zEXP} (mGal)	g_{zANL} (mGal)	Module of Deviation of Experimental Data from Theoretical Data (mGal)	Deviation from the Current Value (%)
1	2	3	4	5	6
1	0	981,100.375	981,100.376	0.001	0
2	10	966,195.234	966,195.257	0.023	1.52
3	20	921,932.665	921,932.784	0.119	6.03
4	30	849,658.072	849,657.849	0.223	13.39
5	40	751,566.893	751,566.491	0.402	23.40
6	50	630,639.662	630,639.161	0.501	35.72
7	60	490,549.470	490,550.188	0.718	50.01
8	70	335,556.981	335,556.091	0.890	65.79
9	80	17,365.725	17,364.818	0.907	98.23
10	90	0	0	0	100

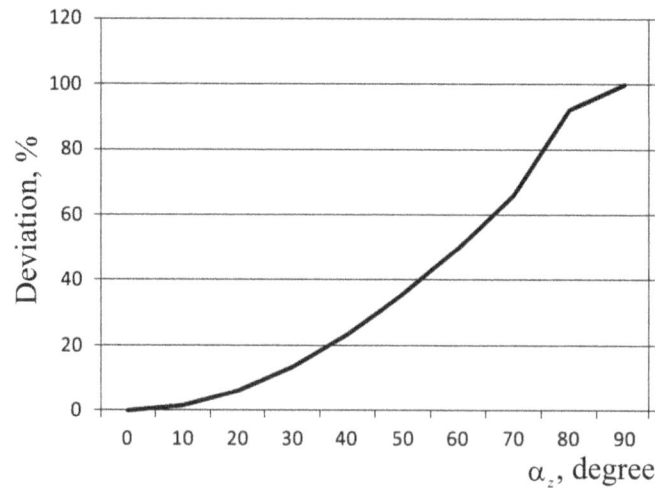

Figure 11. Graph of the dependency of the SE α_z measurement axis deviation on the deviation from the current value of the acceleration due to gravity.

As can be seen from the Table 3, the difference between deviations of the measurement axis of the automated WSS SE at angle α_z, calculated analytically and obtained experimentally, is less than 1 mGal. The angle of rotation of its measurement axis relative to the reference vertical directly affects its initial values and the value of its error.

Given that gravimetric measurements are made on the base moving in space, the coincidence of the SE WSS measurement axis with the reference vertical should always be ensured. To put this into practice, it has been proposed to build a stabilization system which provides a level of acceptable error of the SE sensitivity axis stabilization in the vertical position within 0.5–15 arcmin [5].

4. Determination of Basic Model Parameters

The device is located at the following GPS coordinates: Longitude: 28.637409°; Latitude: 50.244460°. According to these coordinates, and the equation [23]:

$$\gamma_0 = \gamma_{0e}(1 + 0.0052884 \ \sin^2 \ \phi - 0.0000059 \ \sin^2 \ 2\phi),$$

we find a reference value of the acceleration due to gravity γ_{pos}:

$$\gamma_{pos} = 9.78049(1 + 0.0052884 \ \sin^2 \ (50.244460) - 0.0000059 \ \sin^2 \ (2 \times 50.244460)) = 9.81100376 \ \text{m/s}^2. \quad (9)$$

The investigated SE was installed vertically. Tests were conducted on the vibration table shown in Figure 4.

5. The Experiment

Data recorded and processed on the computer are presented in Table 4. Data were processed on the PC in about 50 s intervals.

Systematic error Δ_g was calculated for each case:

$$\Delta_g = \left| \overline{g_{\text{EXP}}(t)} - \gamma_{pos} \right|, \quad (10)$$

where $\overline{g_{EXP}(t)}$ is the average output SE signal obtained in the experiment during an observation period of 50 s:

$$\overline{g_{EXP}(t)} = \frac{1}{N_{EXP} + 1} \sum_{i=0}^{N_{EXP}} \overline{g(t_i)}, \quad (11)$$

where N_{EXP} is the number of measurements during 50 s; $g(t_i)$ is i-th value of the output SE signal.

The absolute error of experimental measurements $\Delta_{g_{EXP}}$ is:

$$\Delta_{g_{EXP}} = \frac{\sigma_{\overline{g_{EXP}}}}{\sqrt{N_{EXP}}}\, t_p;$$

$$\sigma_{\overline{g_{EXP}}} = \sqrt{1/N_{EXP} \sum_{i=0}^{N_{EXP}} \left[\overline{g(t_i)} - \overline{g_{EXP}} \right]^2};$$

$$t_p = qt(p,d), \tag{12}$$

where $\sigma_{\overline{g_{EXP}}}$ is the standard deviation $\overline{g_{EXP}}(t)$; $t_p = qt(p,d)$ is the ratio of Student's inverse distribution according to confidence probability p, and the number of degrees of freedom.

Table 4. SE errors caused by vibrations of the base at $p = 0.90$.

ω (Rad/s)	$\overline{g_{EXP}}(t)$ (mGal)	$\Delta_{g_{EXP}}$ (mGal)	Δ_g (mGal)
0	981,100.3761	0.001136	0.00006001
0.5	981,103.2946	0.006184	2.91861022
1.0	981,103.4298	0.048067	3.05381611
5.0	981,105.7721	0.581020	5.39611120
10.0	981,108.9362	0.851001	8.89863610
30.0	981,113.4471	2.764100	13.0710563

It has been found from the data summarized in Table 4 that:

- the output SE signal coincides with the reference value of the acceleration due to gravity $\Delta_g = 0.00006$ mGal at the zero setting of the vibration table;
- the SE provides measurement accuracy of $\Delta_{g_{EXP}} = 1$ mGal for table translational vibration up to 10 rad/s.

According to [6,12] and the modelling results, a resonant mode can occur at frequencies: $\omega_0 = 0.033; 0.05; 0.1; 0.2;$ and 0.3 rad/s.

The spectrum of perturbing translational vibration accelerations in AV has a maximum at a frequency of 1640 rad/s. Therefore, the amplitudes of perturbing translational vibration accelerations are smaller at lower resonant frequencies. Estimation methods of experimental results were not changed. The results are shown in Table 5.

Table 5. SE errors caused by resonant modes.

ω (Rad/s)	$\overline{g_{EXP}}(t)$ (mGal)	$\Delta_{g_{EXP}}$ (mGal)	Δ_g (mGal)
0	981,100.3761	0.001136	0.000060
0.033	981,100.5046	0.191160	0.128636
0.05	981,100.5798	0.378130	0.203863
0.1	981,101.1799	0.962309	0.803863
0.2	981,101.5961	0.411891	1.220125
0.3	981,102.4886	0.384961	2.112581

According to Table 5 we conclude that SE provides an accuracy of $\Delta_{g_{EXP}} = 1$ mGal even in the most unfavourable resonant modes.

It has been established that experimental results are consistent with the results of digital modelling.

6. Determination of Metrological Characteristics of Piezoelectric SE

In the absence of linear and angular vibrations, the SE can function as a gravimeter sensor. Theoretical and experimental errors of the SE in the absence of disturbances is 0.00006 mGal = 0.00006×10^{-5} m/s^2.

Thus, the static characteristic of the piezoelectric SE as a gravimetric accelerometer is:

$$\overline{g}_{SE} = \overline{g_{EXP}} \mp 6 \times 10^{-5} \text{mGal}. \tag{13}$$

If there are dynamic disturbances, SE operates as WSS accelerometer with the accuracy of:

$$\overline{g}_{SE_{WSS}} = \overline{g_{EXP}} \mp 1 \text{ mGal}. \tag{14}$$

The relative error of the SE WSS is:

$$\delta_g = \frac{\Delta_g}{g_{EXP}} \times 100\% = \frac{1}{981100.37556} \times 100\% = 1.019 \times 10^{-4}\%. \tag{15}$$

The SE in the WSS has a real-time response, and is limited only by the capabilities of modern computers. Therefore, its speed is high enough.

Operational terms of SE WSS:

- Ambient temperature: -20 to $+50\,^{\circ}\text{C}$;
- Atmospheric pressure: 90,000—110,000 Pa;
- Relative humidity of 50% $\pm$ 25%.

The accuracy class of the SE WSS, i.e., the absolute error of the SE under quasi-static laboratory conditions, is Δ_g = 0.00006 mGal, whereas on mobile objects the absolute error is 1 mGal, which meets the highest accuracy class.

7. Conclusions

As a result of experimental investigations we have obtained dependencies of amplitudes of the SE output voltage and the induction transducer on the vibration table oscillation frequency. It has been established that the maximum amplitude of the output voltage SE U_{SE} takes place when values of the vibration table oscillation frequency are equal to the values of the SE natural oscillation frequency. This is a case of the so-called "main resonance". Additionally, voltage U_{SE} decreases with increasing vibration table oscillation frequency.

We have investigated the calibration characteristics of the SE WSS and found that the rotation angle of the SE WSS measurement axis relative to the reference vertical directly impacts on its output values and the value of its error.

SE error has been determined experimentally in the laboratory. It is 0.00006 mGal = 6×10^{-10} m/s^2.

The experiment has shown that the SE provides a measurement accuracy of 1 mGal = 10^{-5} m/s^2 in the most adverse resonant conditions, $\omega = \omega_0 = 0.1$ rad/s, $\omega = 2\omega_0$, $\omega = 3\omega_0$, $\omega = \omega_0/2$, $\omega = \omega_0/3$.

It has been found that systematic error of the SE is at its maximum at $\omega = 3\omega_0 = 0.3$ rad/s and does not affect measurement accuracy.

The main sources of the sensor's error are the temperature coefficient of piezoelectric modulus for given material, mass and dimension variations between individual sensors, and supply voltage stabilization errors. It is thoroughly analysed in [24]. Another source of error are the hysteresis factors of piezoelectric elements, modelling, and measurement of which are described in [25,26]. The identification and compensation of hysteresis effects should be included in future developments of the sensor, if greater accuracy is required.

The new sensor is implemented in the SVU-500 WSS, and produced by the G. Petrovsky Kyiv Automatics Plant for a modernised version of the BMP-2 APC. Newly-produced units are currently (late-2016) being investigated in military trials. Preliminary results indicate an order of magnitude improvement of the modernised system accuracy.

Acknowledgments: This work was supported by the Ministry of Education and Science of Ukraine (grant No. 0115U000210).

Conflicts of Interest: The author declares no conflict of interest.

References

1. Aleksandrov, Y.Y.; Aleksandrova, T.Y. Parametric synthesis of digital stabilization system of tank gun. *J. Autom. Inf. Sci.* **2015**, *47*, 1–17. [CrossRef]
2. Nakashima, A.M.; Borland, M.J.; Abel, S.M. Measurement of noise and vibration in Canadian forces armoured vehicles. *Ind. Health* **2007**, *45*, 318–327. [CrossRef] [PubMed]
3. Korobiichuk, I.; Bezvesilna, O.; Tkachuk, A.; Nowicki, M.; Szewczyk, R. Stabilization system of aviation gravimeter. *Int. J. Sci. Eng. Res.* **2015**, *6*, 956–959.
4. Liu, F.-F.; Rui, X.-T.; Yu, H.-L.; Zhang, J.-S. Study on launch dynamics of the self-propelled artillery marching fire. *J. Vib. Eng.* **2016**, *29*, 380–385.
5. Liu, J.-H.; Shan, J.-Y.; Liu, Y.-S.; Liu, S.-D. Guidance precision factors of terminal correction mortar projectile using pulse jets. In Proceedings of the 2014 IEEE Chinese Guidance, Navigation and Control Conference (CGNCC), Yantai, China, 8–10 August 2014; pp. 991–996.
6. Korobiichuk, I. Mathematical model of precision sensor for an automatic weapons stabilizer system. *Measurement* **2016**, *89*, 151–158. [CrossRef]
7. Hu, H.-J.; Ma, J.-G. Linear accelerometer-assisted high accurate stabilizing and tracking. *High Power Laser Part. Beams* **2006**, *18*, 769–772.
8. Sokolov, A.V.; Krasnov, A.A.; Starosel'tsev, L.P.; Dzyuba, A.N. Development of a gyro stabilization system with fiber-optic gyroscopes for an air-sea gravimeter. *Gyroscopy Navig.* **2015**, *6*, 338–343. [CrossRef]
9. Korobiichuk, I.; Bezvesilna, O.; Tkachuk, A.; Szewczyk, R.; Nowicki, M. Piezoelectric gravimeter of the aviation gravimetric system. *Adv. Intell. Syst. Comput.* **2016**, *440*, 753–763.
10. Korobiichuk, I.; Bezvesilna, O.; Tkachuk, A.; Chilchenko, T.; Nowicki, M.; Szewczyk, R. Design of piezoelectric gravimeter for automated aviation gravimetric system. *J. Autom. Mobile Robot. Intell. Syst.* **2016**, *10*, 43–47. [CrossRef]
11. Panteleev, V.L. Measuring the gravity on a moving basis. In *Manual for "Theory of Gravity Measurement (Additional Chapters)"*; Panteleev, V.L., Bulychev, A.A., Eds.; University of Moscow: Moscow, Russia, 2003; p. 80. (In Russian)
12. Bezvesilna, O.M. *Acceleration Measurement*; Lybid: Kyiv, Ukraine, 2001; p. 261. (In Ukranian)
13. Chudzik, S. Low-costinertialsensors in the system stabilization of balancing robot. *Prz. Elektrotech.* **2015**, *91*, 177–180.
14. Popelka, V. A self stabilizing platform. In Proceedings of the 2014 15th International Carpathian Control Conference, Velke Karlovice, Czech Republic, 28–30 May 2014; pp. 458–462.
15. Villela, T.; Fonseca, R.A.; De Souza, P.; Alves, A.; Mejia, J.; Correa, R.; Braga, J. Development of an attitude control system for a balloon-borne gamma ray telescope. *Adv. Space Res.* **2000**, *26*, 1415–1418. [CrossRef]
16. Eckelkamp-Baker, D.; Sebesta, H.R.; Burkhard, K. Magnetohydrodynamic inertial reference system. *Proc. SPIE* **2000**, *4025*, 99–110.
17. Pathak, A.; Brei, D.; Luntz, J.; LaVigna, C.; Kwatny, H. Design and quasi-static characterization of SMASH: SMA Stabilizing Handgrip. *Proc. SPIE* **2007**, *6523*, 652304.
18. Bezvesilna, O.M. *Aircraft Gravimetric Systems and Gravity Meters: Monograph*; Zhytomyr State Technological University: Zhytomyr, Ukraine, 2007; p. 604. (In Ukranian)
19. Wei, Y.; Xu, J.; Ma, H. Servo control technique of gyro stabilized platform for gravimeter. In Proceedings of the 2014 17th International Conference on Electrical Machines and Systems, Hangzhou, China, 22–25 October 2014; pp. 2238–2241.
20. Xu, J.; Zhu, T.; Bian, H. Gyro-stabilized platform for aerial photography. *Chin. J. Sci. Instrum.* **2007**, *28*, 914–917.
21. G. Petrovsky Kyiv Automatics Plant. Available online: http://www.kza.com.ua/ (accessed on 2 September 2016).
22. A Short Catalog—Vibron. Available online: http://vibronplus.ru/d/116655/d/kratkiy-katalog.pdf (accessed on 2 September 2016).
23. Grewal, S.M. *Global Positioning Systems, Inertial Navigation, and Integration*; Grewal, S.M., Weill, L.R., Andrews, A.P., Eds.; John Wiley & Sons Inc.: Hoboken, NJ, USA, 2007; p. 525.

24. Korobiichuk, I. Analysis of errors of piezoelectric sensor of weapons stabilizer. *Metrol. Meas. Syst.* **2017**. [CrossRef]

25. Liu, L.; Tan, K.K.; Teo, C.S.; Chen, S.L.; Lee, T.H. Development of an approach toward comprehensive identification of hysteretic dynamics in piezoelectric actuators. *IEEE Trans. Control Syst. Technol.* **2013**, *21*, 1834–1845. [CrossRef]

26. Liu, L.; Tan, K.K.; Chen, S.-L.; Huang, S.; Lee, T.H. SVD-based Preisach hysteresis identification and composite control of piezo actuators. *ISA Trans.* **2012**, *51*, 430–438. [CrossRef] [PubMed]

13

A Cu²⁺-Selective Probe Based on Phenanthro-Imidazole Derivative

Dandan Cheng [1,†], **Xingliang Liu** [1,†], **Hongzhi Yang** [1], **Tian Zhang** [2], **Aixia Han** [1,*] and **Ling Zang** [3,*]

[1]   Chemical Engineering College, Qinghai University, Xining 810016, China; hanax@qhu.edu.cn (D.C.); liuxingliang@qhu.edu.cn (X.L.); yhz2011yj@163.com (H.Y.)

[2]   Qinghai Heavy Industry Vocational School, Xining 810101, China; 15897128508@163.com

[3]   Department of Materials Science and Engineering, University of Utah, Salt Lake City, UT 84112, USA

*   Correspondence: hanaixia@tsinghua.org.cn (A.H.); lzang@eng.utah.edu (L.Z.)

†   These two authors contributed equally.

Academic Editors: Jong Seung Kim and Min Hee Lee

Abstract: A novel fluorescent Probe **1**, based on phenanthro-imidazole has been developed as an efficient chemosensor for the trace detection of copper ions (Cu²⁺). Probe **1** demonstrated sensitive fluorescence quenching upon binding with Cu²⁺ through 1:1 stoichiometric chelation. The detection limit for Cu²⁺ ions was projected through linear quenching fitting to be as low as 2.77×10^{-8} M (or 1.77 ppb). The sensing response was highly selective towards Cu²⁺ with minimal influence from other common metal ions, facilitating the practical application of Probe **1** in trace detection.

Keywords: copper ion; fluorescent probe; phenanthro-imidazole

1. Introduction

Copper (Cu²⁺), the third most abundant transition metal ion after Fe²⁺ and Zn²⁺ in the human body, plays a critical role in various fundamental physiological processes, such as those involving mitochondrial, cytosolic and vesicular oxygen-processing enzymes, which need copper as a redox cofactor [1,2]. Therefore, it is of great importance to develop simple, rapid and precise sensor methods to detect and monitor the concentration of Cu²⁺. Currently, there are many analytical methods for detecting Cu²⁺, such as atomic absorption spectroscopy (AAS) [3], inductive coupled plasma-mass spectroscopy (ICP-MS) [4], and fluorescence and surface plasmon resonance sensor methods [5–7]. Among these methods, fluorescence sensing remains one of the most promising approaches due to its high sensitivity, rapid response, and high selectivity through molecular binding design, as well as its simple solution assay processing [8–12]. Particularly, fluorescence sensors are suited for being embedded within tissues or cells for in situ imaging of Cu²⁺ ions and the associated physiological processes.

To date, numerous studies have been performed on the rational design of fluorescent chemosensors (probes) for the detection of ions and neutral analytes [13–19]. Many of these sensors have been proven effective for detecting Cu²⁺ ions [20–28], though in most of the cases the detection limit is not low enough to afford Cu²⁺ monitoring in blood and other biological systems. Moreover, the synthesis of fluorescence sensors often requires multiple step reactions, thus making the final product higher in cost, limiting the commercial use. To overcome these problems, we report herein on a novel fluorescent Probe **1**, which responds to the presence of Cu²⁺ with sensitive fluorescence quenching. Probe **1** is composed of a 1H-phenanthro [9, 10-d] imidazole moiety connected to a N,N-bis(pyridin-2-ylmethyl) benzeneamine unit, and can be synthesized in just one step. The selection of 1H-phenanthro [9, 10-d] imidazole dye is based on the consideration that it can function both as a fluorophore and an electron donor in an electron charge transfer (CT) system [12]. The *N,N*-bis(pyridin-2-ylmethyl) benzenamine moiety was chosen as the binding group for Cu^{2+} ions [29], and it can then become an efficient electron acceptor, resulting in fluorescence quenching through the CT process. Our study showed that the fluorescence quenching of Probe **1** was fast and highly selective towards Cu^{2+} over other common metal ions, implying great potential for using this probe for quick, trace-level detection of Cu^{2+} ions.

2. Experimental

2.1. Materials and Methods

All chemicals and reagents except for Probe **1** were used as purchased without further purification. For the synthesis of Probe **1** 300–400 mesh silica gel was used for column chromatography for the compound purification. ^{1}H NMR and ^{13}C NMR spectra were recorded on an Agilent DD2 NMR spectrometer (Agilent Technologies, Santa Clara, CA, USA) at 600 MHz, using DMSO-d_6 as the solvent. Mass spectra were recorded on an Agilent Technologies 622 spectrometer (Agilent Technologies, Santa Clara, CA, USA). UV/vis spectra were acquired on a Shimadzu UV-2550 spectrophotometer (Shimadzu, Beijing, China). Fluorescence measurements were performed on an Agilent Cary Eclipse fluorescence spectrophotometer (Agilent Technologies, Santa Clara, CA, USA).

2.2. Synthesis

The synthetic route to Probe **1** is shown in Scheme 1. 1.5 g (4.93 mmol) Compound **2**, synthesized following the literature procedures [30], was mixed with phenanthrene-9, 10-dione (1.03 g, 4.93 mmol), and ammonium acetate (7.4 g, 98 mmol) in 63 mL acetic acid, and heated to reflux under nitrogen atmosphere for 16 h. The mixture was then cooled to room temperature and poured into H_2O (100 mL), and the precipitate thus formed was filtered, washed with water and then dried under vacuum. The crude product obtained was purified via column chromatography (300–400 mesh silica gel), CH_2Cl_2/AcOEt, 4/1, v/v) to produce the desired product (1.263 g, 53% yield). ^{1}H NMR (DMSO-d_6, 600 MHz): δ = 8.77 (d, J = 8.4 Hz, 3H), 8.22 (d, J = 6 Hz, 2H), 7.91 (d, J = 7.2 Hz, 1H), 7.64 (t, J = 7.2 Hz, J = 7.2 Hz, 4H), 7.59–7.55 (m, 4H), 7.52 (t, J = 7.2 Hz, J = 7.8 Hz, 1H), 7.31–7.25 (m, 4H), 6.99 (d, J = 7.2 Hz, 1H), 4.08 (s, 1H), 2.49 (d, J = 1.8 Hz, J = 1.8 Hz, 4H). (Figure S1, Supplementary Information). ^{13}C NMR (DMSO-d_6, 150 MHz) δ 141.34, 130.39, 130.02, 129.50, 128.79, 128.44, 127.40, 127.28, 127.22, 126.99, 126.93, 125.01, 124.95, 123.73, 121.52, 48.49 (Figure S2, Supplementary Information). MALDI-TOF MS: m/z calculated for $C_{33}H_{25}N_5$: 491.2110; found: 491.2141 (Figure S3, Supplementary Information).

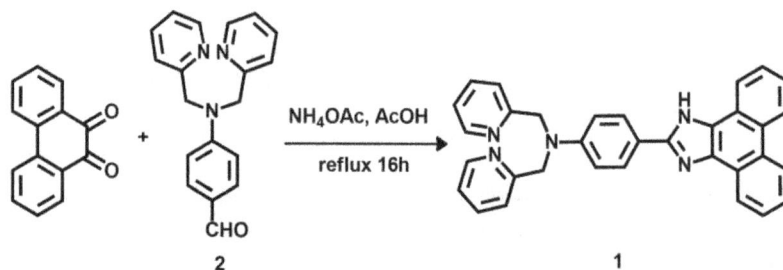

Scheme 1. Synthetic route for fluorescent Probe **1**.

2.3. Spectral Measurements

Distilled water was used for preparing solutions throughout the experiments. Solutions of all the metal ions used (Cu^{2+}, Hg^{2+}, Ca^{2+}, Ba^{2+}, Cd^{2+}, Zn^{2+}, Pb^{2+}, Mg^{2+}, Co^{2+}, Fe^{2+} and Mn^{2+}) were prepared from their nitrate salts. A stock solution (0.5 mM) of Probe **1** in ethanol was prepared, which was then diluted to 10 μM with ethanol. In the spectral titration experiments, 2 mL of Probe **1** solution (10 μM) was placed in a 1 cm quartz cuvette and Cu^{2+} solution was added gradually by micro-pipette; UV-vis and/or fluorescence spectra were measured before and after the addition of Cu^{2+}. Since the volumes of Cu^{2+} solution added were minimal (in μL), the slight change in concentration of Probe **1** can be ignored. For fluorescence measurement, the excitation wavelength was set at 270 nm, and the slit widths for excitation and emission were 5 nm/5 nm.

3. Results and Discussion

3.1. UV-Vis Spectral Response of the Binding between Probe **1** and Cu^{2+}

Probe **1** binds effectively with Cu^{2+} ions through the chelation with *N,N*-bis(pyridin-2-ylmethyl) benzenamine (Scheme 2). The same chelation was previously reported in a crystalline study of the complex of Cu^{2+} [29]. The strong complexation affects the original conjugation between the lone pair of electrons on the aniline amine and the π-orbital of 1H-phenanthro [9, 10-d] imidazole. This can be seen from the significant change in the absorption spectrum of Probe **1** as shown in Figure 1a. Upon the addition of a Cu^{2+} ion, a significant absorption increase was observed for the wavelength region below 260 nm and in the region between 300 and 325 nm, whereas the absorption in the range of 270–300 nm and 325–345 nm was decreased. Clear isosbestic points can be identified at 269, 300, 328 and 343 nm between the increasing and decreasing bands, indicating the stoichiometric chelation equilibrium shown in Scheme 2.

Scheme 2. Chelation with Cu^{2+} quenches the fluorescence of Probe **1** via intramolecular CT.

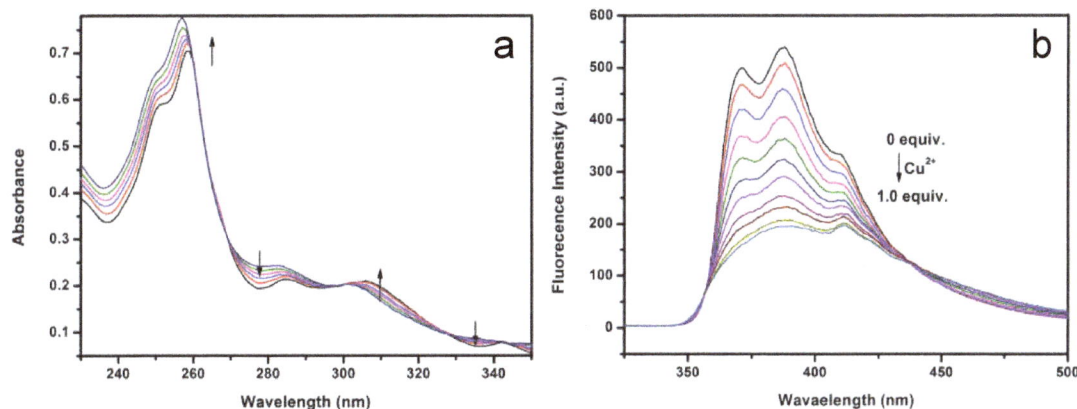

Figure 1. UV/vis absorption (**a**) and fluorescence (**b**) spectra of Probe **1** in ethanol (10 μM) upon addition of varying concentrations of Cu^{2+} ions (0–1 equiv).

Along with the absorption change, the fluorescence spectra recorded accordingly also demonstrated a significant change as shown in Figure 1b. The fluorescence quantum yield of Probe **1** in the absence of Cu^{2+} was determined to be 6.7%, which represents a medium-strength fluorophore suited for being used as a sensor. Upon the addition of a 1:1 molar ratio of Cu^{2+} ions, the fluorescence intensity was quenched by 74%. Interestingly, the fluorescence quenching was dominated by the emission in the shorter wavelength region, while the emission at longer wavelengths (above 437 nm) was actually increased slightly, implying the formation of a charge transfer (CT) transition between the 1H-phenanthro [9, 10-d] imidazole moiety and the Cu^{2+} complex. The fluorescence quenching observed was likely due to the photoinduced electron transfer from the lowest unoccupied molecular orbital (LUMO) of 1H-phenanthro [9, 10-d] imidazole to the Cu^{2+} ion.

3.2. Stoichiometric Ratio of Probe 1-Cu^{2+} Complex

The sensitive fluorescence quenching of Probe **1** by the Cu^{2+} ions provided a way to determine the chelation stoichiometry between the two species simply through a Job plot approach, as shown in Figure 2 [31]. A Job plot is commonly used to determine the stoichiometry of a binding event between two species in a solution. In this method, the total molar concentrations of the two binding species (here Probe **1** and Cu^{2+} ions) are held constant, while their molar fractions are varied. An observable variable (here the fluorescence quenching) that is proportional to the complex formation can be plotted against the molar fractions of the binding species. The maximum of the plot corresponds to the stoichiometry of the complex formed by the two binding species. In this study, by fixing the total concentration of Probe **1** and the Cu^{2+} ions at 10 μM, the molar ratio of the two species was changed from 1:9 to 9:1, and the fluorescence intensity of the mixture was measured at 387 nm under the same conditions. The molar ratio that gives the maximal fluorescence quenching should correspond to the stoichiometry between Probe **1** and Cu^{2+} ions, ca. 1:1 as indicated in Figure 2. The 1:1 ratio is consistent with the previous reports on the same chelation of N,N-bis(pyridin-2-ylmethyl) benzenamine with Cu^{2+} ions [29].

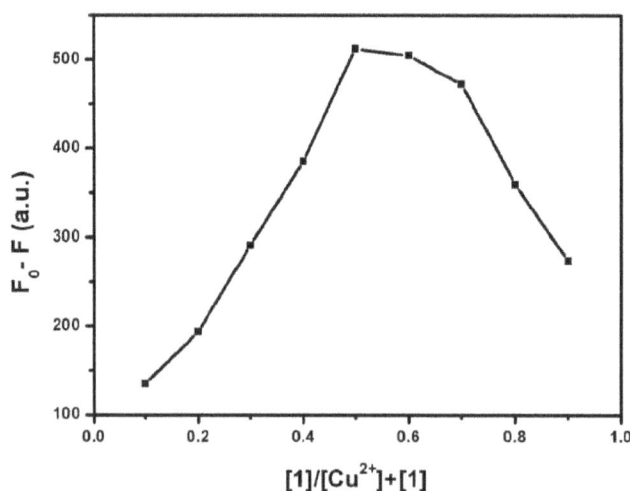

Figure 2. Job plot of the binding between Probe **1** and Cu^{2+} in ethanol, with the total concentration of the two kept constant at 10 μM. F_0 and F are fluorescence intensities measured at 387 nm in the absence and presence of Cu^{2+}, respectively.

3.3. Fluorescence Quenching Selectivity

To examine the fluorescence quenching selectivity of Probe **1** towards Cu^{2+}, comparative experiments were conducted for the same quenching but in the presence of 10 other common metal ions, as shown in Figure 3 and Figures S5 and S6. Compared to the efficient quenching by Cu^{2+} (far left bar in the figure), all other metal ions gave a much lower degree of quenching under the same experimental conditions. Adding the same concentration of Cu^{2+} to each of the 10 solutions containing the different metal ions resulted in dramatic fluorescence quenching at the same level as that observed for the solution containing only Cu^{2+} as the quencher. These results indicate good selectivity for Probe **1** towards Cu^{2+} when used as a fluorescence sensor. The high selectivity is due to the strong chelation interaction between Probe **1** and Cu^{2+} as shown in Scheme 2, as well as the photoinduced electron transfer thus enabled between the two species. Although Probe **1** also binds to other metal ions such as Co^{2+}, Zn^{2+}, Cd^{2+}, these ions do not possess a strong electron-accepting capability as Cu^{2+} does, and thus can hardly induce effective photoinduced electron transfer. The stronger electron acceptability of Cu^{2+} can be seen from its higher standard reduction potential, +0.34 V, much higher than those of Co^{2+}, Zn^{2+}, Cd^{2+}, -0.29, -0.70, -0.40 V, respectively.

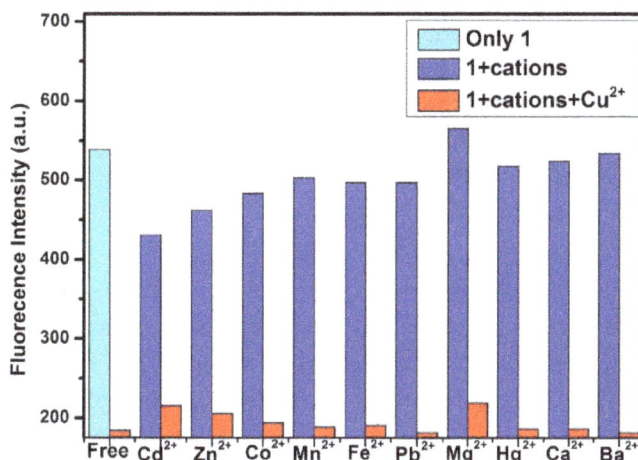

Figure 3. Fluorescence intensity measured at 387 nm for Probe **1** in ethanol (10 μM) in the absence of metal ions (**Green**), and in the presence of 10 μM various metal ions (**Blue**); 10 μM Cu^{2+} was added to each of the 11 solutions and the fluorescence intensity was measured again for comparison (**Red**).

3.4. Detection Limit

Figure 4 shows the fluorescence intensity of Probe **1** (10 μM in ethanol) as a function of the concentration of Cu^{2+} (plotted here as the ratio of $[Cu^{2+}]/[1]$). All the data points can be fitted well into a linear relationship, giving the equation as marked in the plot (with a slope of 368.49). Following the common practice in analytical chemistry, the detection limit can be calculated by defining the lowest detectable signal as three times the standard deviation of the intensity measurement. In this study, the standard deviation of the intensity measurement was 0.34, and three times that gives 1.02. This value represents the minimal detectable change in the fluorescence intensity, which corresponds to the lowest detectable value of $[Cu^{2+}]/[1]$ (calculated as $1.02/\text{slope} = 2.77 \times 10^{-3}$). Since the concentration of Probe **1** was kept at 10 μM, the detection limit of Cu^{2+} was obtained as 2.77×10^{-8} M (or 1.77 ppb). Such a low detection limit is significantly improved, by one to three orders of magnitude, in comparison to the previously reported chemosensors (Figure S7). A low detection limit will be suitable for the trace detection of Cu^{2+} in blood [32].

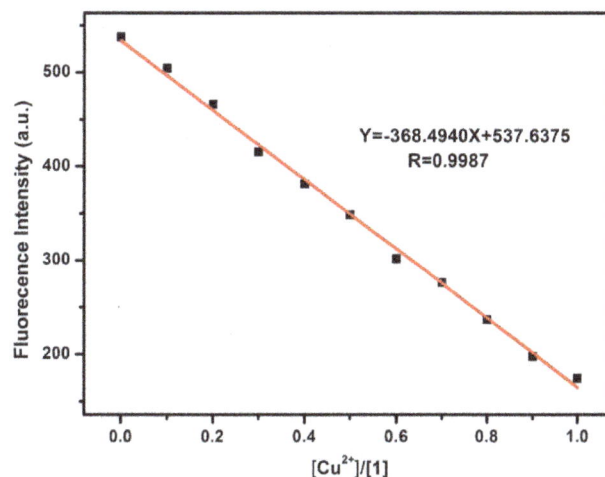

Y=-368.4940X+537.6375
R=0.9987

Figure 4. Fluorescence intensity measured at 387 nm for Probe 1 in ethanol (10 μM) as a function of the concentration of Cu^{2+} (0–1 equiv.). Inset shows the linear fitting result of the data points.

In addition to the high sensitivity and selectivity, a fast sensing response was another feature of Probe **1** regarding the detection of Cu^{2+}. Upon the addition of an equivalent amount Cu^{2+} ions, the fluorescence intensity of Probe **1** (10 μM) in ethanol was quenched rapidly (Figure S4), with a response time estimated to be ca. 10 s (inset of Figure S4). This fast sensing response makes Probe **1** highly suited for real-time monitoring, or portable detection [33], which is not feasible for the traditional analytical methods and many other chemosensors reported before. Moreover, Probe **1** was also proven to have high photostability as shown in Figure S8, wherein the fluorescence of Probe **1** was measured multiple times over 2 h, but no significant decrease in the fluorescence intensity was observed.

4. Conclusions

In conclusion, we have developed an efficient molecular fluorescence sensor, Probe **1**, based on phenanthro-imidazole for quick trace-level detection of Cu^{2+} ions in aqueous solutions. Probe **1** demonstrated sensitive fluorescence quenching upon binding with Cu^{2+} ions through 1:1 stoichiometric chelation. The detection limit was projected through linear quenching fitting to be as low as 2.77×10^{-8} M (or 1.77 ppb), which is improved by one to three orders of magnitude in comparison to the previously reported chemosensors. The fluorescence sensing response was highly selective towards Cu^{2+} ions without significant interference from other common metal ions under the same conditions. The sensing response towards Cu^{2+} ions was also found quickly, on the time scale of seconds. Moreover, high photostability was also proven for Probe **1** by repeatedly measuring the florescence over 2 h. The combination of all these features makes Probe **1** an ideal sensor for the portable, real-time detection of copper ions.

Supplementary Materials
Figure S1: ^{1}H NMR (600 MHz) spectrum of compound 1 in d-DMSO; Figure S2: ^{13}C NMR (150 MHz) spectrum of compound 1 in d-DMSO; Figure S3: MALDI/TOF MS spectrum of compound 1; Figure S4: Fluorescence intensity measured at 387 nm for probe 1 in ethanol (10 μM) as a function of time upon addition of Cu^{2+} (10 μM). Exponential fitting of the fluorescence intensity decrease gives a response time of ca. 10 s; Figure S5: Fluorescence spectra of probe 1 in ethanol (10 μM) in the absence and presence of various metal ions (10 μM); Figure S6: Fluorescence spectra of probe 1 in ethanol (10 μM) in the absence and presence of various metal ions (10 μM) plus 10 μM of Cu^{2+}; Figure S7: Comparison of the detection limit of Probe 1 with the literature reported detection limits of other sensors. Reference # marked in the horizontal axis are the same as cited in the main context; Figure S8: Fluorescence intensity measured at the main peak of probe 1 in ethanol (10 μM) for nine consecutive times over 2 h.

Acknowledgments: This work was financially supported by the Qinghai Science & Technology Department of China (Grant No. 2016-HZ-806), the National Natural Science Foundation of China (Grant No. 21362027).

Author Contributions: Xingliang Liu designed and synthesized Probe 1; Dandan Cheng carried out the majority of the experiments and wrote the article; Hongzhi Yang helped with and advised on the experiments; Tian Zhang did part of the experiments with the help of Dandan Cheng; Aixia Han was responsible for the whole work; Ling Zang helped supervise the research design and manuscript editing.

Conflicts of Interest: The authors declare no conflict of interest.

References

1. Trapaidze, A.; Hureau, C.; Bal, W.; Winterhalter, M.; Faller, P. Thermodynamic study of Cu^{2+} binding to the DAHK and GHK peptides by isothermal titration calorimetry (ITC) with the weaker competitor glycine. *J. Biol. Inorg. Chem.* **2012**, *17*, 37–47. [CrossRef] [PubMed]

2. Hötzer, B.; Ivanov, R.; Brumbarova, T.; Bauer, P.; Jung, G. Visualization of Cu^{2+} uptake and release in plant cells by fluorescence lifetime imaging microscopy. *FEBS J.* **2012**, *279*, 410–419. [CrossRef] [PubMed]

3. Ghaedi, M.; Tavallali, H.; Keshavarz, M.; Niknam, K. Determination of Copper and Zinc Ions by Flame-AAS After Preconcentraction Using Sodium Dodecyl Sulfate Coated Alumina Modified with 3-((1H-Indol-3-yl)-3,4,5-trimethyl)-1H-indole. *Chin. J. Chem.* **2009**, *27*, 2066–2072. [CrossRef]

4. Yang, X.; Wang, E. A Nanoparticle Autocatalytic Sensor for Ag^+ and Cu^{2+} Ions inAqueous Solution with High Sensitivity and Selectivity andIts Application in Test Paper. *Anal. Chem.* **2011**, *83*, 5005–5011. [CrossRef] [PubMed]

5. Jiang, X.C.; Yu, A.B. Silver Nanoplates: A Highly Sensitive Material toward InorganicAnions. *Langmuir* **2008**, *24*, 4300–4309. [CrossRef] [PubMed]

6. Jiang, X.C.; Yu, A.B. Low Dimensional Silver Nanostructures: Synthesis, Growth Mechanism, Properties and Applications. *J. Nanosci. Nanotechnol.* **2010**, *10*, 7829–7875. [CrossRef] [PubMed]

7. Carter, K.P.; Young, A.M.; Palmer, A.E. Fluorescent Sensors for Measuring Metal Ions in Living Systems. *Chem. Rev.* **2014**, *114*, 4564–4601. [CrossRef] [PubMed]

8. Masilamany, K.; Ramaier, N. L-Cysteine-capped ZnS quantum dots based fluorescence sensor for Cu^{2+} ion. *Sens. Actuators B Chem.* **2009**, *139*, 104–109.

9. Shyamaprosad, G.; Debabrata, S.; Nirmal, K.D. A New Highly Selective, Ratiometric and Colorimetric Fluorescence Sensor for Cu^{2+} with a Remarkable Red Shift in Absorption and Emission Spectra Based on Internal Charge Transfer. *Org. Lett.* **2010**, *12*, 856–859.

10. Liu, X.J.; Zhang, N.; Bing, T.; Shangguan, D.H. Carbon Dots Based Dual-Emission Silica Nanoparticles as a Ratiometric Nanosensor for Cu^{2+}. *Anal. Chem.* **2014**, *86*, 2289–2296. [CrossRef] [PubMed]

11. Chen, W.B.; Tu, X.J.; Guo, X.Q. Fluorescent gold nanoparticles-based fluorescence sensor for Cu^{2+} ions. *Chem. Commun.* **2009**. [CrossRef] [PubMed]

12. Guo, Z.Q.; Zhu, W.H.; Tian, H. Hydrophilic Copolymer Bearing Dicyanomethylene-4H-pyran Moiety As Fluorescent Film Sensor for Cu^{2+} and Pyrophosphate Anion. *Macromolecules* **2010**, *43*, 739–744. [CrossRef]

13. Hariharan, P.S.; Anthony, S.P. Substitutional group dependent colori/fluorimetric sensing of Mn^{2+}, Fe^{3+} and Zn^{2+} ions by simple Schiff base chemosensor. *Spectrochim. Acta A* **2015**, *136*, 1658–1665. [CrossRef] [PubMed]

14. Grynkiewicz, G.; Poenie, M.; Tsien, R.Y. A New Generation of Ca^{2+} Indicators with Greatly Improved Fluorescence Properties. *J. Biol. Chem.* **1985**, *260*, 3440–3450. [PubMed]

15. He, G.J.; Guo, D.; He, C.; Zhang, X.L.; Zhao, X.W.; Duan, C.Y. A Color-Tunable Europium Complex Emitting Three Primary Colors and White Light. *Angew. Chem. Int. Ed.* **2009**, *48*, 6132–6135. [CrossRef] [PubMed]

16. Huang, J.H.; Xu, Y.F.; Qian, X.H. A red-shift colorimetric and fluorescent sensor for Cu^{2+} in aqueous solution: Unsymmetrical 4, 5-diaminonaphthalimide with N-H deprotonation induced by metal ions. *Org. Biomol. Chem.* **2009**, *7*, 1299–1303. [CrossRef] [PubMed]

17. Madhu, S.; Ravikanth, M. Boron-Dipyrromethene Based Reversible and Reusable Selective Chemosensor for Fluoride Detection. *Inorg. Chem.* **2014**, *53*, 1646–1653. [CrossRef] [PubMed]

18. Lin, K.K.; Wu, S.C.; Hsu, K.M.; Hung, C.H.; Liaw, W.F.; Wang, Y.M. A N-(2-Aminophenyl)-5-(dimethylamino) -1-naphthalenesulfonic Amide (Ds-DAB) Based Fluorescent Chemosensor forPeroxynitrite. *Org. Lett.* **2013**, *16*, 4242–4245. [CrossRef] [PubMed]

19. Hu, B.; Hu, L.L.; Chen, M.L.; Wang, J.H. A FRET ratiometric fluorescence sensing system for mercury detection and intracellular colorimetric imaging in live Hela cells. *Biosens. Bioelectron.* **2013**, *49*, 499–505. [CrossRef] [PubMed]

20. Goswami, S.; Maity, S.A.; Maity, A.; Maity, A.K.D.; Saha, P.A. A FRET-based rhodamine–benzimidazole conjugate as a Cu^{2+}-selective colorimetric and ratiometric fluorescence probe that functions as a cytoplasm marker. *RSC Adv.* **2014**, *4*, 6300–6305. [CrossRef]

21. Zhang, X.; Shirashi, Y.; Hirai, T. Cu(II)-Selective Green Fluorescence of a Rhodamine-Diacetic Acid Conjugate. *Org. Lett.* **2007**, *9*, 5039–5042. [CrossRef] [PubMed]

22. Grasso, G.I.; Gentile, S.; Giuffrida, M.L.; Satriano, C.; Sgarlata, C.; Sgarzi, M.; Tomaselli, G.; Arena, G.; Prodi, L. Ratiometric fluorescence sensing and cellular imaging of Cu^{2+} by a new water soluble trehalose-naphthalimide based chemosensor. *RSC Adv.* **2013**, *3*, 24288–24297. [CrossRef]

23. Boiocchi, M.; Fabbrizzi, L.; Licchelli, M.; Sacchi, D.; Vazquez, M.; Zampa, C. A two-channel molecular dosimeter for the optical detection of copper (II). *Chem. Commun.* **2003**, *21*, 1812–1813. [CrossRef]

24. Royzen, M.; Dai, Z.; Canary, J.W. Ratiometric Displacement Approach to Cu (II) Sensing by Fluorescence. *J. Am. Chem. Soc.* **2005**, *127*, 1612–1613. [CrossRef] [PubMed]

25. Martinez, R.; Espinosa, A.; Tarraga, A.; Molina, P. New Hg^{2+} and Cu^{2+} Selective Chromoand Fluoroionophore Based on a Bichromophoric Azine. *Org. Lett.* **2005**, *7*, 5869–5872. [CrossRef] [PubMed]

26. An, R.B.; Zhang, D.T.; Chen, Y.; Cui, Y.Z. A "turn-on" fluorescent and colorimetric sensor for selective detection of Cu^{2+} in aqueous media and living cells. *Sens. Actuators B Chem.* **2016**, *222*, 48–54. [CrossRef]

27. Wang, H.L.; Zhou, G.D.; Chen, X.Q. An iminofluorescein-Cu^{2+} ensemble probe for selective detection of thiols. *Sens. Actuators B Chem.* **2013**, *176*, 698–703. [CrossRef]

28. Huang, J.G.; Liu, M.; Ma, X.Q.; Dong, Q.; Ye, B.; Wang, W.; Zeng, W.B. A highly selective turn-off fluorescent probe for Cu(II) based on a dansyl derivative and its application in living cell imaging. *RSC Adv.* **2014**, *4*, 22964–22970. [CrossRef]

29. Almesåker, A.; Bourne, S.A.; Ramon, G.; Scotta, J.L.; Strauss, C.R. Coordination chemistry of *N*,*N*,4-tris(pyridin-2-ylmethyl)aniline: A novel flexible, multimodal ligand. *CrystEngComm* **2007**, *9*, 997–1010. [CrossRef]

30. Peng, X.J.; Du, J.J.; Fan, J.L.; Wang, J.Y.; Wu, Y.K.; Zhao, J.Z. A Selective Fluorescent Sensor for Imaging Cd^{2+} in Living Cells. *J. Am. Chem. Soc.* **2007**, *129*, 1500–1501. [CrossRef] [PubMed]

31. Zhao, W.Q.; Liu, X.L.; Lv, H.T.; Fu, H.; Yan, Y.; Huang, Z.P.; Han, A.X. A phenothiazine–rhodamine ratiometric fluorescent probe for Hg^{2+} based on FRET and ICT. *Tetrahedron Lett.* **2015**, *56*, 4293–4298. [CrossRef]

32. Huang, L.; Cheng, J.; Xie, K.; Xi, P.; Hou, F.; Li, Z.; Xie, G.; Shi, Y.; Liu, H.; Bai, D.; et al. Cu^{2+}-selective fluorescent chemosensor based on coumarin and its application in bioimagin. *Dalton Trans.* **2011**, *40*, 10815–10817. [CrossRef] [PubMed]

33. Han, A.X.; Liu, X.H.; Prestwich, G.D.; Zang, L. Fluorescent sensor for Hg^{2+} detection in aqueous solution. *Sens. Actuators B Chem.* **2014**, *198*, 274–277. [CrossRef]

14

Robust Functionalization of Large Microelectrode Arrays by Using Pulsed Potentiostatic Deposition

Joerg Rothe [1,*], Olivier Frey [1], Rajtarun Madangopal [2,3], Jenna Rickus [2] and Andreas Hierlemann [1]

[1] ETH Zurich, Department of Biosystems Science and Engineering, Bio Engineering Laboratory, Mattenstrasse 26, CH-4058 Basel, Switzerland; olivier.frey@insphero.com (O.F.); andreas.hierlemann@bsse.ethz.ch (A.H.)

[2] Agricultural and Biological Engineering, Biomedical Engineering, Physiological Sensing Facility at the Bindley Bioscience Center and Birck Nanotechnology Center, Purdue University, West Lafayette, IN 47907, USA; rajtarun.madangopal@nih.gov (R.M.); rickus@purdue.edu (J.R.)

[3] Intramural Research Program of the National Institute on Drug Abuse, National Institutes of Health, Baltimore, MD 21224, USA

* Correspondence: joergenmarode@gmail.com

Academic Editors: Andrew J. Mason, Haitao Li and Yuning Yang

Abstract: Surface modification of microelectrodes is a central step in the development of microsensors and microsensor arrays. Here, we present an electrodeposition scheme based on voltage pulses. Key features of this method are uniformity in the deposited electrode coatings, flexibility in the overall deposition area, i.e., the sizes and number of the electrodes to be coated, and precise control of the surface texture. Deposition and characterization of four different materials are demonstrated, including layers of high-surface-area platinum, gold, conducting polymer poly(ethylenedioxythiophene), also known as PEDOT, and the non-conducting polymer poly(phenylenediamine), also known as PPD. The depositions were conducted using a fully integrated complementary metal-oxide-semiconductor (CMOS) chip with an array of 1024 microelectrodes. The pulsed potentiostatic deposition scheme is particularly suitable for functionalization of individual electrodes or electrode subsets of large integrated microelectrode arrays: the required deposition waveforms are readily available in an integrated system, the same deposition parameters can be used to functionalize the surface of either single electrodes or large arrays of thousands of electrodes, and the deposition method proved to be robust and reproducible for all materials tested.

Keywords: electrodeposition; microelectrode array; pulse potential waveform; voltage pulses; pulsed potentiostatic deposition; complementary metal-oxide-semiconductor (CMOS); platinum; gold; poly(phenylenediamine) PPD; poly(ethylenedioxythiophene) PEDOT

1. Introduction

Microelectrode arrays (MEAs) constitute a widely used platform that enables spatially highly resolved parallel sensing [1–6]. Owing to the advances in microfabrication technology, established methods for manufacturing microelectrodes from a large variety of materials over a wide range of geometries and spatial arrangements are available [7–10]. In most cases, arrays of electrodes are fabricated on the same substrate using the same materials due to the parallel and planar nature of microfabrication techniques. Desired surface properties of microelectrodes include: (I) large active surface area and/or low impedance for effective charge-transfer, i.e., high signal-to-noise ratio during recordings and high stimulation efficacy [11,12]; (II) good adhesion or chemical binding properties for biochemical recognition units or matrices that encage biomolecules such as enzymes [13,14], or function as a size discrimination or anti-interference layers to reduce cross-sensitivity [15]; and (III) robustness against electrode fouling [16]. Many of these properties can be obtained by specific surface functionalization of the electrodes with the aim of attaining biosensor systems that are capable of performing parallel measurements through many electrodes or that enable measuring different analytes at the same time. Multi-analyte sensing with the same system requires that various subsets of electrodes on an array can be functionalized with different layers.

Micrometer dimensions of the electrodes and electrode distances (<100 μm) obviate the application of manual coating methods, such as depositing drops of coating solutions onto microelectrodes. Also, dipping methods do not enable selective and local deposition on planar electrode arrangements and arrays. Good spatial control of the deposition process and high reproducibility of the quality and morphology of coating over several sensor electrodes are prerequisites to fabricate electrode arrays with different specific sensing properties and to obtain low sensor–signal crosstalk. This holds particularly true for large arrays of densely packed microelectrodes.

Electrodepostion-based surface modification techniques satisfy several of the criteria listed above, as the deposition process relies on a chemical reaction coupled with a charge transfer through the connected electrode(s). Traditional electrodeposition schemes rely on galvanostatic methods, in which constant currents are applied to obtain a constant deposition rate. To improve the quality of the deposited layers, several protocols for applying pulsed currents instead of a constant current have been established [17]. The current has then to be adjusted with respect to the deposition area in order to obtain the desired current density. In most cases, however, the exact area of the target electrode may be unknown and to find a robust deposition protocol for the respective functional layer is challenging. This challenge holds particularly true if the deposition should be done on electrodes of different size. The deposition protocol needs to be adapted for every coating material and coating scenario. Pulsed potentiostatic deposition, in contrast, relies on voltage pulses, which allows for applying the same parameters independent of the overall electrode area. Publications reporting on the usage of voltage pulses for electrodeposition can be found for applications that are largely different from those described in this paper. Horkans and Romankiw reported on pulsed potentiostatic deposition on a large gold electrode [18]. Plyasova et al. performed electrostatic depositions of platinum on gold and glassy carbon substrates, showing differences in the nanostructure of the surface upon varying the deposition potential [19]. Using square-wave potential pulses, Liu et al. 2014 grew nanodendrites with a different morphology by altering the anodic potential [20].

A robust deposition protocol and method is particularly important for large microelectrode arrays, in which electrode size and the number and arrangement of electrodes to be functionalized can vary. Further, for developing a protocol for a new material or a new electrode design, it would be highly desirable that deposition parameters be tuned and determined using a single electrode and subsequently be applied to a larger subset of the microelectrode array without modification.

In this paper we present a universal electrodeposition method that relies on voltage pulses rather than on the more commonly used current pulses to achieve spatially confined growth of both metallic and polymer layers on electrode surfaces. The voltage-pulse-based deposition method presented in this paper enables the use of the same parameters for different electrode sizes (diameters of 5–50 μm). Moreover, this method is highly scalable in that it enables deposition on different and large numbers of electrodes in parallel. We demonstrate the deposition method for electrodeposition of two key classes of materials, (i) the metals Pt black and gold, and (ii) the polymers poly(phenylenediamine), also known as PPD, and poly(ethylenedioxythiophene), also known as PEDOT. Deposition is carried out on an array of 32×32 platinum microelectrodes, placed on a fully developed complementary metal-oxide-semiconductor (CMOS) microelectronic chip. Adjustment of the material-specific deposition parameters allows for controlling the roughness of the surface, the layer morphology, and the capacitive properties or permselectivity of the deposited films.

2. Materials and Methods

2.1. CMOS Microelectrode Array

The electrodeposition and functionalization protocol was developed on a fully integrated CMOS potentiostat chip [21] featuring an integrated 32×32 array of platinum electrodes (cf. Supplementary Figure S1a). The system formed part of a standard three-electrode setup including an on-chip platinum counter electrode and an external Ag/AgCl/KCl (3 M) reference electrode. The platinum working electrodes were disk electrodes with a standard diameter of 25 μm or featuring different diameters ranging from 5 to 50 μm. The electrodes were arranged on a 32×32 grid configuration at 100 μm pitch. The platinum layer (270 nm) for the microelectrodes was deposited and structured through ion-beam deposition and ion-beam etching processes. A passivation stack (1.6 μm), consisting of alternating SiO_2 and Si_3N_4 layers was deposited on the chips and was structured by reactive ion etching to open the active areas of the electrodes. As a result, the platinum surface of the microelectrodes was located in a shallow recess (approx. 1.5-μm deep) of the passivation stack.

The electronic circuits on the chip allowed for generation of arbitrary signal waveforms at a resolution of 2.8 mV. The voltages were generated on chip, and the measured currents were digitized on chip as well. The chip was connected to a field programmable gate array (FPGA, Xilinx Spartan 6), for signal processing. The experiments were controlled, and the data recorded on a computer through a USB connection. To perform experiments, a reservoir holding approximately 5 mL liquid volume was placed atop the electrode array (cf. Supplementary Figure S1b). The reservoir could be closed by means of a lid with holes for inserting the reference electrode and a nitrogen gas supply line.

2.2. Electrodeposition Protocol

Before usage, the microelectrodes were electrochemically treated by varying the electrode potential between –0.2 and +1.2 V vs. Ag/AgCl for 32 cycles in a de-aerated 0.5 M H_2SO_4 solution at a rate of 100 mV/s. The chips were then rinsed with deionized water (DI water) and blown dry with a nitrogen gun.

Before deposition, the open-circuit potential (OCP) was determined with the help of the Ag/AgCl reference electrode and a platinum wire. The deposition scheme was performed by applying alternating potential pulses (see Figure 1). All depositions were performed by using 1 mL of deposition solution. The solutions were de-aerated by nitrogen bubbling for 5 min before deposition started. A plastic pipette tip mounted to a sonicating electric toothbrush was used for agitating the solution and for reducing diffusion limitations.

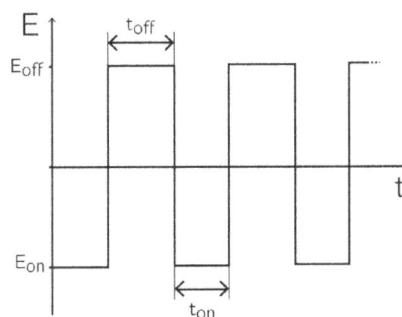

Figure 1. Voltage-pulse-based deposition scheme. Pulse shape for cathodic deposition of, e.g., Au or Pt black. During t_{on}, the potential E_{on} is applied at the selected deposition sites; during t_{off}, the electrodes are brought back to the initial open-circuit potential E_{off}. For oxidative deposition reactions, the polarities have to be inverted.

For deposition, a pulse of length t_{on} and potential E_{on} was applied to the working electrodes with respect to the reference electrode. E_{on} is the potential at which the deposition occurred. During t_{off}, the electrode was brought back to the OCP by applying E_{off}, so that the ionic species to be deposited was replenished in the diffusion layer. The relaxation time t_{off} was defined as the time after which no significant current could be measured anymore.

The parameter windows for the different materials are listed in Table 1. Metals like platinum or gold could be electrodeposited by reduction of their salts in aqueous solutions [22]. In these cases, E_{on} was lower than E_{off} and the OCP. Compounds like phenol and phenylenediamine (PPD) polymerize upon electrochemical oxidation of the aromatic amine portion of the complex [13,23]. In that case E_{on} was larger than E_{off}.

Table 1. Electrodeposition parameters for the different materials.

Material	t_{on}	E_{on}	t_{off}	E_{off}	Cycles
Pt black	0.1–0.8 s	0 V	0.4 s	0.68 V	100–250
Au smooth	0.25 s	−0.43 V	0.25 s	0.12 V	220–360
Au granular	0.1 s	−0.45 V	0.1 s	0.12 V	2000
PEDOT	0.2–0.3 s	1.05 V	0.5 s	0.275 V	20–50
PPD	0.25 s	0.5–0.9 V	0.5 s	0.05 V	60–500

PEDOT: poly(ethylenedioxythiophene); PPD: poly(phenylenediamine).

2.3. Electrochemical Characterizations

Electrochemical characterizations using cyclic voltammograms (CVs) were performed in de-aerated 0.5 M H_2SO_4 at a sweep rate of 100 mV/s, unless otherwise noted. A continuous nitrogen stream was maintained over the sample solution during the measurements.

2.4. Pt Black Deposition

A solution of hexachloroplatinic acid (17.5 mM) and lead(II) acetate trihydrate (0.03 mM) (from Sigma-Aldrich, Buchs, Switzerland) dissolved in DI water was used for electrodeposition of Pt black. Platinum black was deposited by using the parameters listed in Table 1. The pulse-widths t_{on} were varied between 0.1 and 0.8 s, t_{off} was set to 0.4 s in all cases. The total t_{on} times (t_{on}* number of pulses) were kept constant for each set of pulse-widths by adapting the number of pulses. The depositions with different parameter sets were performed sequentially without renewal of the solution.

The roughness factors were determined coulometrically from the hydrogen desorption region of the CVs measured in H_2SO_4 solution at a sweep rate of 100 mV/s [24].

2.5. Gold Deposition

Neutronex 309 solution (Enthone Inc., West Haven, CT, USA) was used for electrodeposition as provided. Two different parameter sets were selected (see Table 1), which allowed for the production of gold surfaces with different surface morphologies.

2.6. Deposition of Conducting Polymers

PEDOT deposition solution was prepared by adding 20 mM 3,4-ethylenedioxythiophene (EDOT, 97%, Sigma Aldrich, Buchs, Switzerland) to a 1 wt % aqueous solution of poly(sodium 4-styrenesulfonate), PSS, MW ~70,000 g/mol, Sigma Aldrich, Buchs, Switzerland.

2.7. Deposition of Non-Conducting Polymers

For the PPD layer, o-phenylenediamine (100 mM, Sigma Aldrich, Buchs, Switzerland) was dissolved in de-aerated phosphate-buffered saline (PBS, Sigma Aldrich, Buchs, Switzerland). Amperometric calibration series for dopamine and ascorbic acid (both from Sigma Aldrich) were performed at 650 mV vs. an Ag/AgCl reference electrode in PBS.

3. Results

3.1. Pt Black Deposition

A layer of amorphous platinum black on top of the bare platinum electrodes increases the effective surface area while preserving a small geometric sensor area. The increase in effective surface area lowers the impedance of the electrode and allows for better charge transfer, which is beneficial for both sensing and stimulation scenarios [12].

Figure 2 shows the current response during a deposition of Pt black (t_{on} = 0.1 s, E_{on} = 0 V, t_{off} = 0.4 s, E_{off} = 0.68 V, 160 cycles). Single pulses (every 13th pulse) were extracted at several time points and overlaid (1–7). The first negative section of the pulse is a combination of non-faradaic (i.e., charging of the double layer) and faradaic (i.e., metal deposition) currents. The second (positive) part of the pulse is only due to non-faradaic charging of the double layer. Both pulses increase over time, as the active surface area is becoming larger, which entails an increase of the double layer capacitance.

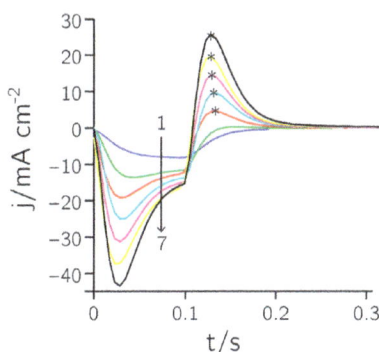

Figure 2. Deposition of Pt black. The curves marked 1 to 7 show the current current density versus time response of every 13th pulse as recorded during the deposition procedure (t_{on} = 0.1 s, E_{on} = 0 V, t_{off} = 0.4 s, E_{off} = 0.68 V, 160 cycles). The asterisks indicate the current density peak values for charging the double layer capacitance.

In Figure 3a,b two electrodes with diameters of 50 μm (a) and 10 μm (b) are shown. The Pt-black layers were deposited by using the deposition parameters mentioned above. In both cases, the electrode surface is homogeneously covered, no overgrowth of the layers can be seen, and the layers show the same morphology upon optical inspection. The dendritic structure of the Pt black surface, which is the reason for the increase in active electrode area, is shown in Figure 3c. The effect of increasing the number of pulses can be seen in Figure 3d. Pt black was deposited in the recessed electrode area during a total t_{on} time of 25 s, 37.5 s, 50 s and 62.5 s (from top right clockwise); at 50 s the layer started to overgrow the recessed electrode opening.

We further determined the roughness factors from the hydrogen desorption region of cyclic voltammograms (CVs) measured in a 0.5 M H_2SO_4 solution at a sweep rate of 100 mV/s [24] (Supplementary Figure S2). Figure 4a shows the results of measurements from all 1024 electrodes on a single chip. It shows the dependence of the roughness factor on pulse width (t_{on}) and total t_{on} time or total deposition time (t_{on} × number of pulses). Each point represents the average of 64 electrodes; the standard deviation is depicted by semi-transparent layers below and above. The relative standard deviation throughout the parameter space is less than 10% (<4.5% for t_{on} = 48 s), which demonstrated the excellent reproducibility of the deposition procedure. The roughness factor (i.e., the ratio of active area to geometric area) is proportional to the total deposition or t_{on}-on time as expected. The roughness factor can be increased to up to 150 $\mu m^2/\mu m^2$ (from the initial 1.6 $\mu m^2/\mu m^2$ on average before the deposition) without overgrowth of the recessed electrodes. Depositions at different pulse widths (t_{on}) cannot be distinguished optically (using SEM). Reducing the relaxation time t_{off} down to 0.25 s did not visibly change the structures or the electrochemical response of the resulting deposited layer.

However, the relaxation time may need to be increased for larger electrodes or closer spacing to allow for replenishing the depleted liquid volume around the deposition electrodes through diffusion in the liquid phase.

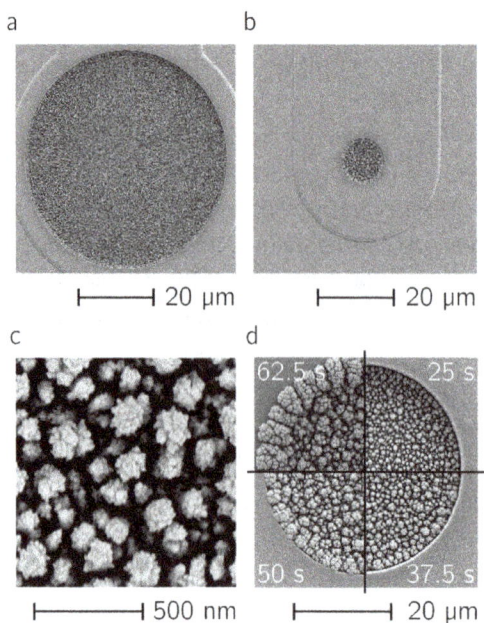

Figure 3. Deposited Pt black. SEM micrographs of deposited Pt black layers. The same deposition parameters have been used for a 50-μm (**a**) and a 10-μm-diameter (**b**) electrode; (**c**) Close up of the Pt black morphology and structure; (**d**) Different total t_{on} times, clockwise from top right: 25 s, 37.5 s, 50 s, 62.5 s for deposition on a 10-μm-diameter electrode.

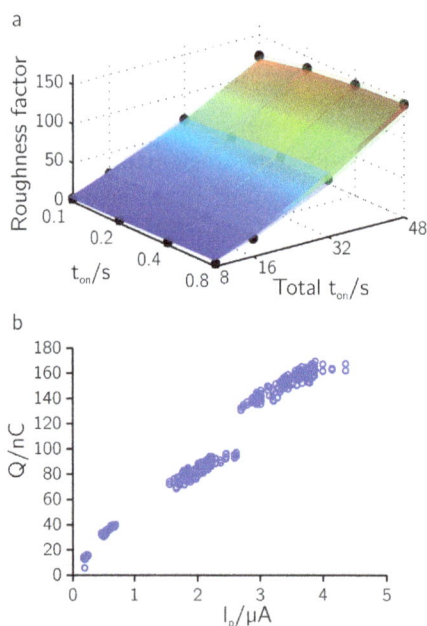

Figure 4. Pt black characteristics. (**a**) Roughness factor determined from integrating cyclic voltammograms (CVs) in H_2SO_4 from all electrodes of one chip for different deposition parameters (n = 64 electrodes per data point). Standard deviations are represented by semitransparent layers. (**b**) Integrated current from the hydrogen desorption region vs. maximum positive pulse height of the pulse train during deposition.

In Figure 4b the charge in the hydrogen desorption region, used for determining the roughness factor, is plotted against the maximum anodic pulse height of the pulse train during deposition (cf. asterisks in Figure 2). The plot reveals a clear linear dependence between pulse height and integrated charge (R^2 = 0.985). The active area can thus be controlled and predicted during the deposition process. This may prove useful to fine tune the active area and impedance of the electrodes in order to achieve uniform electrode characteristics over a large array.

3.2. Gold Deposition

Gold electrodes are widely used to attach biomolecules by using thiol chemistry. The strong affinity of the thiol groups for noble metal surfaces enables the formation of covalent bonds between the sulfur and gold atoms [14]. This biofunctionalization is a common approach used for nucleic-acid biosensors [3,25]. Hence, the voltage-pulse-based deposition protocol was evaluated for the deposition of gold on platinum electrodes.

The surface of the gold deposited on platinum electrodes can be rendered either smooth or granular by altering the deposition time (t_{on}). Smooth gold depositions were achieved by using pulse durations (t_{on}) of 0.25 s or longer, and granular gold depositions by using t_{on} times of 0.1 s. Figure 5a shows a CV in 0.5 M H_2SO_4 performed on 50-µm-diameter gold electrodes, using a sweep rate of 100 mV/s. One of the electrodes has been covered with granular gold and the other one with smooth gold. Complete coverage of the electrode can be deduced from the different potential window of gold with respect to that of platinum. The absence of contamination by other electroactive compounds is indicated by the presence of only the characteristic CV features of gold, including the single oxide reduction peak at approx. 0.9 V [26].

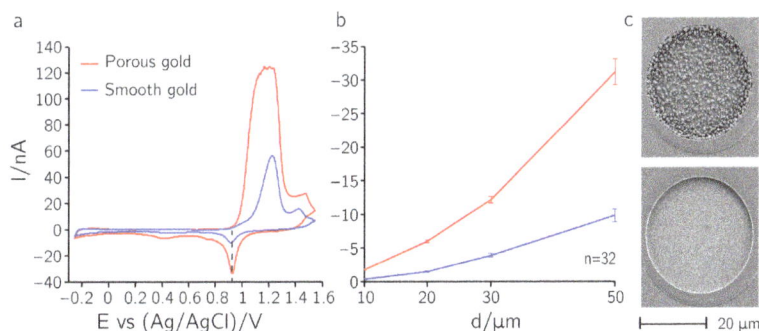

Figure 5. Gold deposition. Cyclic voltammogram (CV) of an electrode with deposited gold layer. (**a**) CV of 50-µm-diameter electrode modified with granular or smooth gold in 0.5 M H_2SO_4 in a voltage range between −0.25 V and 1.55 V vs. Ag/AgCl; the sweeping rate was 100 mV/s. A single sharp reduction current peak can be seen at ~0.9 V. Electrodes were completely covered so that no traces of platinum were visible. (**b**) Reduction current peak heights for different electrode diameters for smooth and granular gold layers; (**c**) scanning electron microscope (SEM) of a 30-µm-diameter electrode after deposition of granular gold (top) and smooth gold (bottom) layers.

The total t_{on} time needed for complete coverage was 50 s for smooth and 150 s for granular gold. If the electrodes were not covered completely, the CV could not be performed in the specific range for gold electrodes (up to 1.6 V vs. Ag/AgCl), because the platinum surface then promoted the generation of gaseous oxygen above 1.2 V vs. Ag/AgCl, which led to excessive currents and overload of the potentiostat. The dependence of the peak height of the oxide reduction peak (marked with a dashed line in Figure 5a) on the electrode diameter is shown for smooth and porous layers in Figure 5b. The low standard deviations within the sets of values obtained from 32 electrodes that have been coated in parallel demonstrate the good reproducibility of the deposition process. Granular gold has an active surface area which is three times larger due to its dendritic nature. Further, the current peak height was proportional to the electrode area. The different surface textures—granular for the top graph and smooth for the lower graph—can clearly be seen in the SEM micrographs shown in Figure 5c. A close-up view of the granular gold layer is provided in Supplementary Figure S3.

3.3. Deposition of the Conducting Polymer PEDOT

PEDOT was chosen as an exemplary system to evaluate the proposed deposition scheme for conducting polymers. PEDOT with embedded enzymes has been used for devising biosensors [27,28], or enzymes have been grafted onto the conducting PEDOT polymer [29]. Furthermore, PEDOT has been used for neurotransmitter sensing [30] and as an anti-fouling layer [16].

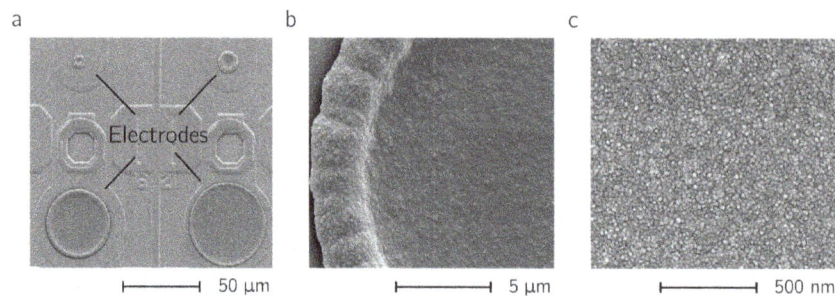

Figure 6. Deposited Poly(3,4-ethylenedioxythiophene) (PEDOT). (**a**) SEM micrographs of 5, 10, 40 and 50 μm diameter electrodes and (**b**,**c**) close-up views of a 25-μm-diameter electrode after PEDOT deposition.

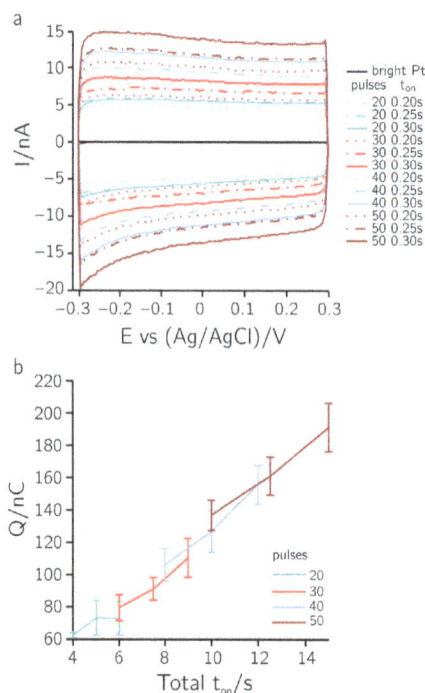

Figure 7. PEDOT. (**a**) CVs in phosphate-buffered saline (PBS) between −0.3 and 0.3 V vs. Ag/AgCl at 100 mV/s sweep rate, on 25-μm-diameter electrodes. The electrode materials included bright Pt and PEDOT deposited with a t_{on} of 0.2 s, 0.25 s, 0.3 s while applying 20, 30, 40, and 50 pulses. (**b**) Integrated total charge for different total-ton times. Colors code the number of pulses and correspond to (**a**). The integrated current over a CV of a bright Pt electrode is 200 pC.

Depositions on electrodes of different sizes are shown in the SEM micrographs of Figure 6a. A slight overgrowth at the rim of the electrodes was observed. The dense structure of the deposited layer can be seen in Figure 6b,c. Figure 7a shows CVs performed in PBS between −0.3 and

0.3 V vs. an Ag/AgCl reference electrode at a sweep rate of 100 mV/s. A large non-faradaic current can be observed as compared to a bare Pt electrode (black trace around the zero line in Figure 7a). In Figure 7b the integrated charge is plotted against the total t_{on} time. The results indicate that the total t_{on} time defines the electrode capacitance. The capacitance (C) can be estimated from the current values of the CV (C = current/sweep rate). For the values in Figure 7, the capacitances range between 50 and 150 nF (10–30 mF/cm^2), which is approx. 1000 times larger than the characteristic capacitance measured using bright Pt electrodes (150 pF or 30 μF/cm^2). In the SEM micrographs, there is no indication for an increase in the roughness of the deposited layers upon applying different total t_{on} times. Therefore, the active area cannot be attributed to an increased capacitance. PEDOT/PSS is known to exhibit a large pseudo-capacitance due to the immobilized anions (PSS$^-$; [31]). Hence, the dependence of the pseudo-capacitance on the total t_{on} time observed in Figure 7 can be attributed to differences in the deposited layer thicknesses.

3.4. Deposition of the Non-Conducting Polymer PPD

The voltage-pulse-based deposition procedure has also been applied for depositing a non-conducting polymer, poly(phenylenediamine (PPD). The non-conducting nature of PPD has been employed to prevent other electro-active molecules, such as dopamine or ascorbic acid, from being oxidized at the electrode surface and to thus avoid interference with the signal of interest [15]. Besides, PPD features perm-selective properties [23] and has been used as a matrix for embedding biosensitive elements [32].

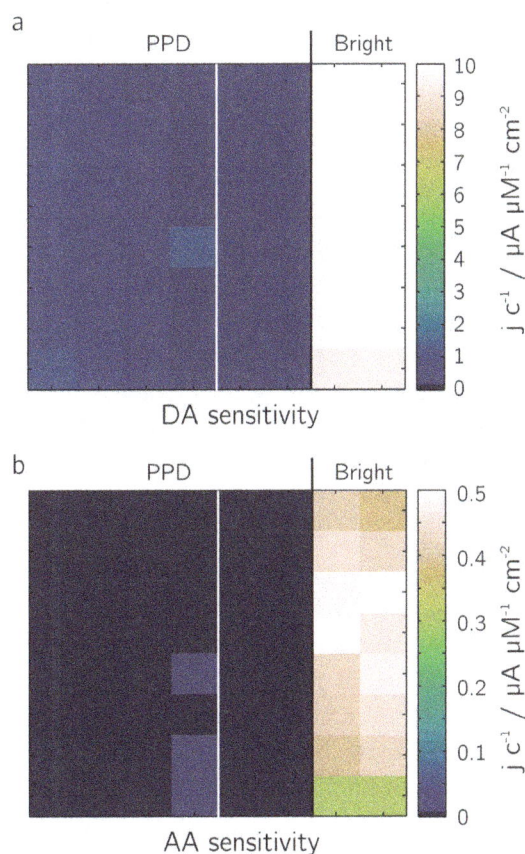

Figure 8. Non-conducting polymer poly(phenylenediamine (PPD). Sensitivity values (in current density per concentration) obtained from measurements of two blocks of 20-μm-diameter electrodes: the left block was completely covered with PPD; half of the right block was covered with PPD and half was left with bright Pt electrodes. Sensitivity was evaluated for (**a**) dopamine and (**b**) ascorbic acid. AA: ascorbic acid; DA: dopamine.

The rejection of the electro-active species dopamine, DA, (Figure 8a) and ascorbic acid, AA (Figure 8b) from the electrode surface has been demonstrated in an amperometric calibration series. Two 4×8 electrode blocks of 20-μm-diameter bright Pt electrodes were evaluated: the electrodes of the left block were completely covered with PPD, those of the right block were half covered with PPD, the other half was left bright Pt for comparison. Sensitivities were determined by measuring the currents upon addition of solutions of different concentrations of DA and AA (0–400 μM). The perm-selective nature of the coating became evident: the signal of the PPD-functionalized electrodes upon DA addition was by a factor of 24 lower, the signal upon addition of AA by a factor of 160, both in comparison to bright Pt electrodes. A variation of the deposition parameters did not yield a clear trend with regard to the suppression of DA or AA redox reactions at the electrodes. This may be due to self-limiting nature of the deposition of the non-conducting polymer whereby additional deposition cycles might not result in further deposition of the polymer, limiting the maximum layer thickness achievable by electodeposition-based growth.

4. Discussion

Galvanostatic electrodeposition allows for controlling and optimizing the deposition rate. When using a voltage scheme, the voltage can be tuned to facilitate specific reactions, instead of forcing reactions to happen at a certain rate, which is the case in applying current-based deposition schemes. Upon forcing electrochemical reactions to happen at a certain rate, the generated overpotentials may promote secondary reactions. These secondary reactions are undesirable, especially for functionalization of microelectrodes. The evolution of hydrogen, for example, can lead to holes in the deposited layers that then render the electrode functionalization useless. Instead, by defining the deposition potential, as has been done in the procedures presented here, secondary reactions could be obviated, and uniform, well-controlled layers were produced.

An advantage of current-based deposition schemes is the adaptation of the voltage during the deposition. This adaptation automatically accounts for a change of the electrode potential when a first layer of deposited material covers the electrode surface so that the deposition rate remains constant. The fact that this adaptation is missing in voltage mode seemed, however, not to be an issue in the depositions performed in this study. Upon entering diffusion-limited regimes or when applying self-limiting deposition processes, the use of current-based schemes will entail the risk of having high electrode potentials and, consequently, having unwanted secondary electrochemical reactions. This issue will not occur in voltage-based schemes, as the voltage is always precisely controlled, so that lower currents will result under voltage-controlled conditions. We found that voltage-controlled conditions were very well suited to achieve homogenous and reproducible deposition of a PPD layer, which features self-limiting characteristics.

Electrodeposition procedures are usually mass-transfer limited by diffusion. Pulsed-current or reverse-pulsed-current protocols are supposed to give better results, not in terms of deposition efficiency, but in uniformity of layer growth and tuning physical layer properties [17]. During the t_{off} time, the voltage-pulse-based deposition scheme presented here produces the same effects as pulse-reverse-current plating techniques: it counteracts surface charging, which may repel further ions, and it counteracts depletion of ions in areas, where deposition occurs [17]. The advantageous effects on diffusion lead to excellent uniformity of the deposited layers, which can be seen in the low standard deviations of the simultaneous parallel depositions of gold and platinum black.

The biggest advantage of the presented deposition scheme becomes evident upon applying it on large microelectrode arrays. In this case it is not necessarily deposition rate efficiency, but the flexibility in selecting the deposition area (which electrodes, how many electrodes, which electrode size), as well as a high yield and deposition uniformity, which are crucial. This flexibility of the deposition scheme is pivotal for depositions on electrodes of different sizes or diameters and for deposition on different numbers of electrodes, as has been shown.

The method can be adapted for other materials to be deposited according to the following procedure: (i) In order to determine E_{off}, the open-circuit potential of the deposition solution has to be measured between an Ag/AgCl reference electrode and the electrode on which the deposition will be done; (ii) E_{on} should then be gradually increased (using the proper polarity for the deposition reaction) until deposition occurs; (iii) t_{on} should be kept small in order not to reach a diffusion limited regime (<1 s for our array dimensions); (iv) t_{off} should be set to a value at which no significant current can be detected anymore.

5. Conclusions

The voltage-pulse-based deposition scheme described in this study constitutes a robust method to achieve spatially controlled and uniform layer depositions of both metallic and polymeric films on large arrays of microelectrodes. In the case of Pt black, the active surface area of the individual electrodes could be increased in a controlled and reproducible manner while keeping the geometric electrodeposition area restricted to the exposed electrode surface. Applying the scheme for gold depositions, the morphology of the resulting gold layer could be varied by changing the pulse parameters. Voltage-pulse-based deposition further allows for fine-tuning the impedance of the commonly used conductive polymer PEDOT through variation of the layer thickness. Even self-limiting depositions (PPD) could be performed without the risk of obtaining large overpotentials. The method is particularly suitable for integrated microelectrode arrays, since the deposition waveforms can be readily generated by using basic electronic circuitry. The method is easily scalable, and the same parameters can be used to functionalize the surface of either single electrodes or large arrays of thousands of electrodes or varying sizes. Scale-up to larger electrode sizes, number or even to whole arrays is simply a matter of switching on the respective electrodes during deposition. Arbitrarily selectable subsets of large arrays of electrodes can be simultaneously functionalized for massive parallel sensing, or individual electrodes or selected subsets of electrodes can be flexibly functionalized for applications in multi-analyte measurements. Finally, the proposed method can be easily adapted to depositing other potential electrode materials by adapting the deposition voltage and pulse length parameters on a single or a few electrodes before then applying the method to larger numbers of electrodes or arrays.

Supplementary Materials:
Figure S1: Fully integrated electrochemical complemenary metal-oxide-semiconductor (CMOS) system: (a) packaged CMOS chip for electrochemical measurements. On the left are the gold contacts for data communication and power supply. In the middle is the opening to the array of 32×32 Pt microelectrodes. (b) The assembled system with a reservoir for the liquids. A standard Ag/AgCl reference electrode was immersed in the electrolyte. Figure S2: Cyclic voltammograms (CV) of H_2SO_4 on Pt black: Pt black characterization via a CV measured in a 0.5 M H_2SO_4 solution at a sweep rate of 100 mV/s. The CV was controlled and recorded by the fully integrated CMOS chip. The hatched area was integrated to determine the active surface area and the roughness factor displayed in Figure 4a,b. Figure S3: Granular Au: Close-up view of the surface of the granular gold in Figure 5c.

Acknowledgments: We acknowledge the ZMB (Zentrum für Mikroskopie) of the University of Basel, Switzerland, for taking the SEM pictures (Figures 3, 5b, 6 and S3). Financial support through the ERC Advanced Grant 267351 "NeuroCMOS" and individual support for Olivier Frey through the Swiss National Science Foundation (Ambizione Grant 142440) is acknowledged. Rajtarun Madangopal was supported by the Indiana Clinical and Translational Sciences Institute funded, in part by Grant # RR025761 from the National Institutes of Health, National Center for Research Resources, Clinical and Translational Sciences Award; and by the National Institutes of Health, National Institute of General Medical Sciences (Grant # 8R21GM103467-03).

Author Contributions: J.Ro., O.F., R.M., J.Ri. and A.H. planned the experiments; J.Ro. and R.M. performed the experiments; J.Ro. performed the data analysis; J.Ro., O.F, and A.H. wrote the manuscript. All authors edited the manuscript.

Conflicts of Interest: The authors declare no conflict of interest.

References

1. Stett, A.; Egert, U.; Guenther, E.; Hofmann, F.; Meyer, T.; Nisch, W.; Haemmerle, H. Biological application of microelectrode arrays in drug discovery and basic research. *Anal. Bioanal. Chem.* **2003**, *377*, 486–495. [CrossRef] [PubMed]

2. Gross, G. The use of neuronal networks on multielectrode arrays as biosensors. *Biosens. Bioelectron.* **1995**, *10*, 553–567. [CrossRef]

3. Drummond, T.G.; Hill, M.G.; Barton, J.K. Electrochemical DNA sensors. *Nat. Biotechnol.* **2003**, *21*, 1192–1199. [CrossRef] [PubMed]

4. Zhang, B.; Adams, K.L.; Luber, S.J.; Eves, D.J.; Heien, M.L.; Ewing, A.G. Spatially and temporally resolved single-cell exocytosis utilizing individually addressable carbon microelectrode arrays. *Anal. Chem.* **2008**, *80*, 1394–1400. [CrossRef] [PubMed]

5. Kalantari, R.; Cantor, R.; Chen, H.; Yu, G.; Janata, J.; Josowicz, M. Label-free voltammetric detection using individually addressable oligonucleotide microelectrode arrays. *Anal. Chem.* **2010**, *82*, 9028–9033. [CrossRef] [PubMed]

6. Jepson, L.H.; Hottowy, P.; Weiner, G.A.; Dabrowski, W.; Litke, A.M.; Chichilnisky, E.J. High-Fidelity Reproduction of Spatiotemporal Visual Signals for Retinal Prosthesis. *Neuron* **2014**, *83*, 87–92. [CrossRef] [PubMed]

7. Voitechovič, E.; Bratov, A.; Abramova, N.; Razumienė, J.; Kirsanov, D.; Legin, A.; Lakshmi, D.; Piletsky, S.; Whitcombe, M.; Ivanova-Mitseva, P.K. Development of label-free impedimetric platform based on new conductive polyaniline polymer and three-dimensional interdigitated electrode array for biosensor applications. *Electrochim. Acta* **2015**, *173*, 59–66. [CrossRef]

8. Frey, O.; Holtzman, T.; McNamara, R.M.; Theobald, D.E.H.; van der Wal, P.D.; de Rooij, N.F.; Dalley, J.W.; Koudelka-Hep, M. Enzyme-based choline and L-glutamate biosensor electrodes on silicon microprobe arrays. *Biosens. Bioelectron.* **2010**, *26*, 477–484. [CrossRef] [PubMed]

9. Kisler, K.; Kim, B.N.; Liu, X.; Berberian, K.; Fang, Q.; Mathai, C.J.; Gangopadhyay, S.; Gillis, K.D.; Lindau, M. Transparent Electrode Materials for Simultaneous Amperometric Detection of Exocytosis and Fluorescence Microscopy. *J. Biomater. Nanobiotechnol.* **2012**, *3*, 243–253. [CrossRef] [PubMed]

10. Müller, J.; Ballini, M.; Livi, P.; Chen, Y.; Radivojevic, M.; Shadmani, A.; Viswam, V.; Jones, I.L.; Fiscella, M.; Diggelmann, R.; et al. High-resolution CMOS MEA platform to study neurons at subcellular, cellular, and network levels. *Lab Chip* **2015**, *15*, 2767–2780. [CrossRef] [PubMed]

11. Gerwig, R.; Fuchsberger, K.; Schroeppel, B.; Link, G.S.; Heusel, G.; Kraushaar, U.; Schuhmann, W.; Stett, A.; Stelzle, M. PEDOT-CNT Composite Microelectrodes for Recording and Electrostimulation Applications: Fabrication, Morphology, and Electrical Properties. *Front. Neuroeng.* **2012**, *5*, 8. [CrossRef] [PubMed]

12. Franks, W.; Schenker, I.; Schmutz, P.; Hierlemann, A. Impedance characterization and modeling of electrodes for biomedical applications. *IEEE Trans. Biomed. Eng.* **2005**, *52*, 1295–1302. [CrossRef] [PubMed]

13. Guerrieri, A.; Ciriello, R.; Centonze, D. Permselective and enzyme-entrapping behaviours of an electropolymerized, non-conducting, poly(o-aminophenol) thin film-modified electrode: A critical study. *Biosens. Bioelectron.* **2009**, *24*, 1550–1556. [CrossRef] [PubMed]

14. Sassolas, A.; Leca-Bouvier, B.D.; Blum, L.J. DNA biosensors and microarrays. *Chem. Rev.* **2008**, *108*, 109–139. [CrossRef] [PubMed]

15. Guerrieri, A.; de Benedetto, G.E.; Palmisano, F.; Zambonin, P.G. Electrosynthesized non-conducting polymers as permselective membranes in amperometric enzyme electrodes: A glucose biosensor based on a co-crosslinked glucose oxidase/overoxidized polypyrrole bilayer. *Biosens. Bioelectron.* **1998**, *13*, 103–112. [CrossRef]

16. Yang, X.; Kirsch, J.; Olsen, E.V.; Fergus, J.W.; Simonian, A.L. Anti-fouling PEDOT:PSS modification on glassy carbon electrodes for continuous monitoring of tricresyl phosphate. *Sens. Actuators B Chem.* **2013**, *177*, 659–667. [CrossRef]

17. Chandrasekar, M.S.; Pushpavanam, M. Pulse and pulse reverse plating—Conceptual, advantages and applications. *Electrochim. Acta* **2008**, *53*, 3313–3322. [CrossRef]

18. Horkans, J. Pulsed Potentiostatic Deposition of Gold from Solutions of the Au(I) Sulfite Complex. *J. Electrochem. Soc.* **1977**, *124*, 1499–1505. [CrossRef]

19. Plyasova, L.M.; Molina, I.Y.; Gavrilov, A.N.; Cherepanova, S.V.; Cherstiouk, O.V.; Rudina, N.A.; Savinova, E.R.; Tsirlina, G.A. Electrodeposited platinum revisited: Tuning nanostructure via the deposition potential. *Electrochim. Acta* **2006**, *51*, 4477–4488. [CrossRef]

20. Liu, J.; Wang, X.; Lin, Z.; Cao, Y.; Zheng, Z.; Zeng, Z.; Hu, Z. Shape-Controllable Pulse Electrodeposition of Ultrafine Platinum Nanodendrites for Methanol Catalytic Combustion and the Investigation of their Local Electric Field Intensification by Electrostatic Force Microscope and Finite Element Method. *Electrochim. Acta* **2014**, *136*, 66–74. [CrossRef]

21. Rothe, J.; Frey, O.; Stettler, A.; Chen, Y.; Hierlemann, A. Fully Integrated CMOS Microsystem for Electrochemical Measurements on 32 × 32 Working Electrodes at 90 Frames Per Second. *Anal. Chem.* **2014**, *86*, 6425–6432. [CrossRef] [PubMed]

22. Pourbaix, B.M.J.N.; van Muylder, J.; de Zoubov, N. Electrochemical Properties of the Platinum Metals. *Platin. Met. Rev.* **1959**, *3*, 47–53.

23. Ohnuki, Y.; Matsuda, H.; Ohsaka, T.; Oyama, N. Permselectivity of films prepared by electrochemical oxidation of phenol and amino-aromatic compounds. *J. Electroanal. Chem. Interfacial Electrochem.* **1983**, *158*, 55–67. [CrossRef]

24. Rodríguez, J.M.D.; Melián, J.A.H.; Peña, J.P. Determination of the Real Surface Area of Pt Electrodes by Hydrogen Adsorption Using Cyclic Voltammetry. *J. Chem. Educ.* **2000**, *77*, 1195. [CrossRef]

25. Wang, J. Electrochemical nucleic acid biosensors. *Anal. Chim. Acta* **2002**, *469*, 63–71. [CrossRef]

26. Hoare, J.P. A Cyclic Voltammetric Study of the Gold-Oxygen System. *J. Electrochem. Soc.* **1984**, *131*, 1808–1815. [CrossRef]

27. Nien, P.-C.; Tung, T.-S.; Ho, K.-C. Amperometric Glucose Biosensor Based on Entrapment of Glucose Oxidase in a Poly(3,4-ethylenedioxythiophene) Film. *Electroanalysis* **2006**, *18*, 1408–1415. [CrossRef]

28. Madangopal, R.; Stensberg, M.; Porterfield, M.; Rickus, J.; Pulliam, N. Directed enzyme deposition via electroactive polymer-based nanomaterials for multi-analyte amperometric biosensors. In Proceedings of the 2012 IEEE Sensors, Taipei, Taiwan, 28–31 October 2012; pp. 1–4.

29. Vidal, J.-C.; Garcia-Ruiz, E.; Castillo, J.-R. Recent Advances in Electropolymerized Conducting Polymers in Amperometric Biosensors. *Microchim. Acta* **2003**, *143*, 93–111. [CrossRef]

30. Samba, R.; Fuchsberger, K.; Matiychyn, I.; Epple, S.; Kiesel, L.; Stett, A.; Schuhmann, W.; Stelzle, M. Application of PEDOT-CNT Microelectrodes for Neurotransmitter Sensing. *Electroanalysis* **2014**, *26*, 548–555. [CrossRef]

31. Bobacka, J.; Lewenstam, A.; Ivaska, A. Electrochemical impedance spectroscopy of oxidized poly(3,4-ethylenedioxythiophene) film electrodes in aqueous solutions. *J. Electroanal. Chem.* **2000**, *489*, 17–27. [CrossRef]

32. Lowry, J.P.; McAteer, K.; el Atrash, S.S.; Duff, A.; O'Neill, R.D. Characterization of Glucose Oxidase-Modified Poly(phenylenediamine)-Coated Electrodes in vitro and in vivo: Homogeneous Interference by Ascorbic Acid in Hydrogen Peroxide Detection. *Anal. Chem.* **1994**, *66*, 1754–1761. [CrossRef]

Permissions

The contributors of this book come from diverse backgrounds, making this book a truly international effort. This book will bring forth new frontiers with its revolutionizing research information and detailed analysis of the nascent developments around the world.

We would like to thank all the contributing authors for lending their expertise to make the book truly unique. They have played a crucial role in the development of this book. Without their invaluable contributions this book wouldn't have been possible. They have made vital efforts to compile up to date information on the varied aspects of this subject to make this book a valuable addition to the collection of many professionals and students.

This book was conceptualized with the vision of imparting up-to-date information and advanced data in this field. To ensure the same, a matchless editorial board was set up. Every individual on the board went through rigorous rounds of assessment to prove their worth. After which they invested a large part of their time researching and compiling the most relevant data for our readers.

The editorial board has been involved in producing this book since its inception. They have spent rigorous hours researching and exploring the diverse topics which have resulted in the successful publishing of this book. They have passed on their knowledge of decades through this book. To expedite this challenging task, the publisher supported the team at every step. A small team of assistant editors was also appointed to further simplify the editing procedure and attain best results for the readers.

Apart from the editorial board, the designing team has also invested a significant amount of their time in understanding the subject and creating the most relevant covers. They scrutinized every image to scout for the most suitable representation of the subject and create an appropriate cover for the book.

The publishing team has been an ardent support to the editorial, designing and production team. Their endless efforts to recruit the best for this project, has resulted in the accomplishment of this book. They are a veteran in the field of academics and their pool of knowledge is as vast as their experience in printing. Their expertise and guidance has proved useful at every step. Their uncompromising quality standards have made this book an exceptional effort. Their encouragement from time to time has been an inspiration for everyone.

The publisher and the editorial board hope that this book will prove to be a valuable piece of knowledge for researchers, students, practitioners and scholars across the globe.

List of Contributors

Jin-Chern Chiou
Department of Electrical and Computer Engineering, National Chiao Tung University, 1001 University Road, Hsinchu City 30010, Taiwan; chiou@mail.nctu.edu.tw (J.-C.C.)
Institute of Electrical Control Engineering, National Chiao Tung University, 1001 University Road, Hsinchu City 30010, Taiwan

Chin-ChengWu and Tse-Mei Lin
Department of Electrical and Computer Engineering, National Chiao Tung University, 1001 University Road, Hsinchu City 30010, Taiwan; chiou@mail.nctu.edu.tw (J.-C.C.)

Yu-Chieh Huang
Institute of Electrical Control Engineering, National Chiao Tung University, 1001 University Road, Hsinchu City 30010, Taiwan

Shih-Cheng Chang
Institute of Biomedical Engineering, National Chiao Tung University, 1001 University Road, Hsinchu City 30010, Taiwan

Norhisam Misron
Faculty of Engineering, Universiti Putra Malaysia, 43400 Serdang, Selangor, Malaysia
Institute of Advance Technology (ITMA), Universiti Putra Malaysia, 43400 Serdang, Selangor, Malaysia

Nor Aziana Aliteh
Faculty of Engineering, Universiti Putra Malaysia, 43400 Serdang, Selangor, Malaysia

Noor Hasmiza Harun
Medical Engineering Section, Universiti Kuala Lumpur-British Malaysia Institute, Batu 8, Jalan Sg Pusu, 53100 Gombak, Selangor, Malaysia

Kunihisa Tashiro, Toshiro Sato and Hiroyuki Wakiwaka
Faculty of Engineering, Shinshu University, Wakasato 4-17-1, Nagano 380-8553, Japan

Yi Du, Xiaowei Xu, Yuanchun Zho and Jianhui Li
Department of Big Data Technology and Application Development, Computer Network Information Center, Chinese Academy of Sciences, Beijing 100190, China; duyi@cnic.cn (Y.D.) zyc@cnic.cn (Y.Z.)

Cuixia Ma
Intelligence Engineering Laboratory, Institute of Software, Chinese Academy of Sciences, Beijing 100190, China

Chao Wu and Yike Guo
Department of Computing, Imperial College London, London SW7 2AZ, UK

Kyu Byung Lee
Department of Nuclear Engineering, Seoul National University, Seoul 08826, Korea
Department of Nuclear Safety, Korea Institute of Nuclear Safety, Daejeon 34142, Korea

Jong Rok Kim
Thermal Hydraulics Safety Research Division, Korea Atomic Energy Research Institute, Daejeon 34057, Korea

Goon Cherl Park and Hyoung Kyu Cho
Department of Nuclear Engineering, Seoul National University, Seoul 08826, Korea

Ryszard Pawlak, Marcin Lebioda and Jacek Rymaszewski
Institute of Electrical Engineering Systems, Lodz University of Technology, 90-924 Lodz, Poland

Witold Szymanski, Lukasz Kolodziejczyk and Piotr Kula
Institute of Materials Science and Engineering, Lodz University of Technology, 90-924 Lodz, Poland

Alessandro Testa
Ministero dell'Economia e delle Finanze, Rome 00187, Italy; alessandro.testa@tesoro.it

Marcello Cinque
Dipartimento di Ingegneria Elettrica e delle Tecnologie dell'Informazione, University of Naples "Federico II", Naples 80125, Italy

Antonio Coronato
CNR-ICAR, Naples 80131, Italy

Juan Carlos Augusto
Department of Computer Science and R.G. on Development of Intelligent Environments, Middlesex University of London, London NW4 2SH, UK

Huijie Zhao, Zheng Ji, Na Li Jianrong Gu and Yansong Li
School of Instrumentation Science & Opto-Electronics Engineering, Beihang University, 37 Xueyuan Road, Haidian District, Beijing 100191, China

Rui Wang
School of Computer and Communication Engineering, University of Science and Technology Beijing, Beijing 100083, China
Beijing Key Laboratory of Knowledge Engineering for Materials Science, Beijing 100083, China

Fei Chang and Suli Ren
School of Computer and Communication Engineering, University of Science and Technology Beijing, Beijing 100083, China

Rosa Ma Alsina-Pagès and Francesc Alías
GTM—Grup de recerca en Tecnologies Mèdia, La Salle—Universitat Ramon Llull, Quatre Camins, 30, Barcelona 08022, Spain

Unai Hernandez-Jayo, and Ignacio Angulo
DeustoTech—Fundación Deusto, Avda. Universidades, 24, Bilbao 48007, Spain
Facultad Ingeniería, Universidad de Deusto, Avda. Universidades, 24, Bilbao 48007, Spain

Hanzhe Li, Shuo Yang and Bo Zhang
College of Mechanical and Electronic Engineering, Northwest A&F University, Yangling 712100, China

Changyuan Zhai
College of Mechanical and Electronic Engineering, Northwest A&F University, Yangling 712100, China
Department of Biosystems and Agricultural Engineering, Oklahoma State University, Stillwater, OK 75078, USA

Paul Weckler and Ning Wang
Department of Biosystems and Agricultural Engineering, Oklahoma State University, Stillwater, OK 75078, USA

Thanh Binh Pham, Huy Bui, and Van Hoi Pham
Institute of Materials Science, Vietnam Academy of Science and Technology, 18 Hoang Quoc Viet Rd, Cau giay District, Hanoi 100000, Vietnam

Huu Thang Le
Small and Medium Enterprise Development and Support Center 1, Directorate for Standards, Metrology and Quality, 8 Hoang Quoc Viet Rd, Cau giay District, Hanoi 100000, Vietnam

Igor Korobiichuk
Industrial Research Institute for Automation and Measurements PIAP, Jerozolimskie 202, 02-486 Warsaw, Poland

Dandan Cheng, Xingliang Liu, Hongzhi Yang and Aixia Han
Chemical Engineering College, Qinghai University, Xining 810016, China

Tian Zhang
Qinghai Heavy Industry Vocational School, Xining 810101, China

Ling Zang
Department of Materials Science and Engineering, University of Utah, Salt Lake City, UT 84112, USA

Joerg Rothe, Olivier Frey and Andreas Hierlemann
ETH Zurich, Department of Biosystems Science and Engineering, Bio Engineering Laboratory,Mattenstrasse 26, CH-4058 Basel, Switzerland

Rajtarun Madangopal
Agricultural and Biological Engineering, Biomedical Engineering, Physiological Sensing Facility at the Bindley Bioscience Center and Birck Nanotechnology Center, Purdue University, West Lafayette, IN 47907, USA
Intramural Research Program of the National Institute on Drug Abuse, National Institutes of Health, Baltimore, MD 21224, USA

Jenna Rickus
Agricultural and Biological Engineering, Biomedical Engineering, Physiological Sensing Facility at the Bindley Bioscience Center and Birck Nanotechnology Center, Purdue University, West Lafayette, IN 47907, USA

Index